Biology and Control of Invasive Fishes

Biology and Control of Invasive Fishes

Editor

Peter W. Sorensen

MDPI • Basel • Beijing • Wuhan • Barcelona • Belgrade • Manchester • Tokyo • Cluj • Tianjin

Editor
Peter W. Sorensen
University of Minnesota
USA

Editorial Office
MDPI
St. Alban-Anlage 66
4052 Basel, Switzerland

This is a reprint of articles from the Special Issue published online in the open access journal *Fishes* (ISSN 2410-3888) (available at: https://www.mdpi.com/journal/fishes/special_issues/invasive_fishes).

For citation purposes, cite each article independently as indicated on the article page online and as indicated below:

LastName, A.A.; LastName, B.B.; LastName, C.C. Article Title. *Journal Name* **Year**, *Volume Number*, Page Range.

ISBN 978-3-0365-2682-9 (Hbk)
ISBN 978-3-0365-2683-6 (PDF)

Cover image courtesy of Todd Koel
The cover photo for this book shows invasive lake trout removed from Yellowstone Lake in the USA using gill nets (see Koel et al. in this volume). This species is of special interest because while it is an exotic and damaging invader in the American west, it is simultaneously a valued native species threated by invasive sea lamprey from the Atlantic Ocean in eastern regions of North America (see Young et al. in this volume). Todd Koel (Yellowstone National Park) graciously provided this photograph.

Contents

About the Editor

Peter W. Sorensen is a professor at the University of Minnesota, where he has led a research and education program on fish physiology and behavior for over 30 years, with a focus on invasive fishes. He is especially interested in how sensory cues such as pheromones and sound might be used to manage invasive fishes without harming native species. During this time, he has studied sea lamprey, common carp, goldfish, brown trout, ruffe and, most recently, bigheaded carp. Peter earned his PhD in Biological Oceanography from the University of Rhode Island, BA from Bates College, and was an Alberta Heritage Fellow for Medical Research before coming to Minnesota.

Preface to "Biology and Control of Invasive Fishes"

The study and management of invasive fish is a rapidly growing discipline. Across the globe, dozens of highly invasive fish are now found in both fresh and marine waters, where they can alter habitats, compete with native fish for food, and prey on native fishes while exerting both indirect and direct effects on ecosystems and economies. Although efforts to understand and control invasive species are in their infancy, a few examples stand out. This book is a collection of 12 insightful articles on freshwater invasive fish and is the first on this discipline. Five topics are addressed: (1) damage caused by invasive fish (1 article on ruffe); (2) techniques to ascertain the presence of invasive fish (1 article on eDNA); (3) techniques to control invasive fish (1 article on CO_2 as a deterrent); (4) strategies to control invasive fish (3 articles on virus control, invasivorism, and modifying lock and dam operation); and (5) lessons learned from ongoing management efforts (5 articles on sea lamprey, lake trout, common carp, and northern pike control). Several commonalities are noted between successful management efforts that include (1) a focus on understanding of species number and distribution, (2) a focus on reducing fish reproductive success, (3) use of multiple complimentary (integrated) control strategies, (4) a deep understanding of the biology of the species in local areas, and (5) a concerted long-term approach. Well over a dozen researchers and managers contributed to this book, and they are thanked for their hard work and insight.

Peter W. Sorensen
Editor

MDPI

Editorial

Introduction to the Biology and Control of Invasive Fishes and a Special Issue on This Topic

Peter W. Sorensen

Department of Fisheries, Wildlife and Conservation Biology, University of Minnesota, St. Paul, MN 55108, USA; soren003@umn.edu; Tel.: +1-612-624-4997

Abstract: Across the globe, dozens of species of invasive fish are now found in fresh as well as marine waters, where they alter habitats, compete with native fish for food, and prey on native fishes, exerting both indirect and direct effects on ecosystems and economies. While efforts to understand and control these species are growing, most are still in their infancy; however, a few examples stand out. This special issue is comprised of 11 notable articles on freshwater invasive fish and is the first to address this topic. This introductory article serves as an introduction to these articles which focus on 5 topics on invasive freshwater fish: (1) the damage they cause (one article); (2) techniques to ascertain their presence (one article); (3) techniques to restrict their movement (one article); (4) strategies to control them (three articles); and (5) lessons learned from ongoing management efforts (five articles). This introduction notes that successful management efforts share a few approaches: (1) they develop and use a deep understanding of local species and their abundance as well as distribution; (2) they focus on reducing reproductive success; (3) they use multiple complimentary control strategies; and (4) they use a long-term approach.

Keywords: integrated pest management; aquatic invasive species; sea lamprey; lake trout; northern pike; common carp; bigheaded carp; eDNA; suppression; eradication

Citation: Sorensen, P.W. Introduction to the Biology and Control of Invasive Fishes and a Special Issue on This Topic. *Fishes* **2021**, *6*, 69. https://doi.org/10.3390/fishes6040069

Received: 21 November 2021
Accepted: 24 November 2021
Published: 30 November 2021

1. Introduction

Rapid increases in human population size and activity over the past century have resulted in an increasing number of species being transported between locations and developing self-sustaining populations that many consider to be invasive from an ecological perspective [1]. Notably, many of these newly self-sustaining populations are now also causing notable ecological and/or economic problems, leading governmental entities to legally recognize them as "invasive species" and deserving of management efforts [2]. Damage caused by aquatic invasive species (AIS), including fish, is often highly problematic because of the high level of interconnectedness of waters. Although both marine and fresh waters are affected by AIS, fresh waters are especially impacted, perhaps because they tend to be smaller, more isolated, as well as evolutionarily younger than marine waters and so have less well-developed ecosystems that less able to fend off invaders. Although many millions (likely hundreds) of US dollars are presently spent each year trying to understand and control invasive fish [1], the topic as a whole remains poorly understood and has not yet been reviewed.

This volume reflects an effort to remediate this deficiency. Work on this volume started in 2018 when the author was invited by the publisher to pursue a volume on invasive fishes and subsequently invited dozens of invasive fish biologists to contribute on any subtopic within this field of their choosing. While many were interested, ultimately 11 groups wrote articles, all of which were then peer-reviewed and published. All happen to be on freshwater fish and most focus on management, a previously poorly reviewed area. This brief article provides an introduction to these 11 articles which I have grouped into five categories: (1) Damage caused by invasive fish (n = 1); (2) Techniques to assess the presence and distribution of invasive fish (n = 1); (3) Techniques to control the abundance of invasive fish (n = 1); (4) Strategies to control invasive fish (a strategy uses multiple techniques to exert a specific effect such as suppression of reproductive success) (n = 3), and (5) Lessons learned from extant invasive fish control programs (n = 5). In each instance, I provide a brief overview of the sub-discipline (category) and then review some of the high points of each article in it and its contributions. A few references are included for context although it is beyond the scope of this paper/special issue to provide a comprehensive review. Common points made by these articles are highlighted to assist readers interested in specific themes. I conclude with a brief summary and a call for future study.

1.1. *Damage Caused by Invasive Fish*

Invasive fish are often a concern because they alter local ecosystem structure, usually in ways that also negatively impact economic activity. Frequently, preservation of bio-diversity (species loss) is a special concern. Invasive fish are usually perceived as "damaging" in several ways that are both interesting and important to understand, especially because invasive species managers often require economic information to justify their efforts. Effects of invasive fishes can be indirect and/ or direct. Important indirect effects include altering an ecosystem structure (for example, the common carp, *Cyprinus carpio*, uproot benthic plants, leading to changes in water clarity/chemistry (see [3])), and/or consuming prey at the base of foodwebs (for example bigheaded carp, *Hypthophthalmichthys* sp., feed on plankton, reducing food availability for gamefishes (see [4])). Direct effects can include aggressive interactions with other species to acquire reproductive territories (lowering reproductive success), consuming or parasitizing native fishes (directly reducing abundance; for example, the sea lamprey, *Petromyzon marinus* preys on lake tout, *Salvelinus namaycush* (see [5]), hybridizing with native fishes (leading to the disappearance of some species), and/or exhibiting behaviors that reduce the value and appeal of sport fishing (for example bigheaded carps jump, reducing the appeal of sport fishing(see [4])). Unfortunately, I am unaware of a comprehensive review of the effects of invasive fish on aquatic ecosystems. In this volume, Newman and colleagues [6] contribute to this field by describing an important experiment in which an invader, the Eurasian ruffe, *Gymnocephalus cernua*, was placed together with a species native to North America, the yellow perch (*Perca flavescens*), in large outdoor mesocosms and found a reduction of the growth rates of perch, which appeared unable to shift their dietary preferences. The latter may be a novel finding.

1.2. *Techniques to Assess the Presence and Distribution of Invasive Fish*

To control a population of invasive fish, managers need to know where members of that population are found and preferably their abundance. More sophisticated control strategies (for example efforts to reduce reproductive success in the common carp (see [7]), also require information on life history stage and reproductive condition. Traditionally, this information has been obtained by various types of netting, electrofishing, and trapping surveys (see [3]). However, these techniques can be of very limited value in deeper waters where fish can often avoid gear, in remote large waters that might be difficult and expensive to reach, or in flowing water that may wash nets away. Also, nets usually only work well for large individuals. Clearly, new sampling techniques and strategies are needed. One recent development has been to measure environmental DNA (eDNA), or DNA released to the water by organisms where it can be sampled and measured using highly sensitive molecular tools (PCR). Sometimes this technique can be conducted together with sex pheromone measurements to acquire more information on reproductive state [8]. eDNA has many distinct advantages (ex. easy to collect, extreme sensitivity, etc.) and some disadvantages (ex. dilution, false positives from dead fish, etc.), but is being actively explored as a means to compliment traditional sampling. There are several reviews of this rapidly developing topic [9]. In this volume, Hayer and colleagues [10] describe how eDNA presence in river waters shows a positive relationship to the presence of invasive grass carp, *Ctenopharyngodon idella*, in North America, especially when these fish are spawning. Because, grass carp, like many invasive fishes, are very difficult to sample, this is an important observation, especially if eDNA concentration could be directly linked to reproductive state.

1.3. Techniques to Control the Abundance of Invasive Fish

The origins of invasive fish control are seemingly found in the responses of the American and Canadian governments to the parasitic sea lamprey, *Petromyzon marinus*, which invaded the Laurentian Great Lakes in the early 1900s and caused a collapse in their fisheries. Efforts initially focused on blocking adult sea lamprey from migrating to their fluvial spawning habitat and adult removal, but soon shifted to chemicals (abiotic) approaches. Eventually, after testing thousands of chemicals, 4-nitro-3-(trifluoromethyl)phenol (TFM) was identified as a compound with unusual potency and specificity (although it does kill other ancient fishes and insects) that will kill most larval lamprey. While TFM presently serves as the primary technique of sea lamprey control, it is now complimented by several other techniques including blocking migratory adults as part of an integrated pest control program (see [5]). Mimicking sea lamprey control, many other invasive fish programs have also attempted to identify a variety of biotic and abiotic techniques to control their target species and assemble them into strategies, and then combine them for use in control programs. Although no program has seemingly been as successful as the sea lamprey control in the Great Lakes, at least on such a large scale, many have developed a variety of effective biotic and abiotic control techniques that have achieved notable levels of success, some of which are described herein. A third approach, genetic engineering/manipulation of populations, is also now being examined but not reviewed. Genetic engineering approaches are still in their infancy and generally attempt to manipulate gender and face challenges with societal acceptance. Biotic control techniques include new techniques to locate/attract and remove adults (ex., "Judas fish" [3,7]); techniques to remove/kill eggs/young using native microbes, etc. [11]; enhanced/targeted fishing [4]; introduction of new predators or management of extant ones, especially predators for vulnerable young [3]; and introduction of novel pathogens [12]. Important abiotic approaches include the strategic use/development of new piscicides (fish poisons) [13], and the development of physical techniques to selectively block movement/migration of invasive fish into new areas where they might breed successfully [14,15]. It has been a major challenge to develop techniques that are simultaneously effective and highly specific, and also acceptable to the public. Although the topic of invasive fish control has never been comprehensively reviewed, it appears that no technique (even TFM) has been developed that meets all desired criteria.

This special issue includes an article by Suski [14] that describes how carbon dioxide, a natural gas that many fishes find inherently repulsive, might be used to control the movement of invasive fishes, especially bigheaded carps that must migrate upstream to reproduce. Suski [14] reviews how and why even small amounts of CO_2 might be introduced into the riverine lock systems to deter carp from entering them. Although the actions of CO_2 are not highly specific, this approach has the advantage that it is not difficult/expensive, only small amounts are needed, and it could be paired with other sensory cues such as sound/air that can be more specific. CO_2 is currently being tested, along with various acoustic and electrical stimuli [16], including multimodal deterrent systems such as ensonified bubble curtains [17].

1.4. Strategies (and Associated Techniques) to Control Invasive Fish

Early in the development of the first invasive fish control programs, including the one for the sea lamprey, it became apparent that techniques were most likely to be effective if targeted to specific life history stages and used in combination with other techniques; in other words, they were part of a strategy. Control programs (see below) now generally employ multiple strategies. Strategies can revolve around either optimizing specific techniques (e.g., blocking adults using different types of barriers using several sensory fields), and/ or using multiple techniques to complement each other to achieve a specific goal (e.g., stopping the successful reproduction by blocking adults while also removing young [3]).

Three articles in this issue describe three new, promising strategic approaches to control invasive fishes. All strategies emphasize the value of targeted approaches that also use local understanding of species and their ecosystems, and include multiple "integrated" techniques.

McColl and Sunarto [12] review the status of a pathogen-based research program that could be used to control common carp in Australia. They describe how cyprinid herpesvirus 3 (CyHV-3; also known as the koi herpesvirus) is highly specific and could be introduced cheaply and easily to kill very large numbers of adult and juvenile common carp across this huge continent. In addition, they describe how numeric models have shown that pathogen release must be extremely strategic, because common carp will eventually develop immunity and their larvae will not be affected. Further, plans must be in place to address large adult die-offs. Fascinating analogies are drawn to a virus-based rabbit control program, and it is emphasized that complimentary control techniques (e.g., fishing-out) will be needed while local conditions must be carefully factored in if this strategy is ever to be employed.

In a study addressing bigheaded carps, Zielinski and Sorensen [15] focus on how the upstream of movement (i.e., invasion) of bigheaded carp through locks and dams in the Upper Mississippi River could be arrested. They show how the velocity of water passing through the locks and dams, which already divide this river into a series of pools, could be exploited to arrest carp movement upstream at a few specific locations (i.e., local conditions are very important) by adjusting dam spillway gate openings to increase water velocities and reduce carp passage. Perhaps most importantly, Zielinski and Sorensen describe a strategy in which, in addition to adjusting lock and dam spillway gate openings, carp removal would also be conducted at key locations while deterrents such as CO_2 and sound could be added to their locks to affect a nearly complete (99%) block at low cost, even in the event of flooding. Zielinski and Sorensen [15] model over 100 scenarios in the Upper Mississippi River to show how this integrated multi-component control scheme could be immediately implemented at low cost at several key sites (pairs of locks and dams that rarely experience "open-river" conditions) without necessarily substantially affecting native fish populations in the river as a whole because particular species and locations can be targeted. Like McColl and Sunarto [13], real biological data are used, meaning that this strategy could be implemented and at low cost. In both cases, few alternatives are presently available to control carps in these vast regions.

Finally, Bouska and colleagues [4] examine the possibility of controlling bigheaded carps in the Illinois River (a tributary of the Mississippi River) by using different types of removal fisheries. This is an important question for many of the invasive species that are edible, because politicians usually wonder if control could be self-funded and/or possibly lead to pressures to maintain an otherwise undesirable fishery. Specifically, Bouska et al. [4] describe how a decade-long contracted harvest program has successfully led to reduced bigheaded carp densities in the upper stretches of the Illinois River, seemingly preventing them from invading the Laurentian Great Lakes, but at great expense. They then describe a previously unpublished study which examined two alternatives ways of removing adult bighead carp with commercial fisheries in this river: (1) one in which a few fishers were offered a "fisher-side" incentive program, in which they were financially rewarded to catch carp for direct consumption if they shared data; (2) another in which many fishers were offered a "market-side" incentive program, which used set-quotas and guaranteed prices set by the industry to provide fish for fertilizer markets. Interestingly, while the market-side program was successful, the fisher-side program was not, suggesting that if demand (and prices) for bigheaded carp could be controlled and maintained, market-side fisheries could become a good alternative and supplement to contract fishing. This study also points out that all fishing favors large fish, but that control requires all size classes of fish to be caught, and so additional complimentary approaches (e.g., deterrent systems in locks and dams) are still required to affect full control if removal fishing is to be used effectively for bigheaded carps in rivers over the long-term.

1.5. Lessons Learned from Invasive Fish Control Programs

Numerous management programs currently attempt to control invasive fishes across the globe. A few of these are reporting success and producing important lessons. This special issue includes five articles which describe success stories from both the southern and northern hemispheres. It is the first such synthesis of management information that I know of. Experiences controlling the common carp, a salmonid, an esocid (northern pike), and sea lamprey are documented. All programs employ multiple techniques optimized to local conditions and use a detailed understanding of invasive fish abundance and distribution generated by some type of numeric model. Interestingly, while all are costly, none rely on expensive state-of-the-art techniques, and all include some type of sustainable adult removal. One example [7] describes eradication. I briefly introduce these studies and their lessons below.

An article by Young and colleagues [5] describe the history of the sea lamprey control in Lake Champlain, a large lake that is just east of, but not directly connected to the Laurentian Great Lakes, whose sea lamprey program (http://www.glfc.org, assessed on 27 November 2021) uses a strategic combination of the larval toxin, TFM (see above), and barriers to block adults to reduce lamprey populations by ~90%. Young and colleagues [5] describe how their multi-agency team instituted a very similar management program in Lake Champlain, but have not observed a similar level of success. Surprising local differences emerged between the two programs; in particular, a few spawning rivers were found to have disproportionate importance in Lake Champlain, and the larvae there appear to survive the TFM treatment better than in the Great Lakes. A system of temporary (and unique) lamprey barriers has thus been implemented to supplement TFM, which is also used more liberally. This activity has required extensive new monitoring and research, along with a more highly coordinated program of control. However, indications that local fishes including lake trout may now be recovering as lamprey numbers also appear to finally be dropping. Comparisons with the Great Lakes' sea lamprey program are very interesting, as they emphasize the importance of adaptive management and an understanding of local conditions to invasive fish control.

A somewhat similar story is described in a large, high-altitude and isolated lake, Yellowstone Lake [11], where lake trout were first observed in the early 1990s after apparently being introduced from the Great Lakes (where they have been decimated by the sea lamprey). Exotic lake trout now prey on the local endemic trout in Yellowstone Lake, the cutthroat (*Oncorhynchus clarkii bouvieri*) and are seriously threatening its survival and biodiversity. Koel and colleagues [11] describe how initial large-scale removal of lake trout by the local agency had little benefit, but how population modeling recommended by an outside review, eventually led to a much more targeted and larger-scale netting effort focused on lake trout spawning beds which is now showing success. Most importantly, they describe how this removal program is now supplemented by a complementary (integrated) program that kills larval lake trout on their benthic spawning beds by sinking the carcasses of extirpated lake trout there, inducing anoxic conditions. Further, private foundation and local angler groups are now assisting with intensive monitoring, and long-term support and planning is now in place. Remarkably, the population of cutthroat trout population is now starting to recover. Integrated sets of strategies with a focus on eliminating reproductive success and which use deep local knowledge have clearly been enabling in this example.

Dunker and colleagues [13] describe an invasive species program designed to protect and preserve migratory Pacific salmonids, *Oncorhynchus* spp., in a large area of southern Alaska which was invaded by northern pike, *Esox lucius*, an apex predator, several decades ago. This esocid (which came from northern Alaska) now preys on native salmon species which it has extirpated from dozens of lakes and greatly reduced in many more. Given the vast size of the area and the limited resources to manage it, the challenge is enormous. Once again, a suite of integrated and targeted approaches has been implemented to consider local conditions carefully and include: (1) population suppression in larger open systems using targeted netting; (2) eradication using rotenone (a natural fish poison) in other systems; and (3) prevention using angler awareness (as in Yellowstone Lake). Restocking native fish has complimented this long-term multi-faceted effort that is now paying hard-won dividends.

Finally, two studies at opposite ends of the world report success controlling the common carp, a species from Eurasia. Yick and colleagues [7] report eradicating this species in a large high-altitude Tasmanian (Australian) lake after a 12-year removal program accompanied (and now followed by) an equally long period of monitoring. The primary goal has been the preservation of biodiversity via eradication in this unique and ecologically-delicate lake. Success was realized using a series of innovative approaches which included containing fish in the lakes; removal of adults using both electrofishing and large-scale seining using radio-tagged "radio-transmitter" or "Judas" fish (tagged fish that lead biologists to other carp so they can be removed); and finally reducing their reproductive success by keeping carp out of known shoreline spawning grounds using kilometer-long fences, and when eggs have been detected, killing them. This might be the only known example of invasive fish eradication that has not used poisons. In another story about common carp and the importance of understanding local ecosystems, Sorensen and Bajer [3] report similar, albeit slightly less dramatic success at controlling and reducing populations of common carp in two chains of Midwestern (USA) lakes. The goal here was to improve water quality, and although success relied on many of the same elements used in Tasmania, some differed because of differences in local ecosystems. Notably, control has been realized using a strategic combination of three approaches: (1) removing aggregating over-wintering adults using large nets and Judas fish; (2) blocking adult migration into spawning areas located in adjoining outlying lakes using simple barriers; and (3) suppressing reproductive success (the production of young) by enhancing the numbers of a native micro-predator, the bluegill sunfish, *Lepomis macrochirus,* which was found capable of consuming very large numbers of carp eggs and larvae. Tasmania does not have similar populations of micro-predators which have reduced the cost of control and made it sustainable without the need for eradication in the Midwest. This study and its control program spanned two decades and involved long-term commitment/ discovery. Importantly, common carp densities have been reduced to levels that are no longer deemed highly environmentally damaging (allowing for treatments to improve water quality) and sustainably controlled by native fish. This example may be the first to control an invasive fish using native predators. Together, these two studies of common carp control show how strategies that include reducing reproductive success (the production of young), adult removal, and blocking movement of adults to key local areas, have been highly effective at controlling a long-lived invasive fish in two different areas of the world with very different ecosystems. Understanding local ecology has in both cases also been key to their success.

2. Summary and Future Study

Great progress is being made understanding how and why fish are invasive in freshwaters and some management programs are registering success, using strategic suites of well-established techniques. However, there is still room for improvement and in particular, a need for new techniques including genetic control, sensory deterrents, and eDNA, especially if these techniques are to be integrated into control plans. Nevertheless, based on the five programs reviewed in this issue, it appears that in order to succeed, control plans need to develop a deep local knowledge and focus on both reducing reproductive success and adult removal in targeted manners supported by numeric models. Long-term financial and administrative support is essential. Although this special issue summarizes some of this progress and many important lessons in science and invasive freshwater fish control, there is still a strong need to review, synthesize, and publish data on other control techniques, strategies and programs we could not cover. Marine fishes in particular need to be considered. I hope that this special issue sparks such efforts and thank its contributors for their many valuable insights and contributions.

Funding: This research received no external funding.

Conflicts of Interest: The author declares no conflict of interest.

References

1. Simberloff, D. *Invasive Species: What Everyone Needs to Know*; Oxford University Press: New York, NY, USA, 2013.
2. Clinton, W.J. Executive Order 13112 of the President of the United States—Invasive Species. 1999. Available online: https://www.invasivespeciesinfo.gov/executive-order-13112 (accessed on 27 November 2021).
3. Sorensen, P.W.; Bajer, P.B. Case studies demonstrate that common carp can be sustainably controlled by exploiting source-sink dynamic in Midwestern lakes. *Fishes* **2020**, *5*, 36. [CrossRef]
4. Bouska, W.A.; Glover, D.C.; Trushenski, J.Y.; Sechchi, S.; Garvey, J.E.; MacNamara, R.; Coulter, D.P.; Coulter, A.A.; Irons, K.; Weiland, A. Geographic-scale harvest program to promote invasivorism of bigheaded carps. *Fishes* **2020**, *5*, 29. [CrossRef]
5. Young, B.; Allaire, B.J.; Smith, S. Achieving sea lamprey control in Lake Champlain. *Fishes* **2021**, *6*, 2. [CrossRef]
6. Newman, R.M.; Henson, F.G.; Richards, C. Competition between invasive ruffe (*Gymnocephalus cernua*) and native yellow perch (*Perca flavescens*) in experimental mesocosms. *Fishes* **2020**, *4*, 33. [CrossRef]
7. Yick, J.L.; Wisniewski, C.; Diggle, J.; Patil, J.G. Eradication of the invasive common carp, *Cyprinus carpio*, from a large lake: Lessons and insights from the Tasmanian experience. *Fishes* **2021**, *6*, 6. [CrossRef]
8. Ghosal, R.; Eichmiller, J.A.; Witthuhn, B.A.; Sorensen, P.W. Attracting Common Carp to a bait site with food reveals strong positive relationships between fish density, feeding activity, environmental DNA, and sex pheromone release that could be used in invasive fish management. *Ecol. Evol.* **2018**, *8*, 6714–6727. [CrossRef] [PubMed]
9. Rourke, M.L.; Fowler, A.M.; Hughs, J.M.; Broadhurst, M.L.; DiBattista, J.D.; Fielder, S.; Walburn, J.W.; Furlan, E.M. Environmental DNA (eDNA) as a tool for assessing fish biomass: A review of approaches and future considerations for resource surveys. *Environ. DNA* **2021**. [CrossRef]
10. Hayer, C.-A.; Bayless, M.F.; George, A.; Thompson, N.; Richter, C.A.; Chapman, D.A. Use of environmental DNA to detect grass carp spawning events. *Fishes* **2020**, *5*, 27. [CrossRef]
11. Koel, T.M.; Arnold, J.L.; Bigelow, P.E.; Brenden, T.O.; Davis, J.D.; Detjens, C.R.; Doepke, P.D.; Ertel, B.D.; Glassic, H.C.; Gresswell, R.E.; et al. Yellowstone Lake ecosystem restoration: A case study for invasive fish management. *Fishes* **2020**, *5*, 18. [CrossRef]
12. McColl, K.A.; Sunrato, A. Biocontrol of the common carp in Australia: A review and future directions. *Fishes* **2020**, *5*, 17. [CrossRef]
13. Dunker, K.; Massengill, R.; Bradley, P.; Jacobson, C.; Swenson, N.; Wizik, A.; DeCino, R. A decade in review: Alaska's adaptive management of an invasive apex predator. *Fishes* **2020**, *5*, 12. [CrossRef]
14. Suski, C. Development of carbon dioxide barriers to deter invasive fishes: Insights and lessons learned from bigheaded carp. *Fishes* **2020**, *5*, 25. [CrossRef]
15. Zielinski, D.P.; Sorensen, P.W. Numeric simulation demonstrates that the upstream movement of invasive bigheaded carps can be blocked at sets of Mississippi River locks-and-dams using a combination of optimized spillway gate operations, lock deterrents and carp removal. *Fishes* **2021**, *6*, 10. [CrossRef]
16. Noatch, M.L.; Suski, C.D. Non-physical barriers to deter fish movements. *Environ. Rev.* **2012**, *20*, 71–82. [CrossRef]
17. Dennis, C.E.; Zielinski, D.P.; Sorensen, P.W. A complex sound coupled with an air curtain blocks invasive carp passage without habituation. *Biol. Invasions* **2019**, *21*, 2837–2855. [CrossRef]

Article

Competition between Invasive Ruffe (*Gymnocephalus cernua*) and Native Yellow Perch (*Perca flavescens*) in Experimental Mesocosms

Raymond M. Newman [1,*], Fred G. Henson [1,2] and Carl Richards [3]

1 Department of Fisheries, Wildlife and Conservation Biology, University of Minnesota, Saint Paul, MN 55108, USA; fred.henson@dec.ny.gov
2 New York State Department of Environmental Conservation, Division of Fish and Wildlife, Albany, NY 12233-4753, USA
3 Center for Water and the Environment, Natural Resources Research Institute, 5013 Miller Trunk Highway, Duluth, MN 55811, USA; crich555@gmail.com
* Correspondence: RNewman@umn.edu

Received: 7 September 2020; Accepted: 13 October 2020; Published: 17 October 2020

Abstract: Ruffe (*Gymnocephalus cernua*) were introduced to North America from Europe in the mid-1980s and based on similar diets and habit use may compete with yellow perch (*Perca flavescens*). To examine competitive interactions between invasive ruffe and native yellow perch, individually marked perch and ruffe were placed in mesocosms in a small lake. Mesocosms allowed fish to interact and feed on the natural prey populations enclosed. In the first experiment, four treatments were assessed: 28 perch, 14 perch + 14 ruffe, 14 perch, and 7 perch + 7 ruffe. Yellow perch growth was significantly lower in the presence of ruffe (ANOVA, $p = 0.005$) than in treatments containing only perch. In a second experiment, an increasing density of one species was superimposed upon a constant density of the other in parallel treatment series. Growth rates of both ruffe and perch declined when ruffe density was increased (*t* test, $p = 0.006$). However, neither ruffe nor perch growth was affected by increasing perch density. Total stomach content mass of perch was significantly decreased by ruffe in both years ($p < 0.02$), but no effects of ruffe on the composition of perch diets were observed. Ruffe growth and food consumption was greater than that of perch for both experiments. Ruffe can outcompete yellow perch when both species depend on a limited benthic food resource. Thus there is reason for concern for the ecological effects of ruffe if they expand their range into Lake Erie or North American inland lakes that contain yellow perch.

Keywords: interference competition; exploitative competition; invasive species; ruffe; yellow perch; growth; diet

1. Introduction

The ruffe (*Gymnocephalus cernua*) is a percid fish, native to southern England, northeastern France, and central Europe eastward through Siberia that was introduced to North America in the mid 1980s, likely via ballast water from ships departing the northern Elbe River or eastern North Sea [1–3]. Ruffe were first collected from the Duluth-Superior Harbor in 1986 and became the most abundant fish in bottom trawl samples by 1991 [4]. While ruffe flourished in the harbor, several native fish populations, including yellow perch (*Perca flavescens*), declined [4,5]. Ruffe subsequently expanded along the North Shore of Lake Superior to Thunder Bay, Ontario, and along the near-shore and tributary waters of southern Lake Superior to Lake Huron [3,6]. Populations developed in Green Bay and Little Bay de Noc in Lake Michigan and in Thunder Bay and the Cheboygan River, Lake Huron, likely the result of inter-lake shipping ballast and natural dispersal [3,7]. Despite concern for their expansion into the southern Great Lakes, ruffe have not been detected in southern Lake Michigan, Lake St. Clair or Lake Erie [7] nor have they been detected in any inland waters not tributary to Lakes Superior, Michigan or Huron [3].

In its native range in Europe and Asia, the ruffe is, at best, of marginal value as a fisheries resource and is widely considered a nuisance species [8]. Concern in North America is for potential negative effects on desired game and forage fishes. Within four years of their discovery in the Duluth Superior Harbor, ruffe became the most abundant fish and declines of other forage fish such as perch and trout perch were noted [4]. The observed declines in native fish abundance since the introduction of ruffe were more likely the consequence of natural population dynamics rather than an effect of ruffe, however, in the case of yellow perch, ruffe were partially responsible for fluctuations in year class strength [4]. Ruffe feed heavily on benthic invertebrates [9–12] and diet overlap suggests that exploitative competition between ruffe, yellow perch, and other native benthivores may occur [11,13–15]. Ruffe are particularly adapted to low light benthic habitats [11,16–18] and will likely be most successful in these conditions

Several laboratory studies have examined the potential for competition between ruffe and yellow perch. Ruffe and yellow perch were found to have similar prey preferences in the laboratory [15], and Fullerton et al. [19] found that perch and ruffe growth decreased with ruffe density in laboratory competition experiments for food between ruffe and yellow perch. However, they noted that overall fish density was more important to fish growth than the presence of ruffe. Savino and Kolar [13] concluded that ruffe compete with yellow perch, but the outcome depended on the situation. Ruffe were more efficient with unlimited food, but perch appeared to do better in food limited situations. However, ruffe were more aggressive than perch and may have an advantage over perch via interference competition. Savino and Kostich [20] noted intraspecific interference competition by ruffe and suggested ruffe will do best at intermediate densities. Fullerton and Lamberti [16] assessed both habitat use and feeding efficiency of ruffe and yellow perch and found no evidence of competition for habitat, but that within shared habitats competition for food may occur when food is limiting. The lower growth and conversion efficiency of ruffe suggests that ruffe will place a greater demand on benthic food resources than an equivalent biomass of perch [21] and thus could increase the potential for competition. Perch may be more effective in macrophyte habitats and ruffe appear also to be less adapted to compete with round gobies [22], particularly at low food densities and sand or macrophyte habitats.

These studies were conducted in aquaria and small tanks (100–280 L), although Bergman and Greenberg [23] documented competition between ruffe and European perch (*Perca fluviatilis*) in larger experimental mesocosms. They showed a decline in European perch growth rate and a diet shift by perch from benthic macroinvertebrates to microcrustaceans when ruffe density was increased. Given the high value of the yellow perch to Great Lakes fisheries [24], the potential for ruffe to invade substantial portions of yellow perch's North American range [24,25], and mixed evidence of competition between these two species, a controlled field experiment, to rigorously test the hypothesis that ruffe will decrease the fitness of yellow perch, was desired. A better assessment of ruffe's competitive potential will further inform modeling efforts and risk assessments that are being regularly updated to predict impacts and concern for invasive species (e.g., [25–27]).

Measuring competition between organisms and understanding its role in communities has been a longstanding challenge to ecologists. To demonstrate that competition exists between two species, one must show that a limited resource is utilized by both species and that the fitness of one species is decreased in the presence of the other [28–30]. The manipulation of the abundances of putative competitors in a carefully controlled field experiment has become a widely accepted means of investigating competition [28]. In previous larger scale tests of competition Bergman and Greenberg [23] manipulated ruffe density to demonstrate competition with European perch and Hanson and Leggett [29,30] manipulated pumpkinseed (*Lepomis gibbosus*) and yellow perch density to assess competition between these species. Hanson and Leggett [29] compared mixed treatments equal in total biomass to single species treatments, whereas Bergman and Greenberg [23] superimposed an increasing density of ruffe on a constant density of perch. Bergman and Greenberg [23] criticized designs of the type used by Hanson and Leggett [29,30] because the results are highly dependent on the densities chosen and because they do not lead to the development of a density response curve. However, the design of Bergman and Greenberg [23] did not allow them to determine which species was the superior competitor or control for the effect of total fish density. Recognizing the advantages and shortcomings of both approaches, we conducted experiments using designs of both types. We also examined the effect of different densities of perch superimposed on a constant density of ruffe to determine whether perch could affect ruffe fitness in addition to assessing the effect of ruffe on yellow perch. This multifaceted approach generated a more comprehensive picture of competition between two freshwater fishes than would have been possible with a more limited set of experiments.

2. Results

2.1. Growth

In the ruffe and fish density experiment (1996), perch consistently lost mass except in the absence of ruffe at low density (14 perch + 0 ruffe) (Figure 1). In contrast, ruffe at both densities displayed positive growth (Figure 2). The degree of perch mass loss was marginally greater at higher overall fish density, whether quantified as a treatment factor (low vs. high, ANOVA, $p = 0.057$) or as a covariate (measured treatment biomass, ANCOVA, $p = 0.050$). At the same overall fish density, perch mass loss was significantly greater in the presence of ruffe than in perch only treatments (ANOVA, $p = 0.005$), and there was also a significant interaction between overall fish density and the presence or absence of ruffe (ANOVA, $p = 0.041$). There was no significant effect of time (ANOVA, $p = 0.44$), block ($p = 0.20$), or interactions between other factors (time and overall fish biomass; all $p \geq 0.44$). The results were similar when gross change in mass was analyzed, except that there were no significant interactions and the effects of density (ANOVA, $p = 0.018$) and time (ANOVA, $p = 0.002$) were stronger. Perch growth became significantly more negative as the experiment progressed. There was no main effect of fish density on the growth of ruffe (ANOVA, $p = 0.320$), but a significant interaction between fish density and time (ANOVA, $p = 0.033$); fish density did affect ruffe growth in the middle of the experiment.

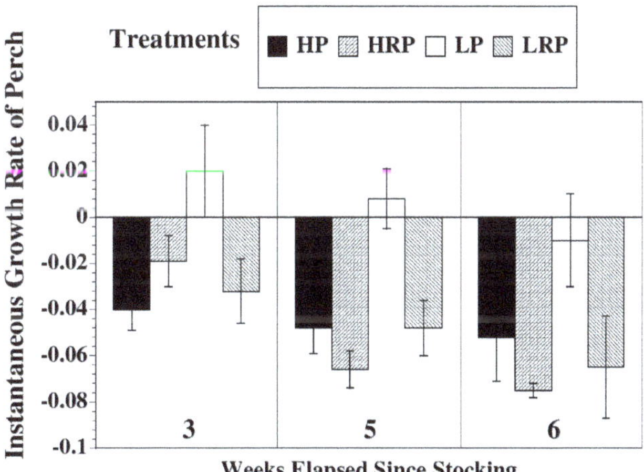

Figure 1. Mean instantaneous growth rate (plus or minus one standard error) by treatment, of yellow perch after three, five, and six weeks in the mesocosms. There were four replicates of each treatment except LRP ($n = 3$). HP = 28 perch, HRP = 14 perch + 14 ruffe, LP = 14 perch, and LRP = 7 perch + 7 ruffe.

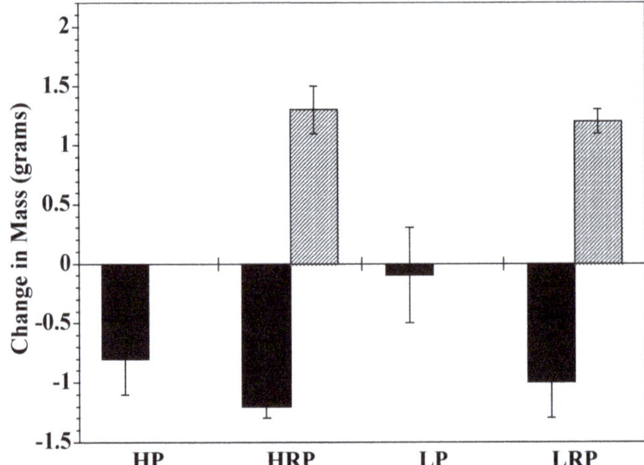

Figure 2. Gross change in mean individual mass (grams) of yellow perch (solid columns) and ruffe (striped columns) by treatment (plus or minus one standard error) during the six-week ruffe and fish density experiment. There are four replicates of each treatment except LRP ($n = 3$). HP = 28 perch, HRP = 14 perch + 14 ruffe, LP = 14 perch, and LRP = 7 perch + 7 ruffe.

As in the first experiment, ruffe growth was generally greater than perch growth in the density gradient experiment (Figures 3 and 4). Growth rate of all fish decreased as the number of ruffe increased (ANOVA, $p = 0.043$) (Figure 3). The two trials (August–September and September–October) were not significantly different (ANOVA, $p = 0.110$) and, because there was no significant interaction between ruffe density and species, the decrease in growth with increasing ruffe density was the same for perch and ruffe. Growth rate of perch and ruffe decreased at $-0.037\% * d^{-1} * \text{ruffe}^{-1}$. This slope was significantly different from zero (t test, $p = 0.006$) and was not significantly different from a linear decrease (ANOVA, $p = 0.552$). The addition of perch to a constant density of ruffe did not have a significant effect on either ruffe or perch growth rate (all $p > 0.1$, Figure 4). There was a significant difference between the mean growth of ruffe and perch (ANOVA, $p < 0.001$); ruffe grew faster than perch, who lost mass.

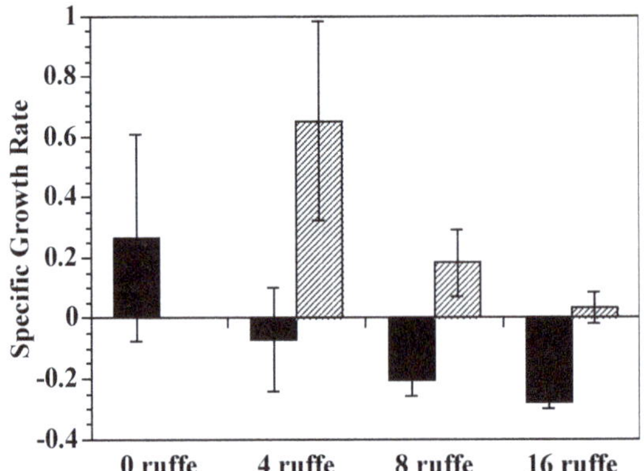

Figure 3. The effect of increasing ruffe density on specific growth rates of ruffe (striped columns) and yellow perch (solid columns) over the two five-week trials of the fish density gradient experiment (August–October 1997). Treatments consist of four replicates each of 0, 4, 8, and 16 ruffe added to 8 yellow perch. Treatment means are shown plus or minus one standard error.

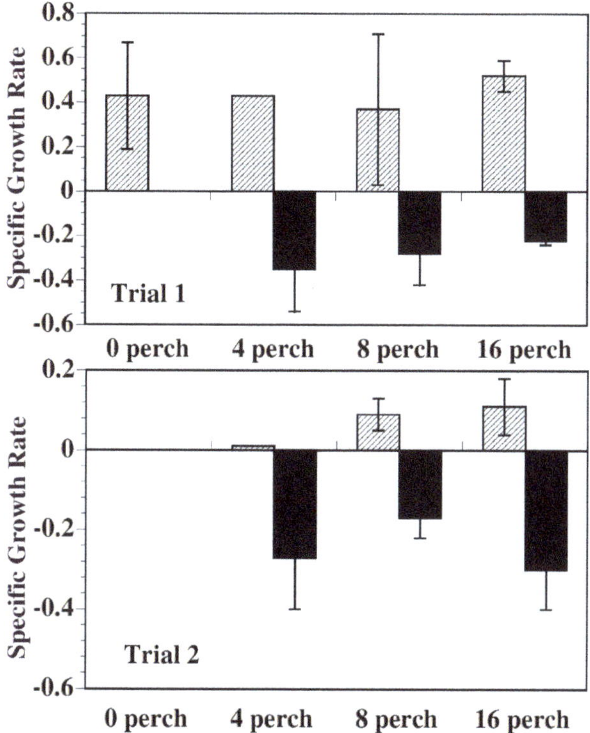

Figure 4. The effect of increasing yellow perch density on specific growth rates of ruffe (striped columns) and yellow perch (solid columns) over the two five-week trials of fish density gradient experiment (August–October 1997). Trials consist of two replicates each of 0, 4, 8, and 16 yellow perch added to 8 ruffe. Trial means are shown plus or minus one standard error. First and second trials are shown separately due to the effect of missing observations (8 ruffe + 0 perch) in second set.

2.2. Diet

In ruffe and fish density experiment, the mean mass of ruffe stomach contents was approximately three times greater than the mean mass of perch stomach contents across treatments. The mean mass of perch stomach contents was not significantly affected by any of the factors that affected perch growth in the repeated measures split-plot ANOVA including presence or absence of ruffe when all dates were considered. However, when diet data from the final (24 October 1996) sample were analyzed separately, mean stomach content mass of perch was significantly reduced in the presence of ruffe (ANOVA, $p = 0.009$; Figure 5). In this analysis, there was also a significant effect of block (ANOVA, $p = 0.024$), but no effect of overall fish density. There was no effect of ruffe on the proportion of microcrustaceans in the perch stomachs; Cladocera and Copepoda composed about 33% of perch diet but only 11% of ruffe diet.

Analysis of the gradient experiment diet data was hindered by low fish recovery from the second trial at the end of the experiment that was due in part to predation by an otter that was first observed in Perch Lake during the final week of the experiment. Therefore, we restricted diet analysis to the first two replicates. There was a significant negative effect of ruffe on the total mass of prey in perch stomachs (t test, $p = 0.014$), but no significant effect of ruffe on ruffe stomach content mass (Figure 6). Perch density did not affect the stomach content mass of either species. As in the first experiment, there was no significant effect of ruffe on the proportion of zooplankton in perch diet. Cladocera and Copepoda composed about 12% of perch diet and <1% of ruffe diet.

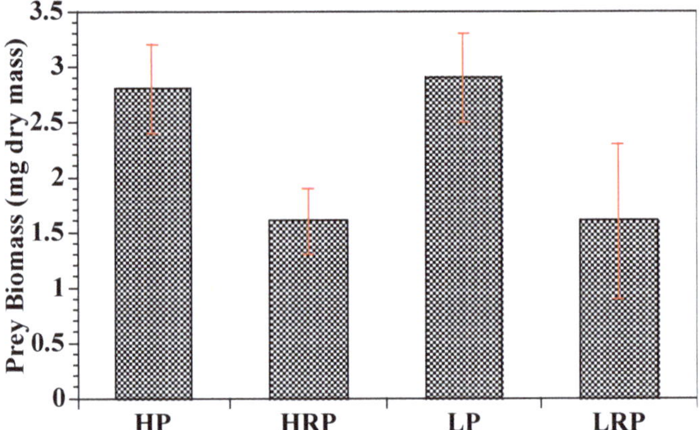

Figure 5. Estimated mass of individual yellow perch stomach contents collected at the conclusion of the ruffe and fish density experiment 1 (24 October 1996). Treatment means are shown plus or minus one standard error. There are four replicates of each treatment except LRP ($n = 3$). HP = 28 perch, HRP = 14 perch + 14 ruffe, LP = 14 perch, and LRP = 7 perch + 7 ruffe.

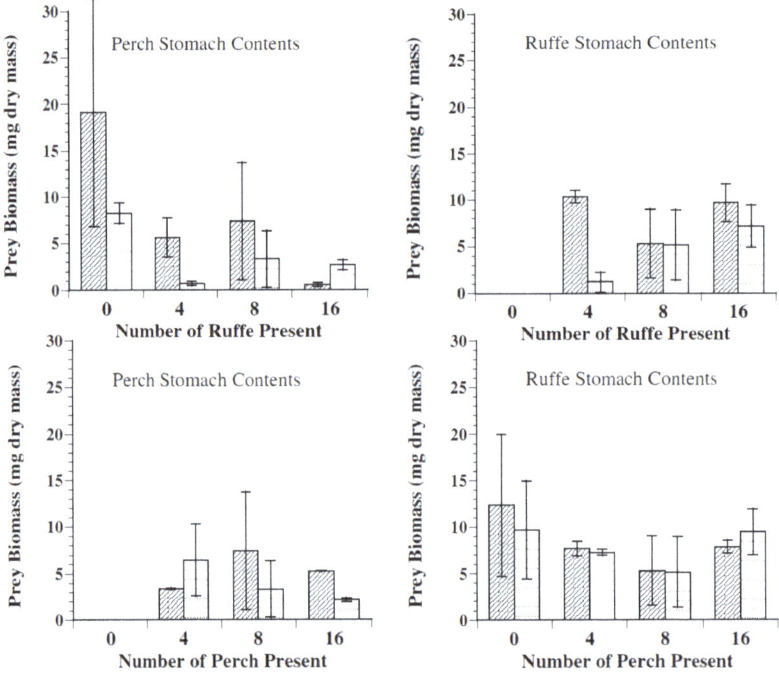

Figure 6. Estimated stomach content mass of yellow perch with increasing ruffe density (**top left**), ruffe with increasing ruffe density (**top right**), ruffe with increasing yellow perch density (**lower right**), and yellow perch with increasing yellow perch density (**lower left**) (experiment 2, 11 August–15 September). Stomach content mass is shown after 2.5 weeks (cross-hatching) and 5 weeks (horizontal stripes). Treatment means are shown plus or minus one standard error. There are two replicates of each treatment. There was a significant effect of ruffe on perch stomach content mass ($p = 0.014$), but no effect of ruffe on ruffe stomach content mass or of perch on either ruffe or perch stomach content mass (all $p > 0.1$).

2.3. *Food Consumption*

Yellow perch daily individual food consumption ranged from 0.14–0.60 g (Tables 1 and 2). Ruffe daily individual food consumption ranged from 0.18–0.93 g. Food consumption by ruffe was generally twice that of perch sharing the same enclosures even though ruffe were generally smaller. In the ruffe and fish density experiment, perch daily ration was reduced in the presence of ruffe (ANOVA, $p = 0.03$) and in the gradient experiment, perch and ruffe daily ration declined with increasing ruffe density (both $p = 0.01$), but no declines in ruffe or perch daily ration were evident with perch density ($p > 0.7$) or total density ($p > 0.08$). Total daily consumption (ruffe and perch combined) increased with both total fish density ($p = 0.01$) and presence of ruffe ($p = 0.04$) in the first experiment. In the gradient experiment, total consumption increased linearly with total fish density and with yellow perch density (ANCOVA, both $p < 0.005$), but not with increasing ruffe density (ANCOVA, $p = 0.14$). The relationship was clearly asymptotic with increasing ruffe density; ln (total consumption) increased significantly with increasing ruffe density (ANCOVA, $p = 0.017$) and the fit was further improved with square root of ruffe density. The fit of the perch density or total density relationship was not improved with a logarithmic transformation of total consumption. Thus, increasing density of perch resulted in a direct increase in consumption, but increasing density of ruffe beyond about 12 ruffe per mesocosm resulted in little increase in total consumption, suggesting either severe intraspecific competition or severe food limitation.

Table 1. Estimates of average daily food consumption by ruffe and yellow perch during the first (1996) experiment, based on bioenergetics modeling. Daily ration (DR) is reported as percent wet mass and daily individual (IC) and total (TC) consumption (means for each treatment) are reported in grams wet mass per day. Standard errors (SE) are based on four replicates (mesocosms) of each treatment, except 7 Perch and 7 Ruffe, where $n = 3$.

Treatment	28 Perch	14 Perch 14 Ruffe	14 Perch	7 Perch 7 Ruffe
Perch DR %	2.09 (0.12)	1.95 (0.14)	2.35 (0.15)	1.69 (0.12)
Perch IC (g)	0.32 (0.02)	0.30 (0.02)	0.39 (0.03)	0.27 (0.02)
Perch TC (g)	9.02 (0.57)	4.21 (0.28)	5.41 (0.37)	1.87 (0.11)
Ruffe DR %		6.05 (0.56)		5.47 (0.36)
Ruffe IC (g)		0.63 (0.05)		0.55 (0.06)
Ruffe TC (g)		8.77 (0.68)		3.88 (0.43)
Ruffe + Perch TC (g)	9.02 (0.57)	12.98 (0.71)	5.41 (0.37)	5.75 (0.51)

Table 2. Estimates of average daily food consumption (mean and 1SE) by ruffe (R) and yellow perch (P) during the second (1997) experiment, based on bioenergetics modeling. Daily ration (DR) is reported as percent wet mass and daily individual (IC) and total (TC) consumption (means for each treatment) are reported in grams wet mass per day. Standard errors (SE) are based on two replicates (mesocosms) of each treatment except for those cases with no SE (-; $n = 1$) and 8R8P in September to October ($n = 4$).

	11 August–15 September						
Treatment	**0R8P**	**4R8P**	**8R8P**	**16R8P**	**8R0P**	**8R4P**	**8R16P**
Ruffe DR %		12.44	7.70	4.33	8.36	8.36	9.28
SE		5.00	3.42	0.87	2.45	0.00	0.71
Ruffe IC (g)		0.93	0.66	0.33	0.64	0.68	0.84
SE		0.41	0.32	0.08	0.18	0.02	0.05
Ruffe TC (g)		3.71	5.29	5.35	5.14	5.47	6.73
SE		1.66	2.53	1.30	1.43	0.18	0.39
Perch DR %	5.23	3.99	3.17	3.13		2.87	3.23
SE	0.98	0.86	0.23	0.16		0.19	0.03
Perch IC (g)	0.60	0.29	0.34	0.33		0.38	0.37
SE	0.04	0.04	0.08	0.07		0.17	0.01
Perch TC (g)	4.77	2.35	2.69	2.62		1.54	5.88
SE	0.31	0.30	0.65	0.58		0.66	0.21
Total C (g)	4.77	6.06	7.98	7.97	5.14	7.00	12.61
SE	0.31	1.95	3.18	1.89	1.43	0.48	0.18

Table 2. *Cont.*

	22 September–28 October						
Treatment	**0R8P**	**4R8P**	**8R8P**	**16R8P**	**8R0P**	**8R4P**	**8R16P**
Ruffe DR %		5.43	3.28	2.67		2.47	3.44
SE		-	0.42	-		0.00	0.66
Ruffe IC (g)		0.54	0.25	0.18		0.19	0.25
SE		-	0.04	-		0.02	0.06
Ruffe TC (g)		2.17	2.03	2.95		1.50	1.97
SE		-	0.32	-		0.16	0.46
Perch DR %	1.59	2.08	1.84	1.61		1.50	1.53
SE	-	-	0.12	0.11		0.33	0.23
Perch IC (g)	0.14	0.18	0.18	0.16		0.18	0.15
SE	-	-	0.01	0.01		0.04	0.02
Perch TC (g)	1.15	1.45	1.46	1.30		0.72	2.45
SE	-	-	0.12	0.04		0.15	0.39
Total C (g)	1.15	3.62	3.49	4.25		2.22	4.42
SE	-	-	0.28	-		0.01	0.85

3. Discussion

3.1. Growth

Competitive interactions are one vehicle for adverse effects of invasive species [31,32], however, the relative importance interspecific completion is unclear, and results of small-scale experiments may not translate to larger scales [33,34]. Relatively few studies have provided experimental evidence of competition between fishes at larger scales. The consistent negative effect of the presence of ruffe on the growth of yellow perch observed in both years of this study provides strong evidence that ruffe compete with yellow perch and that the ruffe is the superior competitor. Whereas the experimental design of Bergman and Greenberg [23] could not identify which species was the superior competitor, our experiments answered that important question. Ruffe was the superior competitor because perch growth in ruffe and the fish density experiment were depressed significantly more by ruffe than by an equivalent biomass of perch and, in the gradient experiment, growth of perch was reduced with increased ruffe density. Mean ruffe mass was less than mean perch mass, so these results may be considered somewhat conservative.

The marginal effect of fish density observed in the first experiment demonstrates intraspecific competition because perch growth was lower at the high-density perch-only treatment than at the low-density perch-only treatment. There was not, however, a similar negative effect of increasing perch density on perch growth in the gradient experiment. One plausible explanation is that because eight ruffe were present in all treatments, intraspecific competition among perch was masked by the stronger effect of interspecific competition with ruffe.

Perch density had no effect on growth of ruffe (Figure 4), suggesting that a dense, established perch population will not deter invading ruffe. However, the results are somewhat less clear concerning intraspecific competition among ruffe. In the gradient experiment, ruffe growth declined at the same rate as perch growth with increasing ruffe density, suggesting intraspecific competition. This observation is in accord with the results of Bergman and Greenberg [23] and Fullerton et al. [19], but the lack of an overall significant difference in ruffe growth between the high and low density mixed species treatments in the first experiment is contradictory. Neither ruffe nor perch growth or ration were significantly different between the high and low density treatments when ruffe were present. The gradient experiment, in combination with the observations of others [19,23], suggests strong intraspecific competition among ruffe; however, our overall results suggest that perch are more consistently affected by competition from ruffe. The degree of intraspecific competition among ruffe is important because, given the lack of a perch effect on ruffe growth, ruffe densities in the wild may ultimately be determined by intraspecific competition rather than by competition with other species. This supports the idea that ruffe should grow best at intermediate densities [20]. There was early evidence that ruffe growth declined as their density increased in the Saint Louis River Harbor [5], but formal analyses of growth related to population density have not been reported for ruffe in North America. Although lengths have been reported along with CPUE (e.g., [35]), aging data, length at age, or growth rates are lacking in population assessments (e.g., [35,36]); these data for ruffe and their competitors would be useful to determine the effects of ruffe in invaded environments.

The negative growth rates of perch in many of the treatments indicates that conditions were not optimal for perch growth. Ruffe and perch were both collected from the wild and the larger perch may have been more stressed, although we held fish for a week before stocking to reduce use of stressed fish. Perch consumed prey throughout the experiments and did grow at low density in the absence of ruffe. The lower water clarity of Perch Lake and the low abundance of larger preferred zooplankton may have also contributed to their lack of positive growth. Finally, the larger size of perch may have resulted in them exceeding the carrying capacity of the mesocosms at total densities above 7 perch (the density at which perch growth was positive). Nonetheless, the consistent reduction of growth with increasing perch and ruffe density suggest that the effects of ruffe and our interpretation of competitive interactions are relevant to field conditions.

3.2. Diet and Consumption

Perch across all treatments were primarily benthivorous, but included substantial proportions of Copepoda and Cladocera in their diet (33% in 1996; 11% in 1997). However, in contrast to the results of similar competition experiments involving yellow perch or European perch [23,29], we found no evidence that perch ate a higher proportion of microcrustaceans in the presence of a superior benthic competitor or in response to increasing density of the superior competitor. Zooplankton densities remained high throughout our experiments and were not influenced by perch or ruffe density (unpublished data). Large zooplankton was scarce and thus zooplankton may not have been a suitable alternative for the larger perch [37]. Indeed, Bergman and Greenberg [23] were able to detect a diet shift in only the first of two trials. They attributed this outcome to a general seasonal decline in food resources during their second trial.

The presence of ruffe or higher densities of ruffe resulted in a reduced stomach content mass for perch. Estimates of food consumption by ruffe and perch calculated from growth and temperature data (Tables 1 and 2) agree with the inferences made from stomach content mass (Figures 5 and 6). Perch daily ration and consumption declined in the presence of ruffe (ruffe and fish density experiment) and declined with increasing density of ruffe (gradient experiment). Several other studies on competition with perch have shown a shift in perch diet associated with competition, but no effect on the total amount of food consumed [23,29]; however, Dieterich et al. [38] found reduced consumption by European perch in the presence of ruffe at low prey density. In our experiments, available benthos was reduced by ruffe (unpublished data) and suitably sized alternative prey (zooplankton) were apparently not available, resulting in lower stomach contents and decreased growth of yellow perch.

Greater food consumption by ruffe could explain the reduced growth of perch if food availability was limiting. Ruffe food consumption was consistently higher than that of perch in the same mesocosms. In the first experiment, total food consumption in treatments containing half perch and half ruffe was 30% greater than total food consumption in the treatments containing perch alone. In the gradient experiment, total food consumption increased asymptotically with ruffe density and in both experiments, total consumption by ruffe was higher than perch in all but one treatment (16 perch and 8 ruffe in the gradient experiment) even when more perch were present. This pattern suggests that ruffe exert a greater demand on the benthic prey resource than perch (see also [21]), but that intraspecific competition among ruffe begins to limit overall demand as ruffe density increases. Both ruffe and perch daily rations decreased with ruffe density, but no changes in ration were found with perch density.

3.3. Competitive Mechanisms and their Implications

There are two common mechanisms by which one population can reduce the food consumption and growth of another [28]. The first mechanism is exploitative (consumptive) competition for food, whereby ruffe eat sufficient numbers of benthic organisms to reduce the quantity or quality of benthic prey available to perch. The second mechanism is interference (encounter) competition, whereby some aspect of ruffe behavior or a behavioral interaction between ruffe and perch decreases the foraging efficiency of perch. For example, if ruffe chase perch while perch are trying to feed, then ruffe decrease the time perch actually spent feeding. In this scenario, not only would the energy intake of perch decrease, but the energetic costs of responding to ruffe aggression may well be large enough to result in a biologically significant increase in energy expenditure. Differences in activity rate can result in substantial differences in growth rate of yellow perch and slower growth may be associated with higher activity rates [39].

The literature is inconclusive on the likelihood that ruffe interfere with perch. Dieterich et al. [38] inferred that interference rather than exploitation was the mechanism of competition between European perch and ruffe. Savino and Kolar [13] described ruffe as more aggressive than yellow perch, based on observations of behaviors such as the initiation of chases and prey stealing. Ruffe aggression was observed towards perch and among themselves [20]. Logically, the energetic cost of agonistic behavior would increase for both ruffe and perch as ruffe density increased. Interference competition would have the same result as resource competition insofar as growth is concerned. Savino and Kostich [20] suggested that ruffe intraspecific competition should be greatest at low and high ruffe densities and growth should be best at intermediate ruffe densities. In the gradient experiment, ruffe growth decreased linearly with increasing ruffe density; however, our densities were much lower than Savino and Kostich [20] used. The turbid waters of Perch Lake precluded any behavioral observations during the experiment.

The observed effects of ruffe on yellow perch growth and diet could be explained entirely by interference competition. However, in 1996, total abundance of benthic macroinvertebrates declined in the presence of ruffe and declines in Oligochaeta and Ceratopoginidae abundance and biomass and size of Chironomidae and Ceratopogonidae were related to presence of ruffe (unpublished data), suggesting consumptive (resource) competition. The decreasing abundance of benthic macroinvertebrates with increasing consumption by fish in the mesocosms reinforces this interpretation.

Bergman and Greenberg [23] did detect the suppression of several benthic taxa by ruffe, but the detection of effects of epibenthic predators on the benthic prey resource has proved elusive in other mesocosm studies and the factors that tend to obscure the link are extensively discussed by those authors (e.g., [30,40]. Our study suggests that consumptive competition partially explains the observed effect of ruffe on perch growth. However, because the growth pattern is consistent even when patterns of prey depletion are more complex, interference competition may be the more important mechanism.

Few studies have been able to definitively separate exploitative from interference competition directly. One approach for future studies would be to add an additional set of treatments whereby the competitor species would be allowed to feed alone for an extended period before being replaced by the other species. The effects of resource suppression by the first species would be apparent in the second species as consumptive competition without interspecific interference. The effects of interference competition could be obtained by subtracting the isolated (consumptive) effect from the competitive effect in sympatry.

The results of our experiments contribute to the understanding of the role of competition in the interactions between an introduced non-indigenous fish and the native fish community by clearly demonstrating competition between ruffe and yellow perch in a natural setting. The extent of harm done to economically important native fishes such as the yellow perch by invading ruffe will depend on a host of modifying environmental influences. For example, Brazner et al. [41] noted that ruffe avoided heavily vegetated littoral areas; ruffe may have less impact in vegetated and more complex systems. Also, Bronte et al. [4] indicated that observed declines in native fish abundance since the introduction of ruffe were more likely the consequence of natural population dynamics rather than an effect of ruffe. However, they did find that in the case of yellow perch, ruffe were partially responsible for fluctuations in year class strength. Given the importance of North America's freshwater fisheries and the results reported here, the threat posed by ruffe and other exotic fishes to our aquatic ecosystems should be taken very seriously.

After their initial introduction to and expansion within the St. Louis River Estuary and Duluth Superior Harbor ruffe remained the most common fish found in assessment bottom trawls from 1990 through 2003 [35]. Since then, invasive round goby (*Neoogobius melanostomus*) have become abundant and alternated with ruffe as the most abundant fish between 2004 and 2011. Although ruffe are no longer the most abundant fish species in the St. Louis River, Duluth-Superior Harbor system, they remain quite abundant and have continued to increase in Chequamegon Bay [42]. Gunderson et al. [43] downplayed the impact of ruffe on native communities and Bronte et al. [4] found effects limited to yellow perch year class strength, however, the analysis of longer term (>20 yr) catch records in the St. Louis River Estuary and Duluth Superior Harbor indicates that native fish declined in the presence of ruffe after 1989 and did not begin to rebound until 20 years later [35]. Invasive round gobies are more likely to have negative impacts in hard bottom systems, whereas ruffe may be more impactful in soft bottom benthic communities [22,35] and low light habitats [11,17]. Efforts to control ruffe by enhancing native predators were largely ineffective because native predators prefer to consume native fish rather than ruffe [5], even when ruffe have been present for a number of years and are the most abundant prey [6,44]. Local reductions of ruffe may be accomplished with bottom trawling removal but physical removal is not an effective tool to control ruffe [45]; currently, there are no effective tools to selectively control ruffe.

Although ruffe have remained contained to the Great Lakes and have not spread beyond Lakes Superior, Michigan and Huron, there is considerable concern for their spread to Lake Erie and the Mississippi River Basin [7,25] and inland lakes and river systems [46] across the upper Midwest and northeastern North America [47]. Their expansion and success in inland lakes in Europe [6,48] indicates the potential for success in a variety of inland lakes. Stepien et al. [3] suggest that low genetic diversity might be limiting the success and expansion of ruffe beyond their current range in North America, however, it is likely that education and possession and transport restrictions have restricted the introduction of ruffe to inland waters. Once introduced to interconnected inland waters ruffe could become established in a number of systems and be of particular concern in lower light productive systems. The recent declines in water clarity in Lake Erie (e.g., [49]) could favor ruffe over yellow perch, particularly if macrophytes decline [16].

4. Materials and Methods

4.1. General Procedures

We conducted two sets of mesocosm experiments, the first in 1996 to assess the effect of ruffe (with and without) and total fish density (high and low) on yellow perch growth and diet and the second in 1997 to examine the effects of an increasing density of ruffe on perch and the effects of an increasing density of perch on ruffe. The mesocosm experiments were conducted in Perch Lake, a shallow backwater of the St. Louis River approximately 20-km upstream from Duluth Harbor (46°45′ N, 92°06′ W). Sixteen enclosures, arranged in four blocks of four units each, were established in the lake to house the experimental treatments. Enclosures were open-ended cylinders, approximately 1.7 m in height and 4 m in diameter, made of 12 mil polyethylene. The bottom margin of each cylinder, a weighted semi-rigid collar, was forced into the soft-muck sediments by SCUBA divers. Approximately 12.6 m^2 of benthic surface was enclosed. The cylinder walls were secured at the top to a ring of PVC pipe that was attached to a floating wooden platform. The platforms held the open ends of the mesocosms above the waterline and served as working platforms from which to access the mesocosms. Monofilament netting was suspended across the tops of the mesocosms when unattended to prevent avian predation. Prior to the experiments, mesocosms were electrofished to remove fish trapped during installation. The dominant open sediment benthic macroinvertebrates were Oligochaeta, Chironomidae, Ceratopogonidae, and *Chaoborus*. The zooplankton community included *Bosmina*, *Chydorus*, *Daphnia*, *Ceriodaphnia*, and Calanoida and Cyclopoida Copepoda.

Ruffe were collected from the St. Louis River Duluth-Superior Harbor by bottom trawl. Depth, tow duration, and location varied between collections. Since sufficient numbers of suitable size perch could not be obtained from the St. Louis River, perch for the experiment were beach seined from Oak Lake near Duquette, MN. Perch and ruffe were selected to be as similar in length and mass as possible, but perch were about 5g heavier than ruffe in 1996 (15 g vs. 10 g) and 3g heavier in 1997 (11 g vs. 8 g). The average total length ± 95% confidence interval of ruffe used in the 1996 experiment was 99 ± 1.0 mm; perch were 116 ± 0.4 mm. In 1997, ruffe mean total length was 91 ± 0.7 mm and perch mean total length was 101 ± 1.5 mm. Perch and ruffe were held in 1.9 m^3 flow-through tanks, positioned in Perch Lake, for up to one week following capture to allow for mortality associated with capture and handling stress. Individuals of both species were anaesthetized with a 0.5 mL/L solution of 2-phenoxyethanol, measured (total length ± 1 mm), and individually marked with passive integrated transponder (PIT) tags. After marking, fish were held for up to one week in the flow-through tanks before stocking into the mesocosms. Fish were then anaesthetized, identified, weighed to the nearest 0.1 g, and assigned to treatments (mesocosms) in a stratified random manner. Mean fish mass at the beginning and end of each experiment are given in Tables S1–S4.

Perch and ruffe were recovered from the mesocosms at the midpoint and end of each experiment by angling. Previous experience with electrofishing and purse seines indicated that angling was the most effective means of retrieving fish. Furthermore, angling minimized disturbance to the substrate and other biota when sampling at the midpoint of an experiment. Electrofishing was conducted after angling at the end of each experiment to retrieve any remaining fish.

Captured fish were immediately anaesthetized in individual plastic tubs with a 120 mg/L solution of MS-222. Regurgitated prey were rarely seen in the MS-222 bath, but when found they were preserved along with other stomach contents. Since only a few seconds elapsed between hooking a fish and placing it into the anaesthetic bath, there was little opportunity for regurgitation during angling. Since we could not be so confident about stomach contents of fish captured by electrofishing, we excluded these fish from the diet analysis. Anaesthetized fish were identified, weighed, and measured. Fish captured during mid-experiment sampling had their stomachs flushed with water to remove contents before being returned to their mesocosms. Stomach contents were preserved in 80% ethanol.

At the conclusion of the experiment in 1996, fish were immediately euthanized with MS-222, sealed in plastic bags, and embedded in ice. Fish were later identified, weighed, and measured before being frozen. Frozen fish later were thawed and dissected in the laboratory, at which time stomachs were removed and PIT tags were recovered. In 1997, fish were weighed, measured, and dissected in the field as they were captured. Their stomachs and PIT tags were immediately placed in 80% ethanol. This change was made because an experiment replicating the 1996 handling procedures showed that a mass loss of approximately 3% of live mass occurred during temporary storage on ice. Masses reported for 1996 were adjusted to reflect live mass.

Stomach contents were examined in the laboratory with a dissecting microscope at a magnification of 10×. Prey items were identified based on [50] to a 9 taxonomic groups (Ephemeroptera, Trichoptera, *Chaoborus*, Ceratopogonidae, Chironomidae, Amphipoda, Ostracoda, Cladocera and Copepoda) chosen for feasibility of identification and because body-part, length-biomass (dry mass of intact specimens) relations were available in the literature [51–53]. Ingested bait was not counted as prey. Occasionally, prey items were found in fish stomachs, such as winged adult insects, that could not be readily classified into any of these categories. If, however, the unusual prey item or items were large enough to obtain a dry weight, then the mass was included with the other estimates. The total mass of stomach contents of each fish was calculated, and the proportion, by mass, of the diet composed of microcrustaceans was determined. Ostracoda, Copepoda, and Cladocera were considered microcrustaceans [29]. Although

Copepoda and Cladocera were not systematically identified to greater taxonomic detail, some of these organisms were probably benthic rather than planktonic. The diet composition of perch and ruffe are given in Tables S5 and S6.

4.2. Ruffe and Fish Density Experiment Design and Analysis

This experiment, conducted in 1996, used a factorial design (high and low density, with or without ruffe) to examine the effects of ruffe on perch growth and diet at two overall fish densities. Four treatments in each of four blocked replicates were established as follows: 28 perch, 14 perch + 14 ruffe, 14 perch, and 7 perch + 7 ruffe. To account for potential spatial variation in Perch Lake, mesocosms were arranged in blocks; each block contained a replicate of each treatment. Fish were added to the mesocosms on 12 September and the experiment was concluded after 6 weeks on the 24th of October. The mean biomass stocked into each treatment is given in Table 3. These densities are similar to those used in other ruffe and perch experiments (e.g., [23,40]) and were representative of localized densities that were seen in the St. Louis River Duluth Superior Harbor.

Table 3. Mean total biomass ± 95% C.I. of ruffe (R) and yellow perch (P) in treatments of the 1996 ruffe and fish density experiment. There were four replicates of each treatment.

Treatments	High Perch Alone	High Perch and Ruffe	Low Perch Alone	Low Perch and Ruffe
Species Numbers	0R 28P	14R 14P	0R 14P	7R 7P
Ruffe (g)		152 ± 15		74 ± 7
Perch (g)	435 ± 12	226 ± 30	229 ± 25	116 ± 10

At 3 and 5 weeks after stocking, samples of ruffe and perch were captured, weighed, and stomach contents preserved. Fish not returned at week 3 were replaced, but fish removed at week 5 were not. At the end of the experiment all fish captured were identified, weighed, and stomach contents preserved. Mean individual fish mass and total length at the beginning and end of the experiment are given in Tables S1 and S2 [54].

A repeated measures blocked split-plot ANCOVA [55] was used to test the effects of block, overall fish density (whole plot), presence or absence of ruffe (whole plot), time (sub-plot), and overall fish biomass (covariate), on the growth response of perch. Data from all three sampling periods were included. One replicate of perch and ruffe at low density was excluded from the analysis because of poor fish recovery and extremely high growth of the few fish recovered.

Instantaneous growth rate was used as the growth response:

$$IG = \ln(Wt_2/Wt_1) \quad (1)$$

where IG = instantaneous growth rate, Wt_2 = final fish mass, and Wt_1 = initial fish mass.

Because there was little variation in the initial mass of the perch and all fish were in the experiment for the same amount of time, gross change in mass ($Wt_2 - Wt_1$) during the experiment was also analyzed as an alternative response. Replacement fish were not included in the analysis of growth.

Two measures of diet response were analyzed with a blocked split-plot ANCOVA: mean stomach content mass and the proportion (arcsin transformed), by mass, of microcrustaceans in the diet. Replacement fish were included in the analysis of diet response. Diet response was also analyzed for the final (24 October) sample only using a factorial ANCOVA that included block, fish density, biomass (covariate), and the presence or absence of ruffe.

4.3. Ruffe and Perch Density Gradient Experiment Design and Analysis

The intent of the 1997 experiments was to compare the effect of ruffe density on yellow perch growth and diet with the effect of perch density on ruffe growth and diet. To that end, two series of treatments were devised by superimposing an increasing density of one species (0, 4, 8, and 16 fish) on a constant density of the other (8 fish). There were two replicates of each treatment in each trial for a total of 14 experimental units (mesocosms). The experiment was executed twice, resulting in four replicates of each treatment. Although the mesocosms were arranged in groups of four, treatments were assigned randomly to the mesocosms. A block size of 7 mesocosms was not feasible given the practical limitations of construction and wind conditions on Perch Lake.

The duration of each trial was 5 weeks. At 2.5 weeks, samples of ruffe and perch were captured (by angling), identified, weighed, and stomach contents preserved. Fish not returned were replaced. The first trial commenced on the 11th of August and concluded on the 15th of September. The second trial commenced on the 22nd of September and concluded on the 28th of October. Mean biomass stocked into each treatment is given in Table 4. Mean fish mass at the beginning and end of each trial are given in Tables S3 and S4 [54].

Table 4. Mean total biomass ± 95% C.I. of ruffe (R) and yellow perch (P) in treatments of the 1997 fish density gradient experiment. There were four replicates of each treatment except for the 0R + 8P and 4R + 8P treatments for which there were three replicates and 8R + 8P (6 replicates).

		Increasing Ruffe		
Species Numbers	**0R** **8P**	**4R** **8P**	**8R** **8P**	**16R** **8P**
Ruffe (g)		29 ± 18	60 ± 5	118 ± 18
Perch (g)	82 ± 50	64 ± 32	85 ± 25	87 ± 45
		Increasing Perch		
Species Numbers	**0P** **8R**	**4P** **8R**	**8P** **8R**	**16P** **8R**
Perch (g)		53 ± 36	85 ± 25	179 ± 28
Ruffe (g)	58 ± 11	61 ± 11	60 ± 5	60 ± 12

The effect of adding ruffe to a constant density of perch on the growth response of both species was analyzed with a factorial ANOVA. The analysis included the effects of trial, species (ruffe or yellow perch) and number of ruffe present. A linear contrast was used to test the effect of the increasing density of ruffe.

Specific daily growth rate was used as the response in this experiment:

$$G = [\ln(Wt_2/Wt_1) * (t_2 - t_1) - 1] \times 100 \quad (2)$$

where G = specific daily growth rate, Wt_2 = final fish mass, Wt_1 = initial fish mass, and $t_2 - t_1$ = interval (days) between measurements [56].

The specific daily growth rate was used because some mortality occurred early in the experiment necessitating the addition of replacement fish. Due to the smaller numbers of fish used in these experiments relative to the 1996 experiment, these replacements together with fish added at mid-experiment to replace angling mortalities constituted a substantial portion of the available sample and thus could not be excluded as in 1996.

The effect on fish growth response of adding perch to constant ruffe density was analyzed separately for each trial due to missing observations in the second trial. Three observations had to be discarded due to a stocking error in the second trial. A factorial ANOVA was used to analyze diet response (mean stomach content mass per treatment and arcsin transformed proportion, by mass, of microcrustaceans in the diet) of both species to increasing ruffe.

4.4. Bioenergetics Modeling

Food consumption in each mesocosm was estimated from water temperature (recorded continuously in 1996 and weekly in 1997), prey type, and growth of ruffe and perch during the experiments. Mean initial and final mass of ruffe and perch in each mesocosm were used to parameterize the models and estimate mean individual consumption. Since final mean mass was influenced by replacement fish in 1997, final mean individual mass was calculated from specific growth rate. Mean water temperature in 1996 was 13.1 °C. Mean water temperature in 1997 was 19.1 °C during the first trial and 11.8 °C during the second trial.

Food consumption of ruffe was estimated using a relationship between food consumption, temperature (T), and specific daily growth (G) derived from the laboratory results reported in Henson and Newman [21] because formal bioenergetics models are not available for ruffe (e.g., [57]). Daily ration (DR; % of live mass) was estimated via multiple regression as:

$$DR = 10.24\,G + 0.22\,T - 0.225\ (r^2 = 0.96, p < 0.001) \quad (3)$$

The ruffe in the mesocosms and the ruffe used in Henson and Newman [21] were of similar size (9 to 10 g) and consumed exclusively invertebrate prey. Mean ruffe mass was 10.8 g in 1996 and 7.4 g in 1997.

For yellow perch, food consumption was estimated using Fish Bioenergetics 3.0 [57]. This software uses bioenergetic relationships from Kitchell and Stewart [58] to predict yellow perch consumption given growth and temperature. The model provides good agreement with field estimates for age 1–3 yellow perch at water temperatures < 22 °C [59], conditions that were met by our data. Mean initial and final masses from each mesocosm were used, along with the temperature data and the fit p-value to estimate consumption. Prey energy density was entered as 2500 joules/g wet mass to represent the mixed invertebrate diet of the perch. Mean perch mass was 16.0 g in 1996 and 10.9 g in 1997. Mean (and SE) consumption for ruffe and perch was averaged over the replicate mesocosms in each treatment. The intent of the 1997 experiments was to compare the effect of ruffe density on yellow perch.

Supplementary Materials: The following are available online at http://www.mdpi.com/2410-3888/5/4/33/s1, Table S1: Initial and final mean individual live masses and mean individual total lengths of perch in the 1996 mesocosm experiment, Table S2: Initial and final mean individual live masses and mean individual total lengths of ruffe in the 1996 mesocosm experiment, Table S3: Initial and final mean individual live masses and mean individual total lengths of perch in the 1997 mesocosm experiment, Table S4: Initial and final mean individual live masses and mean individual total lengths of ruffe in the 1997 mesocosm experiment, Table S5: Diet composition of ruffe and perch at the conclusion of the 1996 mesocosm experiment, Table S6: Diet composition of ruffe and perch at the conclusion of the first trial of the 1997 mesocosm experiment.

Author Contributions: Conceptualization, R.M.N. and C.R.; methodology, F.G.H. and R.M.N.; formal analysis, F.G.H.; writing—original draft preparation, F.G.H. and R.M.N.; writing—review and editing, F.G.H. and R.M.N. and C.R.; visualization, F.G.H.; supervision, R.M.N.; project administration, R.M.N. and C.R.; funding acquisition, R.M.N. and C.R. All authors have read and agreed to the published version of the manuscript.

Funding: This work was prepared by the authors using federal funds under award USDOC-NA46RG0101 (C. Richards, PI) from Minnesota Sea Grant, National Sea Grant College Program, National Oceanic and Atmospheric Administration, U.S. Department of Commerce. The statements, findings, conclusions, and recommendations are those of the authors and do not necessarily reflect the views of NOAA, the Sea Grant College Program or the U.S. Department of Commerce. Additional support was provided by the Minnesota Agricultural Experiment Station under Project 74 and USDA National Institute of Food and Agriculture, Hatch grant MIN-41-081.

Acknowledgments: This manuscript was originally a chapter of Fred Henson's MS thesis that languished as we all moved on to other endeavors. Jeff Schuldt was instrumental in coordinating and overseeing these experiments. We thank Jeremy Trexel, Jim Gangl, Jeff Schuldt and the NRRI technical staff for their steadfast intellectual and physical labor which made this complex field experiment function. We are also grateful for the occasional assistance of Eric Merten, Geoff Schrag, Vanessa Pepi, and Jessica Gurley. Many thanks to Sanford Weisberg for statistical advice. The U.S. Government is authorized to reproduce and distribute reprints for government purposes, not withstanding any copyright notation that may appear hereon.

Conflicts of Interest: The authors declare no conflict of interest. The funders had no role in the design of the study; in the collection, analyses, or interpretation of data; in the writing of the manuscript, or in the decision to publish the results.

References

1. Pratt, D.M.; Blust, W.H.; Selgeby, J.H. Ruffe, *Gymnocephalus cernua*: Newly introduced in North America. *Can. J. Fish. Aquat. Sci.* **1992**, *49*, 1616–1618. [CrossRef]
2. Stepien, C.A.; Brown, J.E.; Neilson, M.E.; Tumeo, M.A. Genetic diversity of invasive species in the great lakes versus their eurasian source populations: Insights for risk analysis. *Risk Anal.* **2005**, *25*, 1043–1060. [CrossRef] [PubMed]
3. Stepien, C.A. Genetic change versus stasis over the time course of invasions: Trajectories of two concurrent, allopatric introductions of the Eurasian ruffe. *Aquat. Invasions* **2018**, *13*, 537–552. [CrossRef]
4. Bronte, C.R.; Evrard, L.M.; Brown, W.P.; Mayo, K.R.; Edwards, A.J. Fish community changes in the St. Louis river estuary, lake superior, 1989–1996: Is it ruffe or population dynamics? *J. Great Lakes Res.* **1998**, *24*, 309–318. [CrossRef]
5. Ogle, D.H.; Selgeby, J.H.; Saving, J.E.; Newman, R.; Henry, M.G. Predation on ruffe by native fishes of the St. Louis river estuary, Lake Superior, 1989–1991. *N. Am. J. Fish. Manag.* **1996**, *16*, 115–123. [CrossRef]
6. Gutsch, M.; Hoffman, J.C. A review of Ruffe (*Gymnocephalus cernua*) life history in its native versus non-native range. *Rev. Fish. Biol. Fish.* **2016**, *26*, 213–233. [CrossRef]
7. Tucker, A.J.; Chadderton, W.L.; Jerde, C.L.; Renshaw, M.A.; Uy, K.; Gantz, C.; Mahon, A.R.; Bowen, A.; Strakosh, T.; Bossenbroek, J.M.; et al. A sensitive environmental DNA (eDNA) assay leads to new insights on Ruffe (*Gymnocephalus cernua*) spread in North America. *Biol. Invasions* **2016**, *18*, 3205–3222. [CrossRef]
8. Hölker, F.; Thiel, R. Biology of Ruffe (*Gymnocephalus cernuus* (L.))—A review of selected aspects from European literature. *J. Great Lakes Res.* **1998**, *24*, 186–204. [CrossRef]
9. Bergman, E. Effects of roach *Rutilus rutilus* on Two Percids, *Perca fluviatilis* and *Gymnocephalus cernua*: Importance of species interactions for diet shifts. *Oikos* **1990**, *57*, 241. [CrossRef]
10. Jamet, J.-L. Feeding activity of adult roach (*Rutilus rutilus* (L.)), perch (*Perca fluviatilis* L.) and ruffe (*Gymnocephalus cernuus* (L.)) in eutrophic Lake Aydat (France). *Aquat. Sci.* **1994**, *56*, 376–387. [CrossRef]
11. Ogle, D.H.; Selgeby, J.H.; Newman, R.; Henry, M.G. Diet and feeding periodicity of Ruffe in the St. Louis River estuary, Lake Superior. *Trans. Am. Fish. Soc.* **1995**, *124*, 356–369. [CrossRef]
12. Ogle, D.H. A synopsis of the biology and life history of Ruffe. *J. Great Lakes Res.* **1998**, *24*, 170–185. [CrossRef]
13. Savino, J.F.; Kolar, C.S. Competition between Nonindigenous Ruffe and Native Yellow Perch in Laboratory Studies. *Trans. Am. Fish. Soc.* **1996**, *125*, 562–571. [CrossRef]
14. Sierszen, M.E.; Keough, J.R.; Hagley, C.A. Trophic analysis of Ruffe (*Gymnocephalus cernuus*) and White Perch (*Morone americana*) in a Lake Superior coastal food web, using stable isotope techniques. *J. Great Lakes Res.* **1996**, *22*, 436–443. [CrossRef]
15. Fullerton, A.H.; Lamberti, G.A.; Lodge, D.M.; Berg, M.B. Prey Preferences of Eurasian Ruffe and Yellow Perch: Comparison of laboratory results with composition of Great Lakes Benthos. *J. Great Lakes Res.* **1998**, *24*, 319–328. [CrossRef]
16. Fullerton, A.H.; Lamberti, G.A. A comparison of habitat use and habitat-specific feeding efficiency by Eurasian ruffe (*Gymnocephalus cernuus*) and yellow perch (*Perca flavescens*). *Ecol. Freshw. Fish.* **2006**, *15*, 1–9. [CrossRef]
17. Schleuter, D.; Eckmann, R. Competition between perch (*Perca fluviatilis*) and ruffe (*Gymnocephalus cernuus*): The advantage of turning night into day. *Freshw. Biol.* **2006**, *51*, 287–297. [CrossRef]
18. Janssen, J. Comparison of response distance to prey via the lateral line in the Ruffe and Yellow Perch. *J. Fish. Biol.* **1997**, *51*, 921–930. [CrossRef]
19. Fullerton, A.H.; Lamberti, G.A.; Lodge, D.M.; Goetz, F.W. Potential for resource competition between Eurasian Ruffe and Yellow Perch: Growth and RNA responses in laboratory experiments. *Trans. Am. Fish. Soc.* **2000**, *129*, 1331–1339. [CrossRef]
20. Savino, J.F.; Kostich, M.J. Aggressive and foraging behavioral interactions among Ruffe. *Environ. Boil. Fishes* **2000**, *57*, 337–345. [CrossRef]

21. Henson, F.G.; Newman, R.M. Effect of temperature on growth at ration and gastric evacuation rate of ruffe (*Gymnocephalus cernuus L.*). *Trans. Am. Fish. Soc.* **2000**, *129*, 552–560. [CrossRef]
22. Bauer, C.R.; Bobeldyk, A.M.; Lamberti, G.A. Predicting habitat use and trophic interactions of Eurasian ruffe, round gobies, and zebra mussels in nearshore areas of the Great Lakes. *Biol. Invasions* **2006**, *9*, 667–678. [CrossRef]
23. Bergman, E.; Greenberg, L.A. Competition between a Planktivore, a Benthivore, and a species with ontogenetic diet shifts. *Ecology* **1994**, *75*, 1233–1245. [CrossRef]
24. Leigh, P. Benefits and costs of the Ruffe control program for the Great Lakes fishery. *J. Great Lakes Res.* **1998**, *24*, 351–360. [CrossRef]
25. US Fish and Wildife Service (USFWS). *Ruffe (Gymnocephalus cernua) Ecological Risk Screening Summary*; US Fish and Wildlife Service, 2015. Available online: https://www.fws.gov/fisheries/ANS/erss/highrisk/Gymnocephalus-cernua-ERSS-revision-June%202015.pdf (accessed on 9 August 2020).
26. Grippo, M.; Hlohowskyj, I.; Fox, L.; Herman, B.; Pothoff, J.; Yoe, C.; Hayse, J. Aquatic nuisance species in the Great Lakes and Mississippi River Basin-A risk assessment in support of GLMRIS. *Environ. Manag.* **2016**, *59*, 154–173. [CrossRef] [PubMed]
27. Zhang, H.; Rutherford, E.S.; Mason, D.M.; Wittmann, M.E.; Lodge, D.M.; Zhu, X.; Johnson, T.B.; Tucker, A. Modeling potential impacts of three benthic invasive species on the Lake Erie food web. *Biol. Invasions* **2019**, *21*, 1697–1719. [CrossRef]
28. Schoener, T.W. Field Experiments on interspecific competition. *Am. Nat.* **1983**, *122*, 240–285. [CrossRef]
29. Hanson, J.M.; Leggett, W.C. Experimental and field evidence for inter-and intraspecific competition in two freshwater fishes. *Can. J. Fish. Aquat. Sci.* **1985**, *42*, 280–286. [CrossRef]
30. Hanson, J.M.; Leggett, W.C. Effect of competition between two freshwater fishes on prey consumption and abundance. *Can. J. Fish. Aquat. Sci.* **1986**, *43*, 1363–1372. [CrossRef]
31. Gozlan, R.E.; Britton, J.R.; Cowx, I.; Copp, G.H. Current knowledge on non-native freshwater fish introductions. *J. Fish. Biol.* **2010**, *76*, 751–786. [CrossRef]
32. Ricciardi, A.; Hoopes, M.F.; Marchetti, M.P.; Lockwood, J.L. Progress toward understanding the ecological impacts of nonnative species. *Ecol. Monogr.* **2013**, *83*, 263–282. [CrossRef]
33. Britton, J.R.; Ruiz-Navarro, A.; Verreycken, H.; Amat-Trigo, F. Trophic consequences of introduced species: Comparative impacts of increased interspecific versus intraspecific competitive interactions. *Funct. Ecol.* **2017**, *32*, 486–495. [CrossRef] [PubMed]
34. Britton, J.R. Empirical predictions of the trophic consequences of non-native freshwater fishes: A synthesis of approaches and invasion impacts. *Turkish J. Fish. Aquat. Sci.* **2019**, *19*, 529–539. [CrossRef]
35. Leino, J.R.; Mensinger, A.F. The benthic fish assemblage of the soft-bottom community of the Duluth-Superior Harbor before and after round goby invasion (1989–2011). *J. Great Lakes Res.* **2016**, *42*, 829–836. [CrossRef]
36. Peterson, G.S.; Hoffman, J.C.; Trebitz, A.S.; West, C.W.; Kelly, J.R. Establishment patterns of non-native fishes: Lessons from the Duluth-Superior harbor and lower St. Louis River, an invasion-prone Great Lakes coastal ecosystem. *J. Great Lakes Res.* **2011**, *37*, 349–358. [CrossRef]
37. Mayer, C.M.; VanDeValk, A.J.; Forney, J.L.; Rudstam, L.G.; Mills, E. Response of Yellow Perch (*Perca flavescens*) in Oneida Lake, New York, to the establishment of Zebra Mussels (*Dreissena polymorpha*). *Can. J. Fish. Aquat. Sci.* **2000**, *57*, 742–754. [CrossRef]
38. Dieterich, A.; Baumgartner, D.; Eckmann, R. Competition for food between Eurasian perch (*Perca fluviatilis* L.) and Ruffe (*Gymnocephalus cernuus* (L.)) over different substrate types. *Ecol. Freshw. Fish.* **2004**, *13*, 236–244. [CrossRef]
39. AubinHorth, N.; Gingras, J.; Boisclair, D. Comparison of activity rates of 1 + yellow perch (*Perca flavescens*) from populations of contrasting growth rates using underwater video observations. *Can. J. Fish. Aquat. Sci.* **1999**, *56*, 1122–1132. [CrossRef]
40. Cobb, S.E.; Watzin, M.C. Trophic interactions between Yellow Perch (*Perca flavescens*) and their benthic prey in a littoral zone community. *Can. J. Fish. Aquat. Sci.* **1998**, *55*, 28–36. [CrossRef]
41. Brazner, J.C.; Tanner, D.K.; Jensen, D.A.; Lemke, A. Relative abundance and distribution of Ruffe (*Gymnocephalus cernuus*) in a Lake Superior Coastal Wetland Fish Assemblage. *J. Great Lakes Res.* **1998**, *24*, 293–303. [CrossRef]
42. Gutsch, M. The Rise and Fall of the Ruffe (Gymnocephalus cernua) Empire in Lake Superior. Ph.D. Dissertation, University of Minnesota, Duluth, MN, USA, 2017; 207p.

43. Gunderson, J.L.; Klepinger, M.R.; Bronte, C.R.; Marsden, J.E. Overview of the International Symposium on Eurasian Ruffe (*Gymnocephalus cernuus*) biology, impacts, and control. *J. Great Lakes Res.* **1998**, *24*, 165–169. [CrossRef]
44. Mayo, K.R.; Selgeby, J.H.; McDonald, M.E. A bioenergetics modeling evaluation of top-down control of Ruffe in the St. Louis River, Western Lake Superior. *J. Great Lakes Res.* **1998**, *24*, 329–342. [CrossRef]
45. Czypinski, G.D.; Ogle, D.H. Evaluating the physical removal of Ruffe (*Gymnocephalus cernuus*) with bottom trawling. *J. Freshw. Ecol.* **2011**, *26*, 441–443. [CrossRef]
46. Drake, J.M. Risk analysis for species introductions: Forecasting population growth of Eurasian Ruffe (*Gymnocephalus cernuus*). *Can. J. Fish. Aquat. Sci.* **2005**, *62*, 1053–1059. [CrossRef]
47. Drake, J.M.; Lodge, D.M. Forecasting potential distributions of nonindigenous species with a genetic algorithm. *Fisheries* **2006**, *31*, 9–16. [CrossRef]
48. Volta, P.; Jeppesen, E.; Campi, B.; Sala, P.; Emmrich, M.; Winfield, I. The population biology and life history traits of Eurasian Ruffe (*Gymnocephalus cernuus* (L.), Pisces: *Percidae*) introduced into eutrophic and oligotrophic lakes in Northern Italy. *J. Limnol.* **2013**, *72*, 22. [CrossRef]
49. Watson, S.B.; Miller, C.; Arhonditsis, G.; Boyer, G.L.; Carmichael, W.; Charlton, M.N.; Confesor, R.; DePew, D.C.; Höök, T.O.; Ludsin, S.A.; et al. The re-eutrophication of Lake Erie: Harmful algal blooms and hypoxia. *Harmful Algae* **2016**, *56*, 44–66. [CrossRef] [PubMed]
50. Sandridge, P.T.; Thorp, J.H.; Covich, A.P. Ecology and classification of North American freshwater invertebrates. *J. N. Am. Benthol. Soc.* **1991**, *10*, 466–467. [CrossRef]
51. Smock, L.A. Relationships between body size and biomass of aquatic insects. *Freshw. Biol.* **1980**, *10*, 375–383. [CrossRef]
52. Rosen, R.A. Length-dry weight relationships of some freshwater zooplankton a. *J. Freshw. Ecol.* **1981**, *1*, 225–229. [CrossRef]
53. Litvak, M.K.; Hansell, R.I.C. Investigation of food habit and niche relationships in a cyprinid community. *Can. J. Zool.* **1990**, *68*, 1873–1879. [CrossRef]
54. Henson, F.G. Competition between Ruffe (*Gymnocephalus cernuus*) and Yellow Perch (*Perca flavescens*) and the Influence of Temperature on Growth and Gastric Evacuation of Ruffe. Master's Thesis, University of Minnesota, St. Paul, MN, USA, 1999.
55. Maceina, M.J.; Bettoli, P.W.; Devries, D.R. Use of a split-plot analysis of variance design for repeated-measures fishery data. *Fisheries* **1994**, *19*, 14–20. [CrossRef]
56. Busacker, G.P.; Adelman, I.R.; Goolish, E.M. Growth. In *Methods for Fish Biology*; Schreck, C.B., Moyle, P.B., Eds.; American Fisheries Society: Bethesda, MD, USA, 1990; pp. 363–387.
57. Hanson, P.C.; Johnson, T.B.; Schindler, D.E.; Kitchell, J.F. *Fish. Bioenergetics 3.0*; University of Wisconsin Center for Limnology and University of Wisconsin Sea Grant Institute: Madison, WI, USA, 1997.
58. Kitchell, J.F.; Stewart, D.J.; Weininger, D. Applications of a bioenergetics model to Yellow Perch (*Perca flavescens*) and Walleye (*Stizostedion vitreum vitreum*). *J. Fish. Res. Board Can.* **1977**, *34*, 1922–1935. [CrossRef]
59. Schaeffer, J.S.; Haas, R.C.; Diana, J.S.; Breck, J.E. Field test of two energetic models for Yellow Perch. *Trans. Am. Fish. Soc.* **1999**, *128*, 414–435. [CrossRef]

Publisher's Note: MDPI stays neutral with regard to jurisdictional claims in published maps and institutional affiliations.

Article

Use of Environmental DNA to Detect Grass Carp Spawning Events

Cari-Ann Hayer [1], Michael F. Bayless [2], Amy George [3], Nathan Thompson [3], Catherine A. Richter [3,*] and Duane C. Chapman [3]

1 US Fish and Wildlife Service; Green Bay Fish and Wildlife Conservation Office, 2661 Scott Tower Drive, New Franken, WI 54229, USA; cari-ann_hayer@fws.gov
2 Missouri Department of Conservation, 2010 South Second St, Clinton, MO 64735, USA; mbayless1969@gmail.com
3 US Geological Survey; Columbia Environmental Research Center, 4200 New Haven Road, Columbia, MO 65201, USA; ageorge@usgs.gov (A.G.); nthompson@usgs.gov (N.T.); dchapman@usgs.gov (D.C.C.)
* Correspondence: CRichter@usgs.gov; Tel.: +1-573-876-1841

Received: 18 July 2020; Accepted: 21 August 2020; Published: 27 August 2020

Abstract: The timing and location of spawning events are important data for managers seeking to control invasive grass carp populations. Ichthyoplankton tows for grass carp eggs and larvae can be used to detect spawning events; however, these samples can be highly debris-laden, and are expensive and laborious to process. An alternative method, environmental DNA (eDNA) technology, has proven effective in determining the presence of aquatic species. The objectives of this project were to assess the use of eDNA collections and quantitative eDNA analysis to assess the potential spawning of grass carp in five reservoir tributaries, and to compare those results to the more traditional method of ichthyoplankton tows. Grass carp eDNA was detected in 56% of sampling occasions and was detected in all five rivers. Concentrations of grass carp eDNA were orders of magnitude higher in June, corresponding to elevated discharge and egg presence. Grass carp environmental DNA flux (copies/h) was lower when no eggs were present and was higher when velocities and discharge increased and eggs were present. There was a positive relationship between grass carp eDNA flux and egg flux. Our results support the further development of eDNA analysis as a method to detect the spawning events of grass carp or other rheophilic spawners.

Keywords: aquatic invasive species; reservoir ecosystems; ichthyoplankton; Asian carp

1. Introduction

Since the developing semibuoyant eggs and the larvae of grass carp (*Ctenopharyngodon idella*) drift in the current [1–4], ichthyoplankton tows during the spawning season can be used to detect spawning events [2,5]. Grass carp are thought to be cued to spawn by a temperature threshold coupled with discharge pulses or other factors, such as turbidity and flow velocity, which are related to discharge. Ichthyoplankton samples from tows taken during high discharge tend to be debris-laden and are expensive and laborious to process. In addition, grass carp eggs in the drift tend to distribute more densely toward the bottom of the water column [6], where sampling is difficult or dangerous. An alternative technology, environmental DNA (eDNA), has been used in determining the presence of aquatic species from water samples taken from a particular waterbody [7]. Grass carp are mass spawners, and the movement and activity of many fish, and the release of sex products, especially sperm, could be expected to cause a sharp increase in the amount of eDNA in the water. Consequently, eDNA technologies could theoretically be used for the detection of spawning or as part of a two-tiered approach with traditional ichthyoplankton sampling.

Grass carp are efficient at removing vegetation and if they reach high abundance, can have undesirable environmental effects in their invaded range [8], although in Asia they are important and desirable species. In both cases, the detection of spawning events is important for managers [9,10]. Furthermore, bighead carp (*Hypophthalmichthys nobilis*) and silver carp (*H. molitrix*), which also have been extremely problematic invasive species in central North America but are important desirable species in Asia, have essentially similar spawning behaviors [1]. If sampling for eDNA is an efficient way to detect the spawning events of grass carp, it should work similarly for these carp species or for other species which are mass spawners in flowing water.

The objective of the present study was to compare the effectiveness of quantitative eDNA analysis to traditional ichthyoplankton sampling for the detection of grass carp spawning events. Our study area was a reservoir system fed by small flashy tributaries that may support grass carp spawning during spring run-off events.

2. Results

One hundred and eighty-eight grass carp eggs were collected from four of five sampled Truman Reservoir rivers (Figure 1, Table 1, Table S1) over a 5 day period (7 June to 11 June 2014) which corresponded to elevated and peak discharge events (Figure 2). Samples visually identified as containing grass carp eggs were verified to be grass carp by quantitative PCR analysis conducted on paired samples preserved in ethanol; a subset was further confirmed by DNA sequencing.

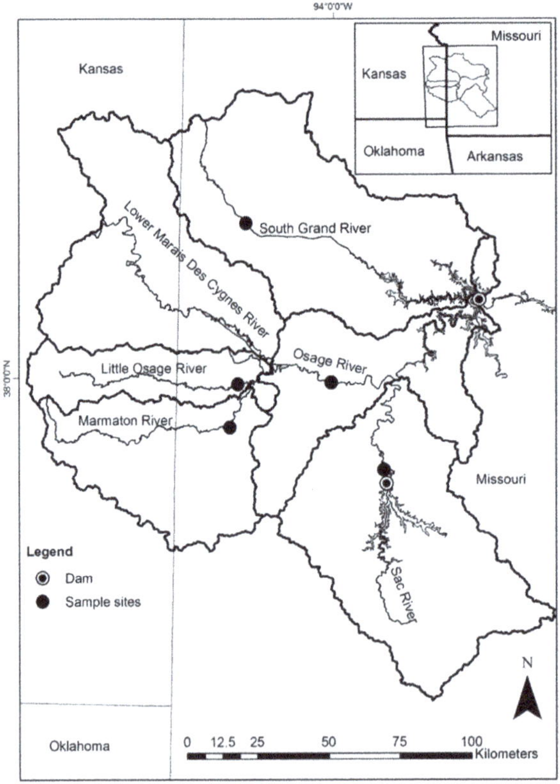

Figure 1. Study area of Truman Reservoir, Missouri, and the major tributary basins (represented by 8-digit hydrologic unit) for the sites sampled during May and June 2014 (map created by K. Anderson, U.S. Geological Survey).

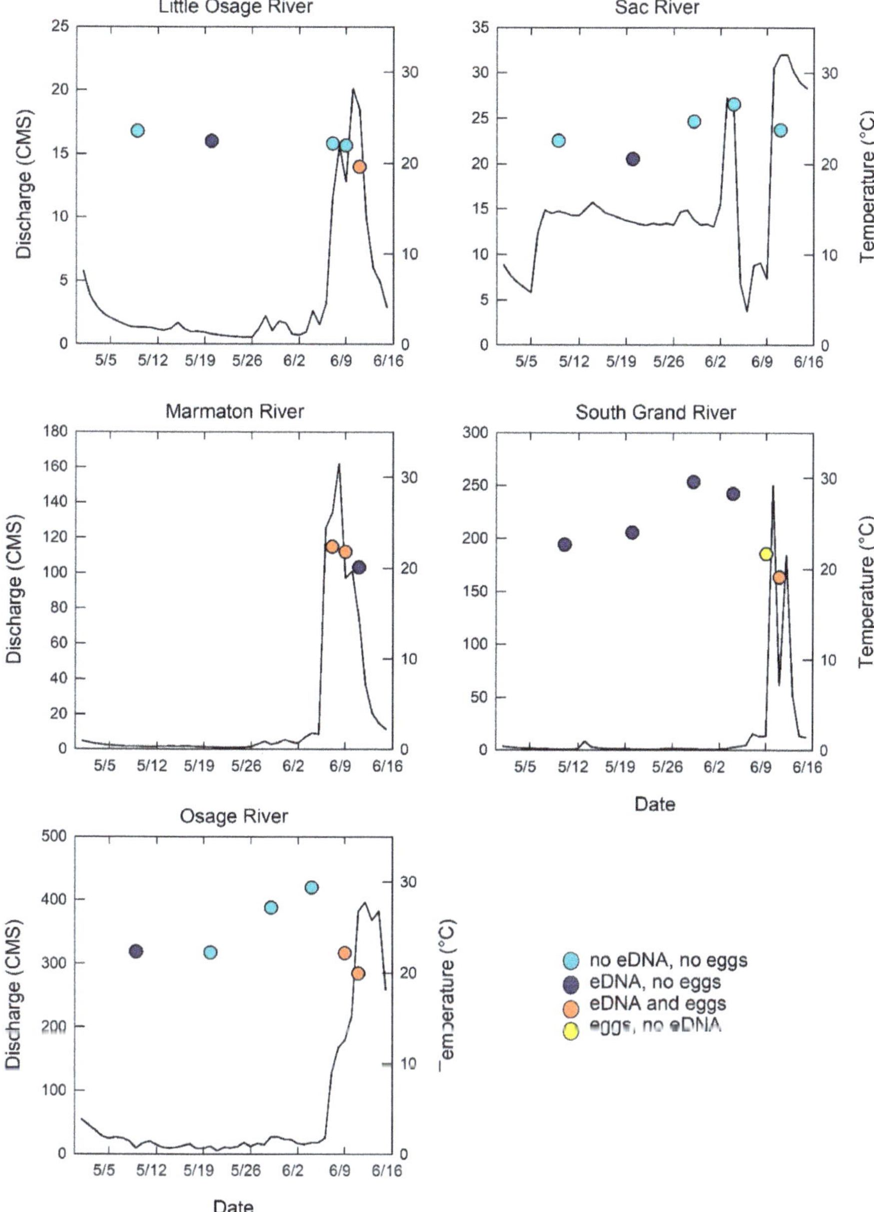

Figure 2. Mean daily discharge (lines, cubic meters per second (CMS)) and measured water temperature (circle symbols, °C) at each sampling event for 5 sampled tributaries to the Truman Reservoir in May and June 2014. The circle symbols are color coded to indicate the grass carp environmental DNA (eDNA) and egg detection results for each sampling event: light blue, no eDNA or eggs detected; dark blue, eDNA detected and eggs not detected; orange, both eDNA and eggs detected; yellow, eDNA not detected but eggs detected. Note that the *y* axis scales for mean daily discharge are not the same.

Grass carp eDNA was detected in water samples for 56% of the sampling occasions and was detected in all five sampled rivers (Figure 3). The flux of grass carp eDNA (copies/h) varied across rivers with the Marmaton, Osage, and South Grand Rivers having the highest grass carp eDNA flux values (Figure 4) and also had the most eggs collected over the duration of the study. The limit of quantification (LOQ) for the eDNA assay was 15.6 copies/reaction, and the effective LOQ for eDNA flux was 5.61×10^{10} copies/h per cubic meter per second (CMS) of river discharge [11]. Only three sampling events, the three highest points in Figure 4, had averages above the LOQ, therefore the quantification presented should be interpreted as the best available estimates, with greater than 35% coefficient of variation at each measurement below the LOQ. The highest grass carp eDNA detection rates occurred in the Marmaton and the South Grand rivers (Figure 3). Concentrations of grass carp eDNA were orders of magnitude higher in June than in May, corresponding to the elevated discharge, egg presence and egg flux (Figure 4). Results of the regression analysis between eDNA flux and egg flux showed a significant positive relationship (Figure 5).

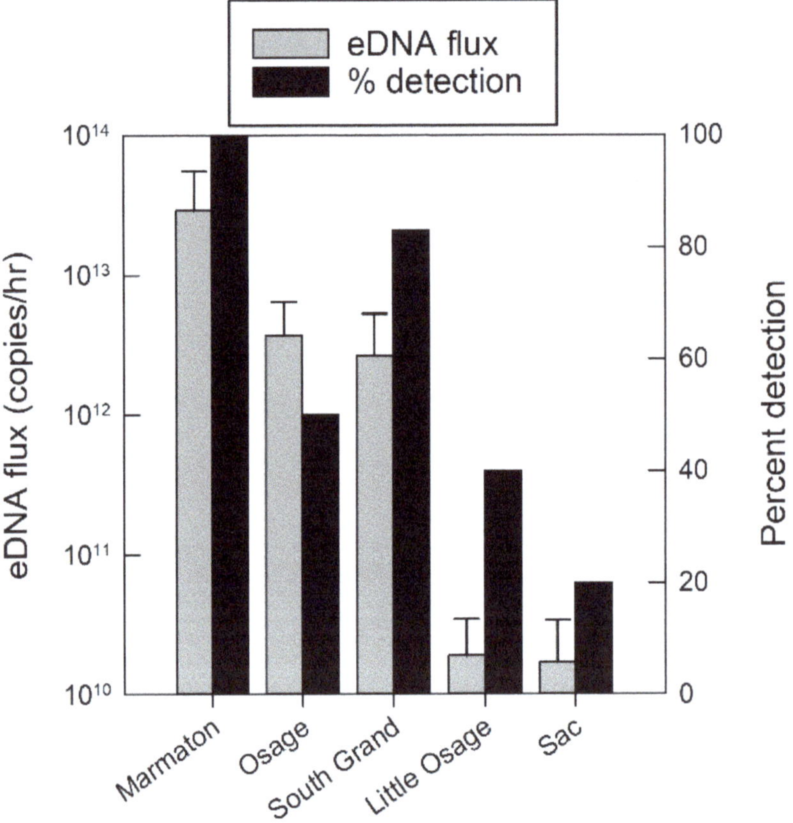

Figure 3. Environmental DNA flux (copies/h) averaged over all the sampling events for each tributary and the percent of sampling events that were positive for eDNA detection in the tributaries to the Truman Reservoir, Missouri, during May and June 2014. Error bars represent one standard error.

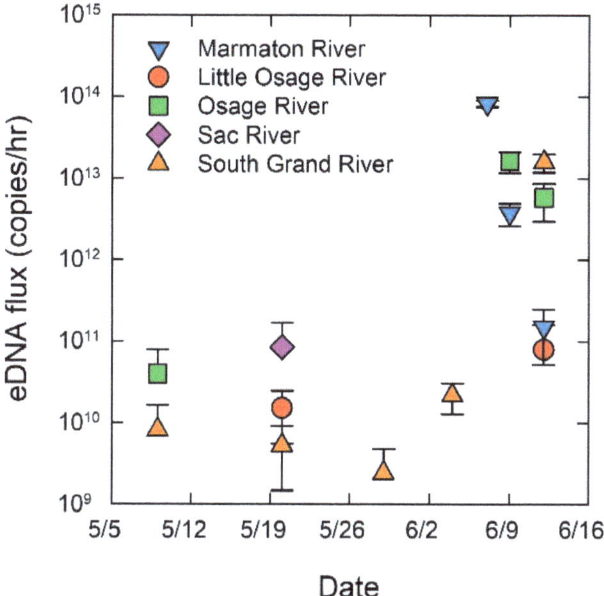

Figure 4. Environmental DNA flux (copies/h) averaged over each sampling event by date for each river sampled in the Truman Reservoir, Missouri, during the period from May to June 2014. Note log scale on *y* axis. Sampling events that did not detect grass carp eDNA are not displayed. Error bars represent one standard error.

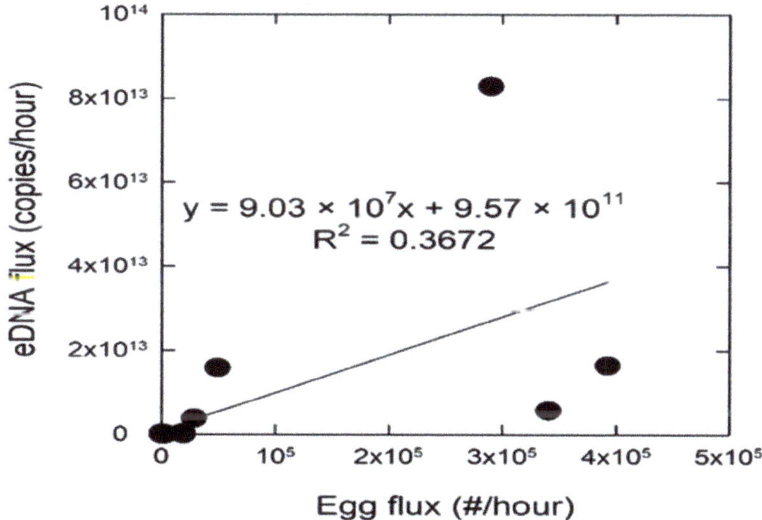

Figure 5. Regression analysis examining the relationship between environmental DNA flux (copies/h) and egg flux (eggs/h) sampled from the Truman Reservoir tributaries, Missouri, during May and June 2014. The regression was significant with an equation of eDNA flux = $9.03 \times 10^7x + 9.57 \times 10^{11}$.

Table 1. U.S. Geological Survey (USGS) water gage and sample site descriptions from five rivers, tributaries to the Truman Reservoir, sampled in May/June 2014. Basin area refers to the basin area for the corresponding 8-digit hydrologic unit. All USGS water gages are upstream of the sampling site, except the Osage River gage, which was located at the sample site. See Figure 1 for a more detailed map of study area.

River	USGS Gage Number	Distance (Gage to Site rkm [1])	Basin Area (km^2)	Sample Site Latitude	Sample Site Longitude	Distance to Truman Reservoir (rkm)	Upstream Barrier Distance (rkm)
Little Osage	06917060	8	1504	38.00658	−94.31944	107	
Marmaton	06918060	26	2956	37.99797	−94.31951	106	
Osage	06918250	0	3117	38.003889	−93.99472	48	
Sac	0691990	26	5104	37.94902	−93.76917	31	Stockton Dam (40)
South Grand	06921760	32	5302	38.45243	−94.00636	28	

[1] rkm refers to river kilometers.

3. Discussion

Grass carp eDNA was found at all sampling locations on at least one sampling occasion. The morphological identification of the collected eggs, eDNA analysis, and DNA sequencing confirm successful spawning (in the form of eggs) of grass carp in four of five sampled rivers in Truman Reservoir in June 2014. This finding represents only the second documentation in the peer-reviewed literature of grass carp spawning or reproduction in a North American reservoir ecosystem [12].

Environmental DNA from grass carp was detected in all five rivers throughout the study period, not only when eggs were detected, suggesting that grass carp were present in each river throughout the study time frame. Grass carp eDNA flux (copies/h) was lower when no eggs were present and was higher when velocities and discharge increased sharply and eggs were present. There was a positive relationship between the grass carp eDNA flux and egg flux. During spawning events, large amounts of sperm are released, up to 35×10^9 mL^{-1} [13]. Spermatozoa contain mitochondria, and thus contain the mitochondrial DNA target sequence for the qPCR assay we used to detect eDNA. The midpiece of a spermatozoon in silver and grass carp is rich in cytoplasmic material, containing spherical mitochondria. This midpiece is further joined with the head, which also contains mitochondria [14]. It follows that spawning events would release large amounts of eDNA into the water. In addition, grass carp spawn aggressively, and additional DNA may be released into the water by the sloughing of mucous, skin, scales, and blood. Increased eDNA in the water could also be indicative of increased biomass of fish resulting from spawning migrations and aggregations. Higher grass carp eDNA fluxes during spawning events suggest that this method could be used to detect spawning events, but the limited number of spawning events that occurred during this study limits interpretation. Other potential causes for the observed increases in eDNA flux could include seasonal changes in eDNA shedding or degradation rates, changes in grass carp ecological or feeding behaviors, or changes in the local abundance of different developmental stages. Furthermore, although the degradation rate of eDNA is an area of substantial research [15–17], and laboratory degradation rates of eDNA from the sperm of bigheaded carps (relatives of the grass carp with similar spawning behavior) have been studied [18], the persistence and degradation rate of eDNA resulting from spawning in flowing systems is inadequately understood, increased flow associated with spawning could affect eDNA detection probability, and these factors could complicate interpretation. Nevertheless, these data support further work in this area.

Grass carp egg flux was much lower relative to the combined bighead, silver, and grass carp egg flux reported in previous studies [19,20], but those studies were performed in much larger rivers. Elevated grass carp eDNA flux (copies/h) coincided with sampling events when eggs were present. We presume that the increase in eDNA is from spawning events; if no spawning had occurred, we would have expected a decrease in eDNA because of the dilution due to the increased river discharge. Further research is needed, but eDNA analysis could be used in addition to traditional sampling to detect spawning events, and to provide an additional, more effective, and less expensive tool to use in the detection of grass carp, or other mass spawners in flowing systems such as bighead carp and silver carp, in novel environments or at invasion fronts. In addition, because of the increased amount of eDNA released during spawning events, timing the collections of eDNA to periods when riverine fishes are spawning could increase detection sensitivity. The use of eDNA to detect spawning events using eDNA would require sampling at times when spawning is not occurring to set a baseline eDNA concentration. Some eDNA would always be present in the system if fish are present, but a spawning event would be expected to coincide with increased eDNA concentration above the baseline. Traditional ichthyoplankton sampling of semi-buoyant eggs is complicated by the distribution of eggs in the drift [6]. Eggs distribute more densely in the lower parts of the water column, especially under conditions of low turbulence, but it is difficult and sometimes dangerous to perform ichthyoplankton tows at depth. Thus, traditional ichthyoplankton tows might fail to detect a spawning event, or underestimate the magnitude of an event, if eggs are distributed deeper than the sampling gear. The detection of spawning events through eDNA would be less prone to this problem because the tiny particles targeted, with an apparent peak in size around 10 μm [21], would likely be more evenly distributed in the water column than the eggs. In addition, because of the laborious sample handling requirements of debris-laden ichthyoplankton tows taken during hydrograph peaks when grass carp are most likely to spawn, the use of eDNA technology to detect spawning events could represent substantial cost savings. Environmental DNA technology could benefit from further investigation of the relationship between the abundance of eDNA and spawning activities in laboratory and field experiments, including the influence of seasonal changes in movement, feeding, eDNA shedding rates, and eDNA degradation rates.

4. Materials and Methods

Harry S. Truman Reservoir (22,510 ha and approximately 196 km in length [22]) is a warmwater impoundment of the Osage River upstream of Lake of the Ozarks (another large impoundment) in west central Missouri (Figure 1). There are four main rivers that converge with the Osage River to flow into Truman Reservoir: Marais des Cygnes, Little Osage, Marmaton, and Sac rivers. The other major river, the South Grand River, flows directly into the Reservoir (Figure 1). To survey for grass carp spawning activity, we chose study sites on the Little Osage, Marmaton, Osage, Sac and South Grand Rivers (Figure 1) because they are small and flashy.

Standardized sampling for grass carp eggs and larvae occurred regularly for approximately one month (9 May to 11 June 2014), at one location, within each of five Truman Reservoir tributaries (N = 25 sampling events; Figure 1). This time frame was selected to incorporate a suite of environmental conditions and site characteristics (e.g., water temperature, base flows, change in hydrograph, and size of river) in which grass carp are known to spawn; we also chose conditions not known for grass carp spawning. At each site, ichthyoplankton tows were collected as paired samples with an ichthyoplankton bongo net (750 μm) for approximately ten minutes at a targeted velocity of 1 m/s relative to the water speed. The bongo net was equipped with a General Oceanics™ mechanical flow meter so that an estimate of the total volume of water sampled could be made. One sample was preserved in formalin for later visual identification, staging, enumeration, and measurement (length), and the other sample was preserved in ethanol for DNA analysis to independently identify the species composition of captured eggs, if necessary. Eggs preserved in ethanol were prepared for DNA extraction by removing ethanol with a pipette, leaving the remaining sample in a biological safety cabinet overnight to allow residual ethanol to evaporate, then suspending the egg in 250 μL of extraction buffer TD-S0 (AutoGen Inc., Holliston, MA, USA) and extracting DNA with the AGP245 automated DNA extraction system according to the manufacturer's instructions (AutoGen Inc., Holliston, MA, USA). The instantaneous flux of eggs was calculated from the number of eggs captured, divided by the volume of water sampled and multiplied by the discharge at the nearest gage, thus obtaining an estimate of the total number of eggs in the drift at a point in time. Egg developmental stages for up to 30 eggs per sampling event were grouped into five categories (i.e., cell division (stage 1–9), blastula (stage 10–12), gastrula (stage 13–15), organogenesis (16–26), and near hatching (stage 27–30); [2,23]). The proportion of eggs within each developmental stage was computed for each sampling occasion.

For eDNA methods, every effort was taken prior to each sampling event to provide a sterile environment and avoid the contamination of water samples from outside sources. This process included the decontamination of all eDNA sampling supplies in 10% bleach for one hour between sampling events. In conjunction with ichthyoplankton sampling, a plastic 1 L water bottle was used to collect river water 2–4 inches below the water surface. A subsample of 50 mL water was extracted from the 1 L bottle with a disposable sterile serological pipette into a 50 mL polypropylene conical bottom tube. This procedure was repeated four times (N = 5 subsamples). Additionally, 1 blank sample was taken from a 1 L bottle of well water brought from the laboratory. Water samples were stored in the dark on ice until returned to the laboratory and stored in a refrigerator for up to 24 h before processing. The water samples were concentrated by centrifugation for 30 min at 5000 RCF at 4 °C, after which water was decanted off. After drying for 10–25 min at room temperature, pellets were suspended in 250 μL of the extraction buffer TD-S0 (AutoGen Inc., Holliston, MA). Total DNA was extracted into 50 μL nuclease-free water and 1 μL DNA sample per reaction was analyzed by qPCR for grass carp DNA quantity (see [24] for more detailed methods). For the qPCR assays, the samples were run using the appropriate species' primer/probe set [25–27]. A qPCR reaction was considered positive for grass carp eDNA detection if there was exponential amplification within 40 cycles, a water sample was considered positive if at least one of three qPCR replicates was positive, and a sampling event was considered positive if at least one of five water subsamples was positive. Units of eDNA concentration are commonly expressed as copies per liter; however, to standardize these measurements among rivers for the detection of spawning events, eDNA concentrations in copies/liter were converted to eDNA flux (copies/h) based on discharge. A linear regression analysis was used to examine the relationships among eDNA flux and egg flux. All analyses were conducted in SAS software (SAS Institute, Cary, NC, USA) with an alpha level of 0.05.

Supplementary Materials: The following are available online at http://www.mdpi.com/2410-3888/5/3/27/s1, Table S1: Data summary per sampling event for use of environmental DNA (eDNA) to detect grass carp spawning events in waters of rivers in the Truman Reservoir, Missouri, USA. Blank cells = not measured; CMS = cubic meters per second.

Author Contributions: Conceptualization, C.-A.H., M.F.B., D.C.C., and C.A.R.; methodology, C.-A.H., A.G., C.A.R. and N.T.; formal analysis, C.-A.H.; data curation, A.G.; writing—original draft preparation, C.-A.H.; writing—review and editing, C.A.R. and D.C.C. All authors have read and agreed to the published version of the manuscript.

Funding: This research was funded by the USGS Ecosystems Mission Area, Invasive Species Program.

Acknowledgments: We would like to thank all those who helped to make this project possible. Specifically, we would like to recognize C. Byrd, J. Deters, B. Howell, T. Kobermann and A. Mueller for assistance in collecting field data, the Missouri Department of Conservation for providing assistance and knowledge of the Truman Reservoir and the development of this project, and K. Anderson for creating the study area map. Use of trade, product, or firm names does not imply endorsement by the U.S. Government.

Conflicts of Interest: The authors declare no conflict of interest. The funders had no role in the design of the study; in the collection, analyses, or interpretation of data; in the writing of the manuscript, or in the decision to publish the results.

References

1. Kolar, C.S.; Chapman, D.C.; Courtenay, W.R., Jr.; Housel, C.M.; Williams, J.D.; Jennings, D.P. *Bigheaded Carps: A Biological Synopsis and Environmental Risk Assessment*; American Fisheries Society: Bethesda, MD, USA, 2007; p. 204.
2. George, A.E.; Chapman, D.C. Embryonic and larval development and early behavior in grass carp, *Ctenopharyngodon idella*: Implications for recruitment in rivers. *PLoS ONE* **2015**, *10*, e0119023. [CrossRef] [PubMed]

3. George, A.E.; Chapman, D.C. Aspects of embryonic and larval development in bighead carp *Hypophthalmichthys nobilis* and silver carp *Hypophthalmichthys molitrix*. *PLoS ONE* **2013**, *8*, e73829. [CrossRef]
4. Murphy, E.A.; Jackson, P.R. *Hydraulic and Water-Quality Data Collection for the Investigation of Great Lakes Tributaries for Asian Carp Spawning and Egg-Transport Suitability*; Report. 2013-5106; U.S. Geological Survey: Reston, VA, USA, 2013; p. 40.
5. Garcia, T.; Murphy, E.A.; Jackson, P.R.; Garcia, M.H. Application of the FluEgg model to predict transport of Asian carp eggs in the Saint Joseph River (Great Lakes tributary). *J. Great Lakes Res.* **2015**, *41*, 374–386. [CrossRef]
6. George, A.E.; Garcia, T.; Chapman, D.C. Comparison of size, terminal fall velocity, and density of Bighead Carp, Silver Carp, and Grass Carp eggs for use in drift modeling. *Trans. Am. Fish. Soc.* **2017**, *146*, 834–843. [CrossRef]
7. Ficetola, G.F.; Miaud, C.; Pompanon, F.; Taberlet, P. Species detection using environmental DNA from water samples. *Biol. Lett.* **2008**, *4*, 423–425. [CrossRef] [PubMed]
8. Cudmore, B.; Mandrak, N.E.; Dettmers, J.M.; Chapman, D.C.; Conover, G.; Kolar, C.S. *Binational Risk Assessment of Grass Carp in the Great Lakes*; Research Document 2016/118; Fisheries and Oceans Canada; Canadian Science Advisory Secretariat: Ottawa, ON, Canada, 2017.
9. Yi, B.; Yu, Z.; Liang, Z.; Sujuan, S.; Xu, Y.; Chen, J.; He, M.; Liu, Y.; Hu, Y.; Deng, Z.; et al. The distribution, natural conditions, and breeding production of the spawning ground of four famous freshwater fishes on the main stream of the Yangtze River. In *Gezhouba Water Control Project and Four Famous Fishes in the Yangtze River*; Yi, B., Yi, Z., Liang, Z., Eds.; Hubei Science and Technology Press: Wuhan, China, 1988.
10. Larson, J.H.; Knights, B.C.; McCalla, S.G.; Monroe, E.; Tuttle-Lau, M.; Chapman, D.C.; George, A.E.; Vallazza, J.M.; Amberg, J. Evidence of Asian Carp spawning upstream of a key choke point in the Mississippi River. *N. Am. J. Fish. Manag.* **2017**, *37*, 903–919. [CrossRef]
11. Klymus, K.E.; Merkes, C.M.; Allison, M.J.; Goldberg, C.S.; Helbing, C.C.; Hunter, M.E.; Jackson, C.A.; Lance, R.F.; Mangan, A.M.; Monroe, E.M.; et al. Reporting the limits of detection and quantification for environmental DNA assays. *Environ. DNA* **2019**. [CrossRef]
12. Hargrave, C.W.; Gido, K.B. Evidence of reproduction by exotic grass carp in the Red and Washita rivers, Oklahoma. *Southwest. Nat.* **2004**, *49*, 89–93. [CrossRef]
13. Bozkurt, Y.; Öğretmen, F. Sperm quality, egg size, fecundity and their relationships with fertilization rate of grass carp (*Ctenopharyngodon idella*). *Iran. J. Fish. Sci.* **2012**, *11*, 755–764.
14. Verma, D.K.; Routray, P.; Dash, C.; Dasgupta, S.; Jena, J.K. Physical and biochemical characteristics of semen and ultrastructure of spermatozoa in six carp species. *Turk. J. Fish. Aquat. Sci.* **2009**, *9*, 67–76.
15. Barnes, M.A.; Turner, C.R.; Jerde, C.L.; Renshaw, M.A.; Chadderton, W.L.; Lodge, D.M. Environmental conditions influence eDNA persistence in aquatic systems. *Environ. Sci. Technol.* **2014**, *48*, 1819–1827. [CrossRef] [PubMed]
16. Strickler, K.M.; Fremier, A.K.; Goldberg, C.S. Quantifying effects of UV-B, temperature, and pH on eDNA degradation in aquatic microcosms. *Biol. Conserv.* **2015**, *183*, 85–92. [CrossRef]
17. Eichmiller, J.J.; Best, S.E.; Sorensen, P.W. Effects of temperature and trophic state on degradation of environmental DNA in lake water. *Environ. Sci. Technol.* **2016**, *50*, 1859–1867. [CrossRef] [PubMed]
18. Lance, R.F.; Klymus, K.E.; Richter, C.A.; Guan, X.; Farrington, H.L.; Carr, M.R.; Thompson, N.; Chapman, D.C.; Baerwaldt, K.L. Experimental observations on the decay of environmental DNA from bighead and silver carps. *Manag. Biol. Invasions* **2017**, *8*, 343–359. [CrossRef]
19. Deters, J.E.; Chapman, D.C.; McElroy, B. Location and timing of Asian carp spawning in the Lower Missouri River. *Environ. Biol. Fishes* **2013**, *96*, 617–629. [CrossRef]
20. Coulter, A.A.; Keller, D.; Bailey, E.J.; Goforth, R.R. Predictors of bigheaded carp drifting egg density and spawning activity in an invaded, free-flowing river. *J. Great Lakes Res.* **2016**, *42*, 83–89. [CrossRef]
21. Turner, C.R.; Barnes, M.A.; Xu, C.C.Y.; Jones, S.E.; Jerde, C.L.; Lodge, D.M. Particle size distribution and optimal capture of aqueous macrobial eDNA. *Methods Ecol. Evol.* **2014**, *5*, 676–684. [CrossRef]
22. Michaletz, P.H.; Siepker, M.J. Trends and synchrony in Black Bass and Crappie recruitment in Missouri reservoirs. *Trans. Am. Fish. Soc.* **2013**, *142*, 105–118. [CrossRef]
23. Yi, B.; Liang, Z.; Yu, Z.; Lin, R.; He, M. A study of the early development of grass carp, black carp, silver carp and bighead carp of the Yangtze River. In *Early Development of Four Cyprinids Native to the Yangtze River*; Chapman, D.C., Ed.; Data Series 239; U.S. Geological Survey: Reston, VA, USA, 2006; pp. 15–51.

24. Klymus, K.E.; Richter, C.A.; Chapman, D.C.; Paukert, C. Quantification of eDNA shedding rates from invasive bighead carp *Hypophthalmichthys nobilis* and silver carp *Hypophthalmichthys molitrix*. *Biol. Conserv.* **2015**, *183*, 77–84. [CrossRef]
25. Coulter, A.A.; Keller, D.; Amberg, J.J.; Bailey, E.J.; Goforth, R.R. Phenotypic plasticity in the spawning traits of bigheaded carp (*Hypophthalmichthys* spp.) in novel ecosystems. *Freshwat. Biol.* **2013**, *58*, 1029–1037. [CrossRef]
26. Wilson, C.; Wright, E.; Bronnenhuber, J.; MacDonald, F.; Belore, M.; Locke, B. Tracking ghosts: Combined electrofishing and environmental DNA surveillance efforts for Asian carps in Ontario waters of Lake Erie. *Manag. Biol. Invasions* **2014**, *5*, 225–231. [CrossRef]
27. Farrington, H.L.; Edwards, C.E.; Guan, X.; Carr, M.R.; Baerwaldt, K.; Lance, R.F. Mitochondrial genome sequencing and development of genetic markers for the detection of DNA of invasive bighead and silver carp (*Hypophthalmichthys nobilis* and *H. molitrix*) in environmental water samples from the United States. *PLoS ONE* **2015**, *10*, e0117803. [CrossRef] [PubMed]

Review

Development of Carbon Dioxide Barriers to Deter Invasive Fishes: Insights and Lessons Learned from Bigheaded Carp

Cory D. Suski

Department of Natural Resources and Environmental Sciences, University of Illinois, 1102 S Goodwin Ave, Urbana, IL 61801, USA; suski@illinois.edu; Tel.: +1-217-244-2237

Received: 2 June 2020; Accepted: 4 August 2020; Published: 13 August 2020

Abstract: Invasive species are a threat to biodiversity in freshwater. Removing an aquatic invasive species following arrival is almost impossible, and preventing introduction is a more viable management option. Bigheaded carp are an invasive fish spreading throughout the Midwestern United States and are threatening to enter the Great Lakes. This review outlines the development of carbon dioxide gas (CO_2) as a non-physical barrier that can be used to deter the movement of fish and prevent further spread. Carbon dioxide gas could be used as a deterrent either to cause avoidance (i.e., fish swim away from zones of high CO_2), or by inducing equilibrium loss due to the anesthetic properties of CO_2 (i.e., tolerance). The development of CO_2 as a fish deterrent started with controlled laboratory experiments demonstrating stress and avoidance, and then progressed to larger field applications demonstrating avoidance at scales that approach real-world scenarios. In addition, factors that influence the effectiveness of CO_2 as a fish barrier are discussed, outlining conditions that could make CO_2 less effective in the field; these factors that influence efficacy would be of interest to managers using CO_2 to target other fish species, or those using other non-physical barriers for fish.

Keywords: invasive species; bigheaded carp; biodiversity; behavior; physiology; toxicity; avoidance

---

1. Background

The transport of species beyond their native range represents a major global problem. The arrival of an invasive species can lead to the suppression of native populations through competition, the introduction of pathogens, predation, hybridization, and disruptions to habitats and ecosystem function [1–3]. Invasive species are therefore believed to be the second most important driver of species extinctions after habitat loss [4], and can lead to billions of dollars in economic costs [1,5]. More importantly, the decrease in biodiversity that invasive species cause can threaten human health and well-being [1,6–8]. Freshwater environments are experiencing declines in biodiversity disproportionately large relative to other biomes [9,10], and invasive species are one reason for this decline [1,11]. Studies have suggested that almost 40% of North American freshwater and diadromous fishes are imperiled [12], and the pace at which freshwater fish are becoming imperiled exceeds other vertebrates, and appears to be accelerating [12,13]; invasive species are a key factor contributing to these declines [4,12]. The rate at which humans have been introducing species beyond their native ranges has also accelerated over the past hundred years, driven primarily by the growth in global trade and mobility [3,14]. More importantly, models suggest that the transport of invasive species around the planet is likely to increase in the future [15,16].

While the eradication of an invasive species is theoretically possible, the unfortunate reality is that, once a species is introduced into an area, its removal is often impossible. For an invasive species to be successfully eradicated, a number of conditions must be met. These conditions include: proper planning and establishing lines of authority, a commitment to complete the eradication effort in terms of resources and enthusiasm, the biology of the target species must be amenable with the entire population of the target species put at risk, the target population must be removed faster than it can reproduce, the target species must be detectable at low densities, and efforts must be made to prevent re-invasion (possibly through restoration activities) [17,18]. These conditions are easiest to meet for isolated, small populations with low reproductive rates and poor dispersal capabilities, often for terrestrial vertebrates, with plants and aquatic species proving more challenging [17,18]. Thus, owing to the challenges associated with eradication efforts, the literature is rich with examples of failed attempts to extirpate invasive species, despite efforts that have extended over many years [18–20]. In some situations, the goal of completely removing an invasive species can be considered controversial as eradication can be costly, unlikely to succeed, and may result in considerable damage to non-target organisms and the environment [17,18,20]. Owing to the obstacles associated with eradication, a common outcome following the invasion of a species is "maintenance management", whereby the goal of elimination is abandoned, and the invader is simply controlled to a density that is deemed tolerable and allowed to persist [18,19]. Therefore, to avoid this sustained presence of an invasive species and perpetual "maintenance management", a more cost-effective, and meaningful approach to invasive species management is to prevent the arrival of an invasive species prior to invasion [1,21,22], or deter the secondary spread of invaders should they arrive [23].

2. Bigheaded Carp

Carps from the family Cyprinidae have been introduced outside of their native range for centuries. Bigheaded carps [24], and particularly bighead carp (*Hypophthalmichthys nobilis*) and silver carp (*H. molitrix*), have been introduced widely as phytoplankton control organisms in commercial aquaculture ponds and sewage lagoons owing to their large size and ability to efficiently filter phytoplankton and zooplankton from the water column [25]. Following transport to the United States for use as a biological control agent, floods allowed bigheaded carp to escape into the Mississippi River where they have spread throughout the basin [25], undeterred by locks, dams or other flood control structures [26], and are currently one of the most abundant species in portions of the Illinois River [27]. More importantly, bigheaded carp have had documented negative impacts on aquatic ecosystems [25]. Silver carp, for example, can consume detritus and bacteria [28], and reduce the size and abundance of both phytoplankton and zooplankton [25,27,29–31]. As a result, studies have shown that populations of bigheaded carps can result in reduced condition and abundance of native planktivorous fishes [32,33], as well as a reduction in the abundance of adult sport fish that compete with bigheaded carps at the larval and juvenile stages [34]. Owing to their abundance, mobility and impacts on receiving ecosystems, a tremendous amount of resources have been devoted to the suppression, removal and eradication of bigheaded carp from the Illinois River for almost a decade [35]. While efforts to date have been successful at reducing population sizes by removing millions of kilograms of fish through contract harvesting and agency collections [35], populations of bigheaded carp still remain throughout the Illinois River, necessitating suppression efforts to prevent the expansion of the population.

3. Chicago Area Waterway System

Bigheaded carp have direct access into the Great Lakes Basin from the Mississippi Basin due to the presence of the Chicago Area Waterway System (CAWS). The CAWS is a series of human-created canals and channels, completed in the early 1900s, that breached the continental divide between the two basins. The CAWS was constructed for the purpose of removing both sewage effluent and stormwater runoff from Chicago, coupled with allowing the passage of commercial shipping vessels to move from the Great Lakes to the Gulf of Mexico [36–38]. At present, the only means of deterring the movement of bigheaded carp through the CAWS from the Mississippi basin into the Great Lakes (beyond extensive suppression/harvest efforts) is a trio of electric barriers near Romeoville, IL, USA, constructed in 2002 [37]. Silver carp and bighead carp currently are over 60 km from Lake Michigan [39], so the effectiveness of these electric barriers at stopping bigheaded carp from passing has not been tested explicitly. However, numerous investigations have documented that these barriers are subject to problems and deficiencies that could allow the passage of bigheaded carp. For example, Dettmers et al. [40] showed that a number of fish confined to cages did not become immobilized when dragged through the barrier alongside steel-hulled barges, Sparks et al. [41] showed that an adult common carp (*Cyprinus carpio*) outfitted with an acoustic telemetry tag was able to traverse the electric barrier (possibly associated with a passing barge), while Evans and Brouder [42] showed that fish can move through the electric barriers if they are trapped between barges. Parker et al. [43] used stationary sonar deployed within the barriers and showed small fish were able to move through the electric fields independent of the presence of barges. In addition, electricity loses effectiveness when applied to small fish [44], the electric barrier is prone to maintenance shut downs, floods and power loss [38], and no non-physical barrier is effective at stopping 100% of fish [45]. Mitigation measures have been proposed to redesign shipping locks to reduce the possibility of the exchange of invasive species between the Mississippi and the Great Lakes basins. The plan to modify locks will cost billions of dollars, take a decade or more to complete, and has yet to start, leaving the Great Lakes vulnerable to the passage of bigheaded carp through the CAWS for the foreseeable future [46]. The consequences, should bigheaded carps traverse the electric barriers and enter the Great Lakes, are not known and are difficult to predict [47–51]. The consensus is that an invasion of bigheaded carps would not be beneficial, however, making the containment of carp within the Mississippi Basin a critical priority for stakeholders. To supplement existing suppression efforts and increase redundancy and effectiveness at preventing movement or spread through the CAWS, additional barrier technologies would be valuable, ideally a technology that will permit the passage of barges and the downstream transport of wastewater through the CAWS.

Based on the above background, the goals of this review are to (1) outline the development of zones of carbon dioxide gas (CO_2) as a non-physical barrier to deter the movement of invasive fishes, with a particular focus on two bigheaded carps: silver carp and bighead carp, and (2) highlight internal and external factors that mediate the performance of CO_2 as a non-physical barrier, either increasing or decreasing its effectiveness as a barrier for invasive fish passage. When taken together, this review will not only share the origins of CO_2 as a fish barrier, but also help researchers think about ways to improve performance and maximize the ability of different barrier technologies to deter the spread of invasive fishes.

4. Carbon Dioxide in the Atmosphere

The idea that CO_2 could be used as a fish barrier is rooted in Earth's history and the evolution of fishes. Billions of years ago, CO_2 levels in the Earth's atmosphere were high, and O_2 was low [52]. As photosynthesizing bacteria on the planet became more abundant, the composition of gasses in the atmosphere changed such that the relative level of O_2 increased and the level of CO_2 declined [52,53]. This change in atmospheric oxygen was concurrent with metabolic evolution that increased reliance on oxidative phosphorylation that uses oxygen as a final electron acceptor resulting in more efficient metabolism, coupled with the production of CO_2 as a waste product [52]. Thus, organisms developed the ability to sense environmental gasses, including CO_2, and respond by either avoiding CO_2-rich areas that might impair energetic processes, or possibly being drawn to CO_2-rich areas if they provide an energetic advantage [53]. Bacterial and fungal pathogens, for example, can sense environmental CO_2 associated with hosts and alter growth or life cycles to maximize virulence [53]. Hawkmoths (*Manduca sexta*, Lepidoptera: *Sphingidae*) use floral CO_2 emissions to quantify food source profitability and the amount of nectar in flowers [54], while honey bees (*Apis mallifera*) actively fan their hives to remove CO_2 wastes, and the number of individuals fanning correlates positively with CO_2 levels inside the colony [55]. Carbon dioxide excreted by vertebrates is used by mosquitoes (*Aedes* spp.) as a signal of a potential host [56,57], while *Drosophila* will avoid CO_2, likely as a signal that rotting fruit is a poor food source [58]. For many terrestrial vertebrates, CO_2 is detected by chemoreceptors in the blood stream and brain stem to regulate breathing [59], while, more specifically, mammals detect of CO_2 in the air with free nerve endings of the trigeminal system [60]. Together, concentrations of environmental CO_2 can be a source of ecologically relevant information, and the ability to detect and respond to CO_2 as a stimulus has persisted across kingdoms.

5. CO_2 and Fish Physiology

Carbon dioxide has a pronounced effect on fishes, resulting in a host of physiological and behavioral responses when concentrations above species-specific set-points are experienced. Fish predominantly sense ambient CO_2 using peripheral chemoreceptors, largely in the gills, that respond to CO_2 tension in the water, not changes in pH; some evidence does exist for the presence of internal CO_2 sensors, but the location of these sensors has not been well-defined [61]. When fish are placed in a high carbon dioxide environment, CO_2 passively diffuses into the fish down its concentration gradient, and arterial CO_2 equilibrates with environmental CO_2 within minutes, resulting in an internal acidosis [62,63]. Over time, this pH imbalance is corrected as fish uptake HCO_3^- from the environment (in exchange for Cl^-) and excrete H^+ (in exchange for Na^+) [63]. Owing to this influx of CO_2, hypercarbic environments cause an elevation of the general stress response [64–66], a drop in blood pH [67], a loss of ions [68], and, ultimately, equilibrium loss and anesthesia (Stage 2 or Stage 3) [64,67,69,70]. At present, the exact mechanism(s) responsible for the loss of equilibrium and the anesthetic impacts of carbon dioxide have not been well defined, but are believed to result from the movement of CO_2 across the blood-brain barrier, which alters brain pH and an impairs brain electrical activity [71,72]; additions of H^+ or HCO_3^- alone will not result in anesthesia for fish [71]. In addition to these physiological changes, studies have documented behavioral changes exhibited by fish in high CO_2 environments including hyperventilation, coupled with a reduction in heart rate, likely to facilitate CO_2 excretion [61]. Together, exposure to water with elevated concentrations of CO_2 has been shown to result in both physiological and behavioral changes to fish.

6. CO_2 and Fish Behavior

A number of past studies have highlighted the propensity for fish to voluntarily swim away from areas of high carbon dioxide, laying the foundation for the use of CO_2 as a fish deterrent. Avoidance reactions are considered the first line of defense for fish that encounter adverse stimuli, and poor water quality can quickly induce a behavioral response that causes fish to depart an area and seek out improved water, presumably to minimize energetic costs [73,74] (Figure 1).

Over a century ago, Shelford and Allee [75] designed a simple experiment to observe the behavior of nine fish species when placed individually in a raceway containing a gradient of CO_2, ranging from approximately 2–88 cubic centimeters of CO_2 per liter of water. Shelford and Allee [75] showed that, upon entering zones of elevated CO_2, some fish started surface ventilations, while others displayed a coughing or yawning reaction coupled with increased ventilation rates. In addition, Shelford and Allee [75] reported that some fish would enter the area of high CO_2, stop suddenly, and then move backwards as if they had "encountered a sheet-rubber wall", and all fish spent less time in areas of high CO_2 relative to areas with lower CO_2. Powers and Clark [76] used a laboratory gradient tank design similar to Shelford and Allee [75] and showed that both brook trout (*Salvelinus f. frontinalis*) and rainbow trout (*Salmo gairdnerii iridus*) also avoided water that had received "very slight" additions of CO_2 (a drop of approximately 0.4 pH units). This pattern was also confirmed by Collins [77] who showed that individual alewife (*Polumbus pseudoharengus*) and glut herring (*P. aestivalis*) (likely river herring) migrating upriver both avoided water with elevated CO_2 that exceeded 0.3 ppm, independent of pH changes. Bishai [78] showed that juvenile brown trout (*Salmo trutta*) and Atlantic salmon (*Salmo salar*) demonstrated a stronger avoidance response to a pH change caused by CO_2 relative to a pH change caused by hydrochloric acid. Jones et al. [79] noted that individual arctic char (*Salvelinus aplinus*) will avoid concentrations of CO_2 that exceed 50 µmol/L, Ross et al. [80] showed that brook trout and blacknose dace (*Rhinichthys atratulus*) would avoid water with ≥2% CO_2, while Clingerman et al. [81] reported that intentional elevations of CO_2 to 60 mg/L in an aquaculture would induce avoidance behavior in groups of rainbow trout (*Oncorhynchus mykiss*), thereby facilitating harvest and collection in recirculating tanks. Finally, both Bernier and Randall [64], as well as Yoshikawa [82], revealed that rainbow trout exhibited a "violent" struggle upon being exposed to water maintained at 36–350 mmHg CO_2, while Clingerman et al. [81] indicated that rainbow trout in aquaculture tanks demonstrated "chaotic" swimming when CO_2 levels were increased to 35–60 mg/L. Thus, for over 100 years, studies have documented that many fish species will avoid areas of elevated CO_2 once a threshold is reached, providing the proof of concept that CO_2 could be a potential non-physical barrier for invasive fishes.

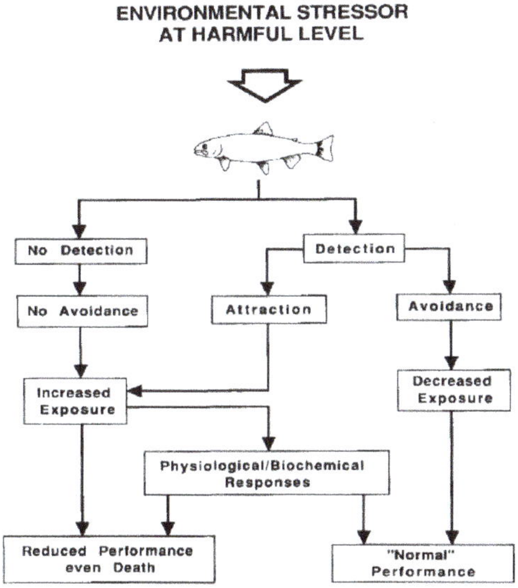

Figure 1. Flow chart showing the possible reactions of fishes to a potentially adverse environmental stimulus, such as an area of elevated carbon dioxide [73].

Despite suggestions from past work that CO_2 could induce avoidance behaviors and act a barrier to the movement of bigheaded carp, a key unknown was the threshold CO_2 level that should be targeted to induce avoidance. While the response of fishes to high concentrations of CO_2 when applied as an anesthetic appeared to be consistent [67,83], and the physiological responses of fishes to general hypercarbia had been well-defined [63], relatively less was known about the thresholds or "inflection points" that cause the onset of disturbances (i.e., a dose-response curve), and if those threshold concentrations were consistent across species and life stages. For example, Ross et al. [80] exposed book trout, slimy sculpin (*Cottus cognatus*) and blacknose dace to four levels of CO_2 (0%, 1.4%, 2.8% and 5.1%) for either one or 24 h and noted differences in physiological responses both across species and across exposure durations, suggesting species-specific responses to CO_2 exposure. To address this need and define concentrations that induced onset of disturbances, Kates et al. [66] exposed bluegill (*Lepomis macrochirus*), largemouth bass (*Micropterus salmoides*), silver carp (>450 mm) and bighead carp (>700 mm) to two different concentrations of CO_2 (30 mg/L and 70 mg/L) for three hours and showed that, 30 mg/L CO_2 (approximately 2000 μatm CO_2) had minimal physiological or behavioral impacts, but a three hour exposure to 70 mg/L CO_2 (approximately 50,000 μatm CO_2) resulted in a drop in ventilation rates, and an increase in irregular behaviors such as erratic swimming, twitching and escape attempts for silver carp and bighead carp [66]. One of the challenges with the study by Kates et al. [66], however, was that adult bigheaded carp were used, which provided little evidence in support of how small fish, those presumably less vulnerable to the existing electric barriers in Romeoville, IL, would respond to CO_2. In an effort to better define the allometric response of fish to CO_2 exposure, Dennis et al. [84] exposed larval and juvenile (73 mm) silver carp and bighead carp to 120 mg/L CO_2 (approximately 42,000 µatm CO_2) for 30 and 60 min. Results from this study were similar to previous work with adult fish, in that exposing larval silver carp and bighead carp to 42,000 µatm CO_2 for 30 min resulted in an increase in the mRNA coding for genes associated with the stress response (*Hsp70* and *c-fos*) [84] (Figure 2).

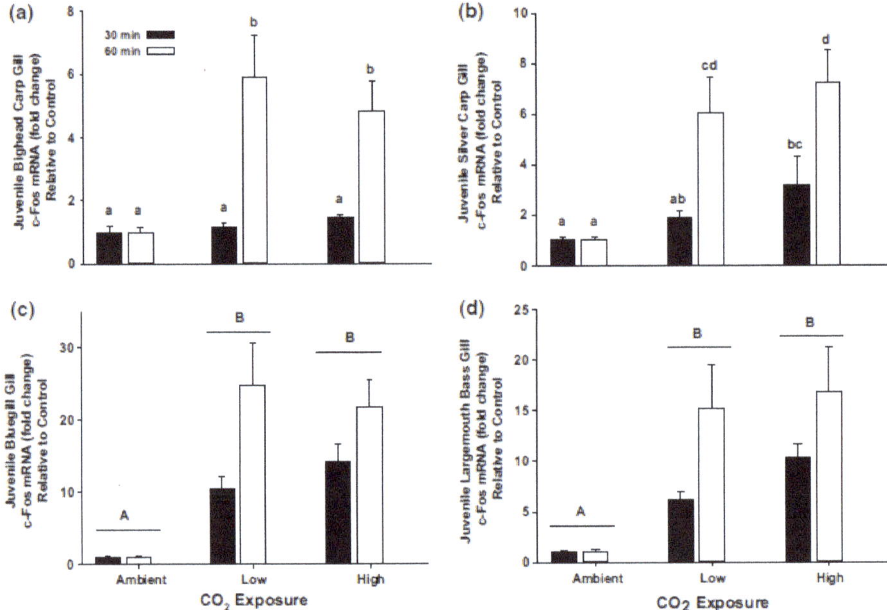

Figure 2. Relative expression of c-fos mRNA extracted from the gill tissue of juvenile bighead carp (**a**), silver carp (**b**), bluegill (**c**), and largemouth bass (**d**) exposed to a two hypercarbic treatments. Relative mRNA expression of juvenile fish that had an exposure duration of 30 min are shown in black bars, while white bars show the mRNA expression of juvenile fish exposed for 60 min. Horizontal lines denote a significant CO_2 concentration effect across exposure durations within a species. Dissimilar letters indicate significant differences between bars within a species. Data are mean ± SE, calculated relative to the expression of the reference gene (i.e., either *18s* or *ef1-a*). For clarity, data are expressed relative to the mean of juvenile fish exposed to ambient water conditions [84].

Thus, when results from these two studies are taken together, data suggest that thresholds of approximately 42,000 µatm CO_2 (70–120 mg/L) induce a suite of physiological and behavioral responses for a range of sizes of silver and bighead carp consistent with stress or discomfort, providing a target in the development of a non-physical barrier for fish.

7. CO_2 and Physiological Responses

Following the identification of putative thresholds that would induce behavioral and physiological disturbances, studies on CO_2 barriers shifted to quantify aspects of avoidance (Figure 1). Despite the research mentioned above that indicated a pattern of fish avoiding zones of elevated CO_2, there were suggestions in the literature that avoidance responses may be variable across species. Ross et al. [80], for example, showed that individual slimy sculpin did not avoid zones of elevated CO_2 and preferred to rest in place when confronted with hypercarbia, while Summerfelt and Lewis [85] noted that CO_2 concentrations from 3.0–9.7 mg/L did not repel green sunfish (*Lepomis cyanellus*) in a graded laboratory tank. Early work with CO_2 avoidance and bigheaded carp was conducted by Kates et al. [66] who used a "shuttle-box" apparatus in a laboratory to show that individual adult silver carp (460 mm) would voluntarily swim away from CO_2 once concentrations were elevated to approximately 120 mg/L CO_2, although there was considerable variation around this mean value (Figure 3) (Table 1).

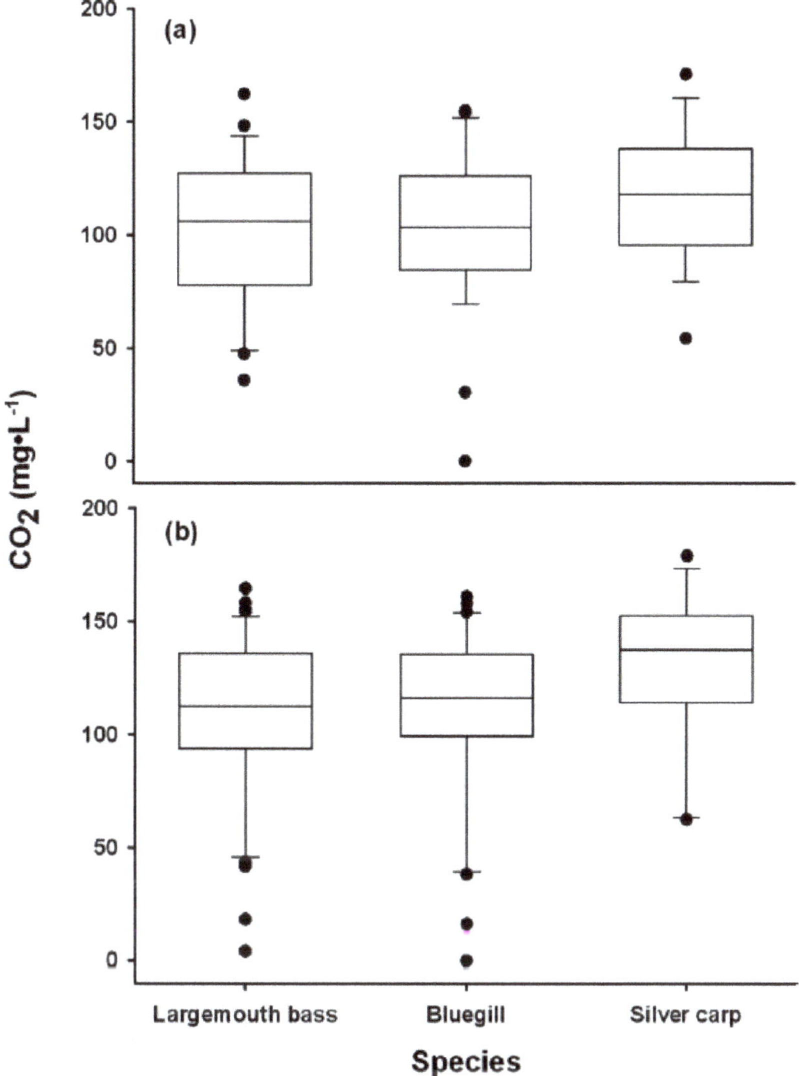

Figure 3. Concentration of CO_2 at which largemouth bass, bluegill, and silver carp displayed either an agitated activity (surface ventilations, twitching, or elevated swimming activity) (**a**) or movement out of high CO_2 environment to a lower CO_2 environment (**b**) during the course of avoidance trials [66].

Dennis et al. [84] later used juvenile silver carp and bighead carp (67 mm and 71 mm, respectively) and the shuttle-box apparatus, and, again, showed that individual fish would voluntarily swim away from zones of elevated CO_2. The concentration of CO_2 required to induce avoidance in this series of tests averaged approximately 180–220 mg/L CO_2, and, again, the variance around the mean was considerable (Figure 4). The success of these laboratory trials led to work at larger settings, including Donaldson et al. [86] who showed that a number of fish species, including silver carp and bigheaded carp, released into a 4000 m^3 outdoor pond in groups of 5–10 avoided zones of elevated CO_2 elevated to approximately 30,000 µatm (60 mg/L), and Cupp et al. [87] who used a two-channel, outdoor raceway (approximately 60 m^3) with flowing water and showed that CO_2 levels of approximately 30,000–40,000 µatm (~75 mg/L) would deter the movement of both silver and bigheaded carp (278 mm and 212 mm, respectively) when tested in groups of 10 (Table 1). Cupp et al. [88] showed that CO_2 deployed at the mouth of an outflow structure draining a backwater lake could reduce the abundance of shoals of mixed fish species by 70–95% at low water flows once a target threshold of 100 mg/L was reached. Finally, Hasler et al. [89] worked in a 12 m long indoor swim flume and showed that bighead carp (145 mm) in shoals of 3 would avoid CO_2 in water flowing at approximately 15 cm/seconds (equivalent to 1 body length per second), and CO_2 levels in this study were approximately 190,000 µatm. When considered together, these studies used a number of environments (indoor, outdoor, static water, flowing water) to demonstrate that a range of sizes of invasive silver and bighead carp, including small fish presumed to be less vulnerable to electricity, would voluntarily swim away from zones of elevated CO_2 once a threshold of approximately 70,000 µatm (100 mg/L) was reached, providing support for the use of CO_2 as a non-physical barrier.

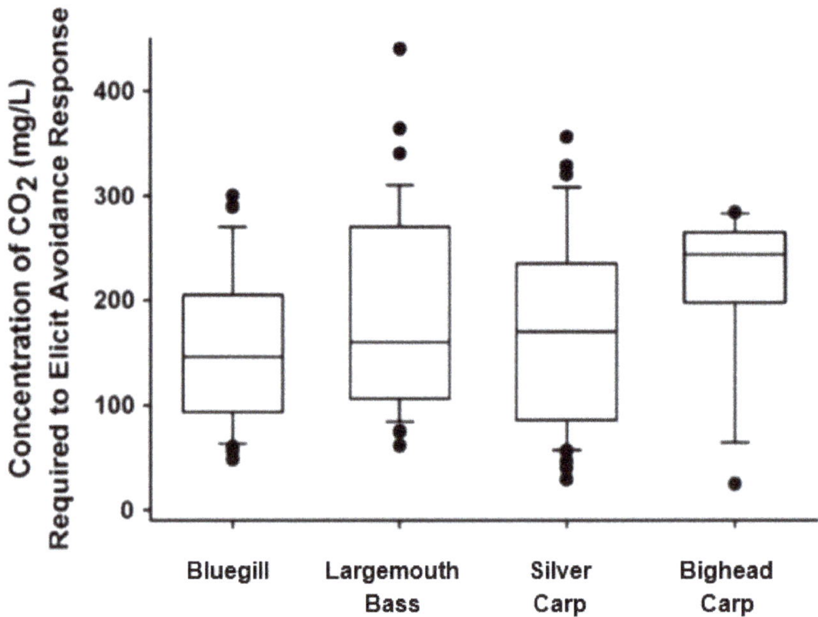

Figure 4. Concentration of CO_2 at which juvenile bluegill, largemouth bass, silver carp and bighead carp displayed avoidance behaviors [84].

Table 1. Summary of studies quantifying CO_2 thresholds that caused avoidance within the framework of generating a non-physical barrier for silver carp and bighead carp. Data have been approximated from figures whe.2e it was not clearly outlined in text of the citation. Units are left in the format that was used during publication.

Species	Avoidance Threshold	Test Environment	Temperature	pH	Mean Fish Size (mm)	Citation
Silver carp	135 mg/L	Shuttle box	18 °C	8.0	460	[66]
	125 mg/L	Shuttle box	16 °C	7.46	67	[65]
	59 mg/L (29,193 µatm)	Outdoor static pond	16 °C	8.25	254	[86]
	75 mg/L (29,532–41,393 µatm)	Outdoor flowing raceway	8–13 °C	7.5	278	[87]
Bighead carp	180 mg/L	Shuttle box	16 °C	7.46	71	[65]
	59 mg/L (29,193 µatm)	Outdoor static pond	16 °C	8.25	205	[86]
	75 mg/L (29,532–41,393 µatm)	Outdoor flowing raceway	8–13 °C	7.5	212	[87]
	160,000–186,000 µatm	Indoor flowing raceway	21 °C	8.3	145	[89]

8. CO_2 as a Potential Fish Barrier

When the general anesthetic properties of CO_2 exposure [67] were combined with results from field and lab avoidance trials, there was a considerable amount of evidence to suggest that zones of elevated CO_2 could deter the spread of invasive bigheaded carps. More specifically, a CO_2 barrier could be deployed in one of two different ways. First, CO_2 could be deployed as a "fence" or wall with the goal of inducing *avoidance* behaviors in fishes, exploiting the fact that fish voluntarily swim away from areas of high CO_2 once a threshold had been reached. For example, CO_2 could be used to confine carp in backwater areas [90] preventing access to turbulent, high-velocity water flowing river environments used for spawning [91], or at a choke-point in a river (e.g., shipping lock) to stop movement. A CO_2 barrier deployed in this way could be temporary (e.g., deployed only during summer or during harvest), or for longer periods of time. Secondly, zones of CO_2 could be deployed to intentionally induce equilibrium loss for fish, taking advantage of the ability of fish to *tolerate* CO_2 as an anesthetic. Again, an application of this kind could be temporary (i.e., deployed at specific times of the year) or longer-term (e.g., added to a shipping lock) [46].

9. Questions from Avoidance Data

While the concept of using CO_2 as a barrier based on either avoidance or a tolerance has support from a number of studies, there were several puzzling trends in the data, which generated questions and presented challenges related to possible deployment. For example, shuttle-box work by Kates et al. [66] showed that avoidance thresholds for individual adult silver carp and bighead carp spanned from approximately 50 mg/L to 160 mg/L (Figure 3) (Table 1). Subsequent shuttle-box work by Dennis et al. [84] with juvenile fishes, showed that avoidance thresholds ranged 6-fold, from approximately 50–300 mg/L (Figure 4). This variation in avoidance is further complicated by work from outdoor ponds showing avoidance occurred for groups of bighead carp, but CO_2 never exceed 60–75 mg/L [86,87]. Questions therefore arose related to the source of this variation (i.e., Is there inter-individual variation? What is the nature of the differing test environments? Is this variation inherent in how animals respond to zones of CO_2?), the potential for inter-individual differences in tolerance, and the effectiveness/consistency of CO_2 across time periods or environments. Thus, it was difficult to make recommendations to managers on target thresholds necessary to achieve an effective CO_2 barrier, or to predict possible changes in barrier effectiveness, without a more thorough understanding of the response of fish to CO_2 barriers. A series of studies were therefore conducted to quantify endogenous and exogenous factors that influenced the avoidance and tolerance of fishes to elevated CO_2 in hopes of refining this technology, providing stronger, more definitive recommendations to managers on target thresholds for CO_2 barriers, and improving the likelihood of long-term performance of CO_2 as a non-physical barrier.

10. Factors Influencing the Avoidance of CO_2

Several different endogenous and exogenous factors have the potential to influence the avoidance response of CO_2 in the context of a non-physical barrier (Table 2). For instance, in recent years, it has become apparent that fish consistently differ from each other in behavior, often termed "personality"; some individuals are more bold than others, some are more active, and some are more likely to explore novel areas [92]. Invasive round goby at the leading edge of their range, for example, were shown to be more bold and willing to explore novel areas than individuals from established, core populations [93], and it is therefore plausible that individual differences in personality could be manifesting in inter-individual differences to CO_2 avoidance [94] (Figures 3 and 4). More importantly, personality differences covary with characteristics such as the response to stressors, life span and growth rate through the pace-of-life continuum [92]. Therefore, if the response to an environmental stress and avoidance thresholds are mediated through behavior (e.g., proactive vs. reactive coping styles [92,95]), due to links between personality, life history and fitness [96], if target levels for an avoidance barrier are too low and fish of a particular behavioral type are able to pass, this could translate to population-level shifts in phenotypes, possibly changing avoidance thresholds for a population. In exploration of this concept, Tucker et al. [97] showed that aspects of personality (e.g., activity and boldness) did not influence CO_2 avoidance in individual bluegill, with fish of all personality types avoiding CO_2 at a threshold of approximately 67,000 µatm. In addition, Tucker and Suski [98] showed that social personality in bluegill (e.g., sociability, clustering with conspecifics and conspecific aggression) also did not influence CO_2 avoidance thresholds or the order that fish avoided CO_2. Related to this, past work has shown that food deprivation can alter the behavior of fish through plastic or flexible changes, with animals deprived of food taking more risks and becoming more active, likely as they search for food [99,100]. Interestingly, Suski et al. [101] showed that nine days of food deprivation did not influence CO_2 avoidance thresholds for individual largemouth bass; fish that had been fed and fish that had been deprived of food, both avoided high CO_2 at thresholds of approximately 70,000 µatm. However, Tucker et al. [97] showed that CO_2 avoidance in individual largemouth bass was influenced by artificial activation of the stress axis as fish that received an intraperitoneal injection of cortisol (hydrocortisone 21-hemisuccinate) required 45% more CO_2 to induce avoidance behavior relative to fish that did not receive an artificial elevation of the stress axis. Many initial studies of CO_2 avoidance [66,84] were conducted on individual fish, but shoals have a number of benefits for fish including predator vigilance and food detection, resulting in a calming effect and a reduced response to environmental stressors [102]. Tucker and Suski [98] showed pronounced differences in CO_2 avoidance thresholds for individual fish relative to shoals, with groups of bluegill choosing to swim away from CO_2 at significantly lower thresholds than individual bluegill; interestingly Tucker et al. [97] also showed that shuttling thresholds were not repeatable within individuals. Allometry is known to influence a number of characteristics of fish including metabolism and survival, but intraspecific differences in CO_2 avoidance thresholds across size categories is not clear. When avoidance thresholds for small and large bighead carp are compared across Kates et al. [66] and Dennis et al. [103], small fish appear to require higher CO_2 thresholds to induce avoidance. These results, however, were obtained in separate studies, not in a single investigation, so inter-study differences may have been a complicating factor. The quantity of CO_2 necessary to induce avoidance in round goby (*Neogobius melanostomus*) [104], silver carp and bighead carp [105] correlated positively with water temperature (range from 5–25 °C), such that more CO_2 was required to induce avoidance at high temperatures for all three species tested. Note that both Cupp et al. [104] and Tix et al. [105] did not acclimate fish at test temperatures for a period of two to three weeks as is common [106,107], with holding times listed at two to six days, which could have influenced these results. Together, a number of factors have been shown to influence the threshold of CO_2 required to induce avoidance behaviors, which have implications for the application of CO_2 as an avoidance barrier to deter the movement of invasive fishes (Table 2).

Table 2. Factors influencing the thresholds of CO_2 required to induce avoidance behaviors in fishes.

	Factor	Outcome	Species	Citation
Factors resulting in **more** CO_2 needed for avoidance	Temperature	Higher concentrations of CO_2 needed to induce avoidance at high temperatures relative to low temperatures.	Bighead carp, silver carp, round goby	[104,105]
	Stress	Fish with artificially-elevated cortisol levels required more CO_2 to induce shuttling than non-stressed controls.	Largemouth bass	[97]
Factors requiring **less** CO_2 for avoidance	Shoals	Shoals of fish avoided CO_2 at lower thresholds than did individual fish.	Bluegill	[97]
Factors **not influencing** CO_2 avoidance	Social personality	Preference for associating with conspecifics did not influence thresholds for CO_2 avoidance.	Bluegill	[97]
	Personality	Activity and boldness did not influence CO_2 avoidance thresholds.	Bluegill	[97]
	Feeding	Fish that had been deprived food for 9 days avoided CO_2 at the same threshold as fed conspecifics	Largemouth bass	[101]

11. Factors Influencing CO_2 Tolerance

Similar to work with avoidance, a number of studies have been carried out to quantify inter-individual differences in CO_2 tolerance, as well as potential mechanisms for any differences (Table 3). Importantly, Hasler et al. [108] showed that, for largemouth bass, CO_2 tolerance not only varied across individuals, with some fish losing equilibrium in high CO_2 sooner than others, but also that the individual response to high CO_2 was repeatable within individuals, suggesting potential for this to be a heritable trait that can be acted upon by natural selection [109]. In general, tolerance to CO_2 is a function of the interaction of exposure time × concentration, further mediated by temperature [64,69,70,110–112]. More specifically, a brief exposure to a high concentration of ambient CO_2, or an extended exposure to lower concentrations of CO_2, will both result in equilibrium loss, provided that the concentration of ambient CO_2 is sufficient to passively diffuse into the bloodstream of the fish [64,70,110,112]. Owing to reduced respiratory and metabolic rates at low temperatures, fish typically require additional time at lower temperatures before anesthetic effects are realized relative to high temperatures [112]. Indeed, this has been demonstrated for CO_2 as both Fish [69] and Gelwicks et al. [110] used study designs where individual fish were transferred to containers of CO_2 at a target concentration and showed decreased time to equilibrium loss at high temperatures, suggesting that fish are more sensitive to CO_2 at high temperatures. Interestingly, both Cupp et al. [104], and Tix et al. [105], showed that, when CO_2 was continually added to a test tank, round gobies [104], silver carp and bighead carp [105] all required higher concentrations of CO_2 before equilibrium loss occurred when animals were at high temperatures relative to low temperatures, suggesting that fish were more tolerant to CO_2 at higher temperatures. There are three potential explanations for the discrepancies across these studies. Firstly, differences across studies could be due to experimental animals, as Fish [69] and Gelwicks et al. [110] worked with salmonids, while Tix et al. [104] and Cupp et al. [105] used round goby and bigheaded carp. Secondly, Tix et al. [105] and Cupp et al. [104] applied CO_2 to fish continuously until equilibrium loss occurred, while Fish [69] and Gelwicks et al. [110] pre-treated tanks of water with CO_2 to a set point and added fish. Finally, both Cupp et al. [104] and Tix et al. [105] did not acclimate animals to each test temperature for extended periods of time, and, rather, animals were first held at 12 °C and then transferred to the test temperatures for 24–144 h prior to testing, which may have influenced their response to CO_2

exposure [106,107]. Clingerman [81] showed that, when CO_2 level was held constant, large rainbow trout were more likely to lose equilibrium than small rainbow trout in aquaculture tanks, suggesting an increased tolerance for smaller fish. Tucker et al. [97] showed that aspects of personality (e.g., activity and boldness) did not influence CO_2 tolerance in bluegill, with fish of all personality types requiring similar durations of time to induce equilibrium loss when exposed to 123,000 μatm. Hasler et al. [108] showed that tolerance to CO_2 was influenced by the metabolic phenotype of largemouth bass, and fish with higher anaerobic performance, quantified as time to become exhausted when burst swimming, required less time to lose equilibrium when exposed to high CO_2, and also that aerobic aspects of metabolic phenotype (i.e., standard metabolic rate, aerobic scope) did not influence tolerance to carbon dioxide. Suski et al. [101] showed that largemouth bass that had been deprived of food for 14 days required 25% longer exposure to high CO_2, relative to fish that had been fed over this 14 day period, thereby demonstrating an increased tolerance to CO_2 from food deprivation. Together, tolerance to high CO_2 can vary due to a number of endogenous and exogenous factors and should be considered should CO_2 be deployed to deter the movement of invasive fishes (Table 3).

Table 3. Factors influencing the tolerance of CO_2, indicated by loss of equilibrium.

	Factor	Outcome	Species	Citation
Factors resulting in a **higher** threshold for equilibrium loss in CO_2	Size	large fish lost equilibrium sooner (were more sensitive) at a given CO_2 concentration than small fish.	Rainbow trout	[81]
	Time × concentration interaction	Equilibrium loss occurs at extended exposure to low CO_2 concentration, or brief exposure to high CO_2 concentration.	Several species of salmonid (steelhead, chinook)	[69,70]
	Temperature	When CO_2 was added to a tank at a constant rate, a higher CO_2 concentration was required to induce equilibrium loss at high temperatures relative to low temperatures.	Silver carp, bighead carp, round goby.	[104,105]
	Food deprivation	Fish that had been deprived food for 14 days required more CO_2 to induce equilibrium loss than fed conspecifics.	Largemouth bass	[101]
Factors resulting in a **lower** threshold for equilibrium loss in CO_2	Temperature	Equilibrium loss occurs faster at higher temperature when CO_2 concentration is held constant.	Several species of salmonid (steelhead, chinook)	[69,110]
	Anaerobic swimming potential	Fish that required longer to become exhausted during burst swimming required less time to lose equilibrium at high CO_2.	Largemouth bass	[108]
Factors **not influencing** equilibrium loss in high CO_2 environments	Standard metabolic rate	Variation in standard metabolic rate did not predict time to equilibrium loss in high CO_2.	Largemouth bass	[108]
	Aerobic scope	Variation in aerobic scope did not predict time to equilibrium loss in high CO_2.	Largemouth bass	[108]
	Personality (activity, boldness)	Variation in activity and boldness did not influence time to equilibrium loss in high CO_2.	Bluegill	[97]

12. Management Implications

There are a number of potential non-physical barriers that can be deployed to prevent the spread of invasive fishes, including bubble screens, sound or electricity, each with particular strengths and weaknesses [45]. A non-physical barrier that uses zones of elevated carbon dioxide to deter fish movements has a number an advantageous as a chemical control tool relative to other technologies as it has few human health concerns, can be applied in a carbon neutral fashion using repurposed CO_2 (i.e., harvesting waste CO_2 destined to be released into the atmosphere), is relatively inexpensive and readily available, can be deployed with relatively little infrastructure, and residual CO_2 does not persist in the environment [113]. Carbon dioxide was recently registered with the United States Environmental

Protection Agency (USEPA) as a pesticide for use as a deterrent of bigheaded carp under the name Carbon Dioxide—Carp (EPA Registration Number 6704-95). Dennis et al. [103] held largemouth bass at 21,000 μatm (13 mg/L) CO_2 for almost two months and showed no decline in avoidance thresholds, suggesting that acclimation to the presence of high CO_2 is not likely. Most important, the avoidance response of fishes to environmental CO_2 appears to be canalized, demonstrated by virtually all fish species tested, while CO_2 tolerance is repeatable and consistent [108], giving CO_2 a number of advantages as a non-physical fish barrier as a tool to deter invasive fishes.

Results from the studies listed above have a number of implications for the deployment of CO_2 as a non-physical barrier and can be used to minimize the likelihood of unintentional fish passage, while also helping minimize waste CO_2 and reduce deployment costs. For example, for a CO_2 barrier deployed with the intention of causing avoidance, it is important for managers to consider the context in which the barrier is deployed. More specifically, although not repeatable within individuals, avoidance of CO_2 has been shown to be consistent across virtually all species tested when CO_2 pressures reach approximately 30,000–60,000 μatm (60–100 mg/L). However, avoidance thresholds will likely be lower for fish in shoals (rather than individual fish) but will increase if fish are experiencing stress (independent of food availability), such as chronic hypoxia or environmental pollution. Finally, studies suggest that higher concentrations of CO_2 may be required to induce avoidance at warmer water temperatures (summer) relative to cooler conditions (Table 2). It should be noted that Schneider et al. [114] showed that CO_2 did not impair either the burst or sprint swimming performance of largemouth bass until thresholds of 100,000 μatm were reached (approximately 150 mg/L), well in excess of thresholds required to induce avoidance, suggesting that, if fish choose to challenge a CO_2 barrier and burst through it, the barrier will likely not impair swimming performance. When considered together, a number of factors should be considered to ensure maximum effectiveness should CO_2 be used in the field to deter the movement of invasive fishes (Table 2).

If a CO_2 barrier is deployed with the intent of stopping fish via equilibrium loss (tolerance), aspects of individual fish need to be considered as these factors can influence effectiveness. At present, the relationship between exposure time × CO_2 concentrations that results in equilibrium loss for most species has not been defined, so these data would need to be collected to help guide management targets, and owing to individual variation in the loss of equilibrium time for fish [108], a large number of fish would need to be assessed to quantify a range of equilibrium loss times. In general, however, small fish, and individuals that had been deprived of food, would be expected to have improved tolerance in high CO_2 relative to larger, well-fed individuals. The role of environmental temperature has not been clearly defined, but studies suggest that a longer exposure time may be required at lower water temperatures and in periods of low food availability (e.g., winter) (Table 3).

13. Future Work

At present, there are five areas that should be the focus of future studies to improve the performance and efficacy of CO_2 as a non-physical barrier. Firstly, additional work should focus on defining differences in both avoidance and tolerance thresholds across fish of different sizes; this is particularly important given the possibility that electricity as a barrier may lose effectiveness against small fish [43]. Currently, work that quantifies avoidance and tolerance thresholds across a range of sizes of fishes, within a single study with consistent methods, has not occurred. Owing to the likelihood that a CO_2 barrier would be encountered by fish of a range of sizes, the ability to confidently predict the response of different sized individuals to either a tolerance or avoidance application of CO_2 is critical. Secondly, the exact parameters of the time × concentration interaction to induce equilibrium loss associated with a tolerance-focused barrier have not been defined extensively, and would need to be parameterized across target species before tolerance barriers could be developed and/or implemented. Ideally, this work would be conducted across a range of temperatures. Thirdly, work should be conducted that pairs CO_2 barriers with additional stimuli (e.g., deploy a sound barrier and CO_2 barrier concurrently as in Ruebush et al. [115], or use CO_2 as part of a bubble curtain rather than compressed air). No non-physical barrier is 100% effective at stopping all fish [45], but a CO_2 barrier paired with a second stimulus (e.g., light or sound barriers) could synergistically improve the overall effectiveness of each barrier, increasing the potential to deter invasive fishes across a range of conditions. Penultimately, efforts need to occur to quantify the logistics of CO_2 deployment, including cost estimates, deployment feasibility and infrastructure requirements to assist with future planning efforts. The design for deploying a CO_2 barrier will vary across sites and situations, but efforts to share costs and strategies across successful applications will help improve deployment efficiency and ensure success across locations. Finally, owing to the unavoidable reductions in pH that occur with zones of elevated CO_2, work should continue to quantify the environmental impacts [116,117], consequences for non-target organisms (e.g., mussels [118,119]; native fishes [120–122]; crayfish [123]) and strategies for CO_2 off-gassing. Work to both mitigate CO_2 applications, coupled with efforts to predict possible impacts to non-target organisms or the receiving environment, will help improve the likelihood of a successful application. Together, work to address these 5 concerns will not only help improve the effectiveness of CO_2 as a non-physical barrier, but also will help minimize unintended environmental consequences and improve the efficiency of CO_2 as a tool to deter invasive fishes.

14. Conclusions

Invasive species represent a significant threat to global biodiversity, and models suggest that the rate of introduction of invasive species will likely accelerate in the future [16]. For fishes in North America, bigheaded carp represent a current threat to the Mississippi ecosystem, and there is potential for them to gain access to waterbodies in the eastern portion of the continent should they pass through the Chicago Area Waterway System (CAWS) into the Great Lakes basin. Carbon dioxide (CO_2) is a naturally occurring compound that provides ecologically-relevant information to a host of taxa. A number of different studies, conducted across a range of conditions, have demonstrated that zones of elevated carbon dioxide gas can be an effective non-physical barrier to deter the spread of invasive fishes. More specifically, fish will voluntarily swim away from zones of high CO_2 once a target threshold has been reached, or else equilibrium loss will occur due to the anesthetic properties of CO_2, providing two different mechanisms by which carbon dioxide can deter fish movement. This response has been documented for a number of taxonomically diverse species of fish, and also across a range of sizes spanning from larvae to adults. In addition, unlike physical barriers, a CO_2 barrier can be deployed without requiring the construction of permanent structures that can modify water flow or boat traffic. Several internal and external factors can influence the response of fishes to CO_2, making them more effective, or less effective (e.g., fish experiencing stress will require additional CO_2 to induce avoidance relative to non-stressed individuals; shoals of fish require less CO_2 to induce avoidance relative to solitary individuals) and need to be considered when defining target thresholds should CO_2 be deployed in the field. Additional studies to define effective deployment strategies at large scales, cost, and impacts to the receiving environment should continue as CO_2 barriers grow in popularity and field applications. Work is currently ongoing to develop other non-physical barriers to deter invasive fishes (e.g., sound, electricity, strobe lights), and the lessons learned and experiences described here from CO_2 can serve as potential considerations to refine the application of other barrier technologies to increase their effectiveness. Together, with continued exploration and testing, it is hoped that barrier technologies can be further developed to prevent the spread of invasive fishes and protect freshwater biodiversity.

Funding: This research was funded by the Illinois Department of Natural Resources, and the United States Geological Survey, through funds provided by the United States Environmental Protection Agency's Great Lakes Restoration Initiative.

Acknowledgments: A number of individuals have contributed to field and laboratory work that have generated the data contained in this review, including Caleb Hasler, Jennifer Jeffrey, Kelly Hannan, John Tix, Eric Schneider, Emi Tucker, Madison Philipp, Ian Bouyoucos, Christa Woodley, Cody Sullivan, Jason Romine, Dave Smith, Steve Midway, Aaron Cupp, Jon Amberg, Mark Gaikowski, Clark Dennis, Matt Noatch, Dan Kates, Shivani Adhikari, Michael Donaldson, and Adam Wright. Jake Wolf fish hatchery provided fish for experiments, and staff at the Illinois River Biological Station facilitated the collection of carp. Caleb Hasler provided comments on an early version of this review.

Conflicts of Interest: The author declares no conflict of interest.

References

1. Lodge, D.M.; Williams, S.; MacIsaac, H.J.; Hayes, K.R.; Leung, B.; Reichard, S.; Mack, R.N.; Moyle, P.B.; Smith, M.; Andow, D.A.; et al. Biological Invasions: Recommendations for US Policy and Management. *Ecol. Appl.* **2006**, *16*, 2035–2054. [CrossRef]
2. Britton, J.R.; Davies, G.D.; Harrod, C. Trophic Interactions and Consequent Impacts of the Invasive Fish Pseudorasbora Parva in a Native Aquatic Foodweb: A Field Investigation in the UK. *Boil. Invasions* **2009**, *12*, 1533–1542. [CrossRef]
3. Gozlan, R.E.; Britton, J.R.; Cowx, I.; Copp, G.H. Current Knowledge on Non-Native Freshwater Fish Introductions. *J. Fish Boil.* **2010**, *76*, 751–786. [CrossRef]
4. Wilcove, D.S.; Rothstein, D.; Dubow, J.; Phillips, A.; Losos, E. Quantifying Threats to Imperiled Species in the United States. *BioScience* **1998**, *48*, 607–615. [CrossRef]
5. Pimentel, D.; Zuniga, R.; Morrison, D. Update on the Environmental and Economic Costs Associated with Alien-Invasive Species in the United States. *Ecol. Econ.* **2005**, *52*, 273–288. [CrossRef]
6. Pejchar, L.; Mooney, H.A. Invasive Species, Ecosystem Services and Human Well-Being. *Trends Ecol. Evol.* **2009**, *24*, 497–504. [CrossRef]
7. Strayer, D.L.; Dudgeon, D. Freshwater Biodiversity Conservation: Recent Progress and Future Challenges. *J. N. Am. Benthol. Soc.* **2010**, *29*, 344–358. [CrossRef]
8. Wu, J. Landscape Sustainability Science: Ecosystem Services and Human Well-Being in Changing Landscapes. *Landsc. Ecol.* **2013**, *28*, 999–1023. [CrossRef]
9. Jenkins, M. Prospects for Biodiversity. *Science* **2003**, *302*, 1175–1177. [CrossRef]
10. Reid, A.J.; Carlson, A.K.; Creed, I.F.; Eliason, E.J.; Gell, P.A.; Johnson, P.T.J.; Kidd, K.A.; MacCormack, T.J.; Olden, J.D.; Ormerod, S.J.; et al. Emerging Threats and Persistent Conservation Challenges for Freshwater Biodiversity. *Boil. Rev.* **2018**, *94*, 849–873. [CrossRef]
11. Dudgeon, D.; Arthington, A.H.; Gessner, M.O.; Kawabata, Z.-I.; Knowler, D.J.; Lévêque, C.; Naiman, R.J.; Prieur-Richard, A.; Soto, D.; Stiassny, M.L.J.; et al. Freshwater Biodiversity: Importance, Threats, Status and Conservation Challenges. *Boil. Rev.* **2005**, *81*, 163–182. [CrossRef] [PubMed]
12. Jelks, H.L.; Walsh, S.J.; Burkhead, N.M.; Contreras-Balderas, S.; Díaz-Pardo, E.; Hendrickson, D.A.; Lyons, J.; Mandrak, N.E.; McCormick, F.; Nelson, J.S.; et al. Conservation Status of Imperiled North American Freshwater and Diadromous Fishes. *Fisheries* **2008**, *33*, 372–407. [CrossRef]
13. Burkhead, N.M. Extinction Rates in North American Freshwater Fishes, 1900–2010. *BioScience* **2012**, *62*, 798–808. [CrossRef]
14. Seebens, H.; Blackburn, T.M.; Dyer, E.E.; Genovesi, P.; Hulme, P.E.; Jeschke, J.M.; Pagad, S.; Pyšek, P.; Winter, M.; Arianoutsou, M.; et al. No Saturation in the Accumulation of Alien Species Worldwide. *Nat. Commun.* **2017**, *8*, 1–9. [CrossRef] [PubMed]
15. Chapin, F.S.; Zavaleta, E.S.; Eviner, V.T.; Naylor, R.L.; Vitousek, P.M.; Reynolds, H.L.; Hooper, D.U.; Lavorel, S.; Sala, O.E.; Hobbie, S.; et al. Consequences of Changing Biodiversity. *Nature* **2000**, *405*, 234–242. [CrossRef] [PubMed]
16. Sardain, A.; Sardain, E.; Leung, B. Global Forecasts of Shipping Traffic and Biological Invasions to 2050. *Nat. Sustain.* **2019**, *2*, 274–282. [CrossRef]

17. Clout, M.N.; Veitch, C.R. Turning the tide of biological invasion: the potential for eradicating invasive species. In *Turning the Tide: The Eradication of Invasive Species*; IUCN SSC Invasive Species Specialist Group: Gland, Switzerland; Cambridge, UK, 2002; pp. 1–3. Available online: http://www.issg.org/pdf/publications/turning_the_tide.pdf (accessed on 1 June 2020).
18. Simberloff, D. Eradication—Preventing Invasions at the Outset. *Weed Sci.* **2003**, *51*, 247–253. [CrossRef]
19. Simberloff, D. We can eliminate invasions or live with them. Successful management projects. In *Ecological Impacts of Non-Native Invertebrates and Fungi on Terrestrial Ecosystems*; Langor, D., Sweeney, J., Eds.; Springer: Dordrecht, The Netherlands, 2008; pp. 149–157, ISBN 978-1-4020-9680-8.
20. Simberloff, D. Eradication: Pipe dream or real option? In *Plant Invasions in Protected Areas*; Foxcroft, L.C., Pyšek, P., Richardson, D.M., Genovesi, P., Eds.; Springer: Dordrecht, The Netherlands, 2013; pp. 549–559, ISBN 978-94-007-7750-7.
21. Leung, B.; Lodge, D.M.; Finnoff, D.; Shogren, J.F.; Lewis, M.A.; Lamberti, G. An Ounce of Prevention or a Pound of Cure: Bioeconomic Risk Analysis of Invasive Species. *Proc. R. Soc. B Boil. Sci.* **2002**, *269*, 2407–2413. [CrossRef]
22. Finnoff, D.; Shogren, J.F.; Leung, B.; Lodge, D. Take a Risk: Preferring Prevention over Control of Biological Invaders. *Ecol. Econ.* **2007**, *62*, 216–222. [CrossRef]
23. Zanden, M.J.V.; Olden, J.D. A Management Framework for Preventing the Secondary Spread of Aquatic Invasive Species. *Can. J. Fish. Aquat. Sci.* **2008**, *65*, 1512–1522. [CrossRef]
24. Kocovsky, P.M.; Chapman, D.C.; Qian, S. "Asian Carp" Is Societally and Scientifically Problematic. Let's Replace It. *Fisheries* **2018**, *43*, 311–316. [CrossRef]
25. Kolar, C.S.; Chapman, D.C.; Courtenay, W.R.J.; Housel, C.M.; Williams, J.D.; Jennings, D.P. *Asian Carps of the Genus Hypophthalmichthys (Pisces, Cyprinidae)—A Biological Synopsis and Environmental Risk Assessment*; American Fisheries Society Special Publication: Bethesda, MD, USA, 2007; Volume 33, ISBN 978-1-888569-79-7.
26. Whitledge, G.W.; Knights, B.; Vallazza, J.; Larson, J.; Weber, M.J.; Lamer, J.T.; Phelps, Q.E.; Norman, J.D. Identification of Bighead Carp and Silver Carp Early-Life Environments and Inferring Lock and Dam 19 Passage in the Upper Mississippi River: Insights from Otolith Chemistry. *Boil. Invasions* **2018**, *21*, 1007–1020. [CrossRef]
27. Sass, G.G.; Hinz, C.; Erickson, A.C.; McClelland, N.N.; McClelland, M.A.; Epifanio, J.M. Invasive Bighead and Silver Carp Effects on Zooplankton Communities in the Illinois River, Illinois, USA. *J. Great Lakes Res.* **2014**, *40*, 911–921. [CrossRef]
28. Kuznetsov, Y.A. Consumption of Bacteria by the Silver Carp (*Hypophthalmichthys molitrix*). *J. Ichthyol.* **1977**, *17*, 398–403.
29. Fukushima, M.; Takamura, N.; Sun, L.; Nakagawa, M.; Matsushige, K.; Xie, P. Changes in the Plankton Community Following Introduction of Filter-Feeding Planktivorous Fish. *Freshw. Boil.* **1999**, *42*, 719–735. [CrossRef]
30. Laws, E.A.; Weisburd, R. Use of Silver Carp to Control Algal Biomass in Aquaculture Ponds. *Progress. Fish-Culturist* **1990**, *52*, 1–8. [CrossRef]
31. Lieberman, D.M. Use of Silver Carp (*Hypophthalmichthys molotrix*) and Bighead Carp (*Aristichthys nobilis*) for Algae Control in a Small Pond: Changes in Water Quality. *J. Freshw. Ecol.* **1996**, *11*, 391–397. [CrossRef]
32. Irons, K.S.; Sass, G.G.; McClelland, M.A.; Stafford, J.D. Reduced Condition Factor of Two Native Fish Species Coincident with Invasion of Non-Native Asian Carps in the Illinois River, USA Is This Evidence for Competition and Reduced Fitness? *J. Fish Boil.* **2007**, *71*, 258–273. [CrossRef]
33. Pendleton, R.M.; Schwinghamer, C.; Solomon, L.E.; Casper, A.F. Competition among River Planktivores: Are Native Planktivores Still Fewer and Skinnier in Response to the Silver Carp Invasion? *Environ. Boil. Fishes* **2017**, *100*, 1213–1222. [CrossRef]
34. Chick, J.H.; Gibson-Reinemer, D.K.; Soeken-Gittinger, L.; Casper, A.F. Invasive Silver Carp is Empirically Linked to Declines of Native Sport Fish in the Upper Mississippi River System. *Boil. Invasions* **2019**, *22*, 723–734. [CrossRef]
35. Asian Carp Regional Coordinating Committee. Asian Carp Monitoring and Response Plan. 2018. Available online: https://www.asiancarp.us/Documents/MRP2018.pdf (accessed on 1 June 2020).
36. Hill, L. *The Chicago River: A Natural and Unnatural History*; Lake Claremont Press: Chicago, IL, USA, 2000.

37. Moy, P.B.; Polls, I.; Dettmers, J.M. The Chicago Sanitary and Ship Canal aquatic nuisance species dispersal barrier. In *Invasive Asian Carps in North America, American Fisheries Society Symposium*; Chapman, D.C., Hoff, M.H., Eds.; American Fisheries Society: Bethesda, MD, USA, 2010; Volume 74, pp. 121–137.
38. Rasmussen, J.L.; Regier, H.A.; Sparks, R.E.; Taylor, W.W. Dividing the Waters: The Case for Hydrologic Separation of the North American Great Lakes and Mississippi River Basins. *J. Great Lakes Res.* **2011**, *37*, 588–592. [CrossRef]
39. Asian Carp Regional Coordinating Committee. Asian Carp Action Plan. 2020. Available online: https://www.asiancarp.us/Documents/2020-Action-Plan.pdf (accessed on 1 June 2020).
40. Dettmers, J.M.; Boisvert, B.A.; Barkley, T.; Sparks, R.E. *Potential Impact of Steel-Hulled Barges on Movement of Fish across an Electric Barrier to Prevent the Entry of Invasive Carp into Lake Michigan*; Aquatic Ecology Technical Report 2005/19; Illinois Natural History Survey Center for Aquatic Ecology: Zion, IL, USA, 2005; Available online: https://www.ideals.illinois.edu/bitstream/handle/2142/10091/inhscaev02005i00019_opt.pdf?sequence=2&isAllowed=y (accessed on 1 June 2020).
41. Sparks, R.E.; Barkley, T.L.; Creque, S.M.; Dettmers, J.M.; Stainbrook, K.M. Evaluation of an electric fish dispersal barrier in the Chicago Sanitary and Ship Canal. In *Invasive Asian Carps in North America, American Fisheries Society Symposium*; Chapman, D.C., Hoff, M.H., Eds.; American Fisheries Society: Bethesda, MD, USA, 2010; Volume 74, pp. 139–161.
42. Evans, N.T.; Brouder, M.J. *Asian Carp Entrainment, Retainment and Upstream Transport by Commercial Barge tows on the Illinois Waterway—2018 Trials*; US Fish & Wildlife Service Report; US Fish and Wildlife Service Carterville Fish and Wildlife Conservation Office: Willmington, IL, USA, 2020. Available online: https://www.fws.gov/midwest/fisheries/carterville/documents/2018-Barge-Entrainment-Study-Report.pdf (accessed on 1 June 2020).
43. Parker, A.D.; Rogers, P.B.; Finney, S.T.; Simmonds, R.L.J. *Preliminary Results of Fixed DIDSON Evaluations at the Electric Dispersal Barrier in the Chicago Sanitary and Ship Canal*; US Fish & Wildlife Service Report; US Fish and Wildlife Service, Carterville Fish and Wildlife Conservation Office: Willmington, IL, USA, 2013. Available online: https://www.fws.gov/midwest/fisheries/carterville/documents/DIDSON.pdf (accessed on 1 June 2020).
44. Reynolds, J.B. Electrofishing. In *Fisheries Techniques*, 2nd ed.; Murphy, B.R., Willis, D.W., Eds.; American Fisheries Society: Bethesda, MD, USA, 1996; pp. 221–253, ISBN 9781888569001.
45. Noatch, M.R.; Suski, C.D. Non-Physical Barriers to Deter Fish Movements. *Environ. Rev.* **2012**, *20*, 71–82. [CrossRef]
46. USACE. *The Great Lakes and Mississippi River Interbasin Study—Brandon Road Final Integrated Feasibility Study and Environmental Impact Statement—Will County, Illinois*; US Army Corps of Engineers, Rock Island and Chicago Districts: Rock Island, TN, USA; Chicago, IL, USA, 2018; Available online: https://usace.contentdm.oclc.org/utils/getfile/collection/p16021coll7/id/11394 (accessed on 1 June 2020).
47. Cooke, S.; Hill, W.R. Can Filter-Feeding Asian Carp Invade the Laurentian Great Lakes? A Bioenergetic Modelling Exercise. *Freshw. Boil.* **2010**, *55*, 2138–2152. [CrossRef]
48. Cuddington, K.; Currie, W.J.S.; Koops, M.A. Could an Asian Carp Population Establish in the Great Lakes from a Small Introduction? *Boil. Invasions* **2013**, *16*, 903–917. [CrossRef]
49. Wittmann, M.E.; Cooke, R.M.; Rothlisberger, J.D.; Rutherford, E.S.; Zhang, H.; Mason, D.M.; Lodge, D.M. Use of Structured Expert Judgment to Forecast Invasions by Bighead and Silver Carp in Lake Erie. *Conserv. Boil.* **2014**, *29*, 187–197. [CrossRef] [PubMed]
50. Lauber, T.B.; Stedman, R.C.; Connelly, N.A.; Rudstam, L.G.; Ready, R.C.; Poe, G.L.; Bunnell, D.B.; Höök, T.O.; Koops, M.A.; Ludsin, S.A.; et al. Using Scenarios to Assess Possible Future Impacts of Invasive Species in the Laurentian Great Lakes. *N. Am. J. Fish. Manag.* **2016**, *36*, 1292–1307. [CrossRef]
51. Zhang, H.; Rutherford, E.S.; Mason, D.M.; Breck, J.T.; Wittmann, M.E.; Cooke, R.M.; Lodge, D.M.; Rothlisberger, J.D.; Zhu, X.; Johnson, T.B. Forecasting the Impacts of Silver and Bighead Carp on the Lake Erie Food Web. *Trans. Am. Fish. Soc.* **2015**, *145*, 136–162. [CrossRef]
52. Cummins, E.P.; Strowitzki, M.J.; Taylor, C.T. Mechanisms and Consequences of Oxygen and Carbon Dioxide Sensing in Mammals. *Physiol. Rev.* **2020**, *100*, 463–488. [CrossRef]
53. Cummins, E.P.; Selfridge, A.C.; Sporn, P.H.S.; Sznajder, J.I.; Taylor, C.T. Carbon Dioxide-Sensing in Organisms and Its Implications for Human Disease. *Cell. Mol. Life Sci.* **2013**, *71*, 831–845. [CrossRef]

54. Thom, C.; Guerenstein, P.G.; Mechaber, W.L.; Hildebrand, J.G. Floral CO_2 Reveals Flower Profitability to Moths. *J. Chem. Ecol.* **2004**, *30*, 1285–1288. [CrossRef]
55. Seeley, T.D. Atmospheric Carbon Dioxide Regulation in Honey-Bee (*Apis mellifera*) Colonies. *J. Insect Physiol.* **1974**, *20*, 2301–2305. [CrossRef]
56. Gillies, M.T. The Role of Carbon Dioxide in Host-Finding by Mosquitoes (Diptera: Culicidae): A Review. *Bull. Entomol. Res.* **1980**, *70*, 525–532. [CrossRef]
57. Takken, W.; Knols, B.G.J. Odor-Mediated Behavior of Afrotropical Malaria Mosquitoes. *Annu. Rev. Entomol.* **1999**, *44*, 131–157. [CrossRef] [PubMed]
58. Faucher, C. Behavioral Responses of Drosophila to Biogenic Levels of Carbon Dioxide Depend on Life-Stage, Sex and Olfactory Context. *J. Exp. Boil.* **2006**, *209*, 2739–2748. [CrossRef]
59. Lahiri, S.; Forster II, R.E. CO_2/H^+ Sensing: Peripheral and Central Chemoreception. *Int. J. Biochem. Cell Biol.* **2003**, *35*, 1413–1435. [CrossRef]
60. Shusterman, D. Individual Factors in Nasal Chemesthesis. *Chem. Senses* **2002**, *27*, 551–564. [CrossRef]
61. Tresguerres, M.; Milsom, W.K.; Perry, S.F. CO_2 and acid-base sensing. In *Carbon Dioxide*; Farrell, A.P., Brauner, C.J., Eds.; Elsevier: San Diego, CA, USA, 2019; Volume 37, pp. 33–68, ISBN 9780128176108.
62. Eddy, F.B.; Lomholt, J.P.; Weber, R.E.; Johansen, K. Blood Respiratory Properties of Rainbow Trout (*Salmo gairdneri*) Kept in Water of High CO_2 Tension. *J. Exp. Boil.* **1977**, *67*, 37–47.
63. Brauner, C.J.; Baker, D.W. Patterns of acid–base regulation during exposure to hypercarbia in fishes. In *Cardio-Respiratory Control in Vertebrates*; Glass, M.L., Wood, S.C., Eds.; Springer: Berlin/Heidelberg, Germany, 2009; pp. 43–63, ISBN 978-3-540-93984-9.
64. Bernier, N.J.; Randall, D.J. Carbon Dioxide Anaesthesia in Rainbow Trout: Effects of Hypercapnic Level and Stress on Induction and Recovery from Anaesthetic Treatment. *J. Fish Biol.* **1998**, *52*, 621–637.
65. Dennis, C.E.; Kates, D.F.; Noatch, M.R.; Suski, C.D. Molecular Responses of Fishes to Elevated Carbon Dioxide. *Comp. Biochem. Physiol. Part A Mol. Integr. Physiol.* **2015**, *187*, 224–231. [CrossRef]
66. Kates, D.; Dennis, C.; Noatch, M.R.; Suski, C.D. Responses of Native and Invasive Fishes to Carbon Dioxide: Potential for a Nonphysical Barrier to Fish Dispersal. *Can. J. Fish. Aquat. Sci.* **2012**, *69*, 1748–1759. [CrossRef]
67. Iwama, G.K.; McGeer, J.C.; Pawluk, M.P. The Effects of Five Fish Anaesthetics on Acid-Base Balance, Hematocrit, Blood Gases, Cortisol, and Adrenaline in Rainbow Trout. *Can. J. Zool.* **1989**, *67*, 2065–2073. [CrossRef]
68. Brauner, C.J.; Seidelin, M.; Madsen, S.S.; Jensen, F.B. Effects of Freshwater Hyperoxia and Hypercapnia and Their Influences on Subsequent Seawater Transfer in Atlantic Salmon (*Salmo salar*) Smolts. *Can. J. Fish. Aquat. Sci.* **2000**, *57*, 2054–2064. [CrossRef]
69. Fish, F.F. The Anaesthesia of Fish by High Carbon-Dioxide Concentrations. *Trans. Am. Fish. Soc.* **1943**, *72*, 25–29. [CrossRef]
70. Post, G. Carbonic Acid Anesthesia for Aquatic Organisms. *Progress. Fish-Culturist* **1979**, *41*, 142–144. [CrossRef]
71. Yoshikawa, H.; Yokoyama, Y.; Ueno, S.; Mitsuda, H. Changes of Blood Gas in Carp, Cyprinus Carpio, Anesthetized with Carbon Dioxide. *Comp. Biochem. Physiol. Part A Physiol.* **1991**, *98*, 431–436. [CrossRef]
72. Yoshikawa, H.; Kawai, F.; Kanamori, M. The Relationship between the EEG and Brain pH in Carp, *Cyprinus carpio*, Subjected to Environmental Hypercapnia at an Anesthetic Level. *Comp. Biochem. Physiol. A Physiol.* **1994**, *107*, 307–312.
73. Beitinger, T.L. Behavioral Reactions for the Assessment of Stress in Fishes. *J. Great Lakes Res.* **1990**, *16*, 495–528. [CrossRef]
74. Tierney, K.B. Chemical Avoidance Responses of Fishes. *Aquat. Toxicol.* **2016**, *174*, 228–241. [CrossRef]
75. Shelford, V.E.; Allee, W.C. The Reactions of Fishes to Gradients of Dissolved Atmospheric Gases. *J. Exp. Zool.* **1913**, *14*, 207–266. [CrossRef]
76. Powers, E.B.; Clark, R.T. Further Evidence on Chemical Factors Affecting the Migratory Movements of Fishes, Especially the Salmon. *Ecology* **1943**, *24*, 109–113. [CrossRef]
77. Collins, B.G. Factors Influencing the Orientation of Migrating Anadromous Fishes. *Fish. Bull.* **1952**, *52*, 375–396.
78. Bishai, H.M. Reactions of Larval and Young Salmonids to Different Hydrogen Ion Concentrations. *ICES J. Mar. Sci.* **1962**, *27*, 181–191. [CrossRef]
79. Jones, K.A.; Hara, T.J.; Scherer, E. Locomotor Response by Arctic Char (*Salvelinus alpinus*) to Gradients of H^+ and CO_2. *Physiol. Zool.* **1985**, *58*, 413–420. [CrossRef]

80. Ross, R.M.; Krise, W.F.; Redell, L.A.; Bennett, R.M. Effects of Dissolved Carbon Dioxide on the Physiology and Behavior of Fish in Artificial Streams. *Environ. Toxicol.* **2001**, *16*, 84–95. [CrossRef]
81. Clingerman, J.; Bebak, J.; Mazik, P.M.; Summerfelt, S.T. Use of Avoidance Response by Rainbow Trout to Carbon Dioxide for Fish Self-Transfer between Tanks. *Aquac. Eng.* **2007**, *37*, 234–251. [CrossRef]
82. Yoshikawa, H.; Ishida, Y.; Ueno, S.; Mitsuda, H. Anesthetic Effect of CO_2 on Fish. I. Changes in Depth of Anesthesia of the Carp Anesthetized with a Constant Level of CO_2. *Nippon. Suisan Gakkaishi* **1988**, *54*, 457–462. [CrossRef]
83. Pirhonen, J.; Schreck, C.B. Effects of Anaesthesia with MS-222, Clove Oil and CO_2 on Feed Intake and Plasma Cortisol in Steelhead Trout (*Oncorhynchus mykiss*). *Aquaculture* **2003**, *220*, 507–514. [CrossRef]
84. Dennis, C.E.; Adhikari, S.; Suski, C.D. Molecular and Behavioral Responses of Early-Life Stage Fishes to Elevated Carbon Dioxide. *Boil. Invasions* **2015**, *17*, 3133–3151. [CrossRef]
85. Summerfelt, R.C.; Lewis, W.M. Repulsion of Green Sunfish by Certain Chemicals. *J. Water Pollut. Control. Fed.* **1967**, *39*, 2030–2038.
86. Donaldson, M.R.; Amberg, J.; Adhikari, S.; Cupp, A.R.; Jensen, N.; Romine, J.; Wright, A.; Gaikowski, M.P.; Suski, C.D. Carbon Dioxide as a Tool to Deter the Movement of Invasive Bigheaded Carps. *Trans. Am. Fish. Soc.* **2016**, *145*, 657–670. [CrossRef]
87. Cupp, A.R.; Erickson, R.; Fredricks, K.T.; Swyers, N.M.; Hatton, T.W.; Amberg, J.J. Responses of Invasive Silver and Bighead Carp to a Carbon Dioxide Barrier in Outdoor Ponds. *Can. J. Fish. Aquat. Sci.* **2017**, *74*, 297–305. [CrossRef]
88. Cupp, A.R.; Smerud, J.; Tix, J.; Schleis, S.; Fredricks, K.; Erickson, R.A.; Amberg, J.; Morrow, W.; Koebel, C.; Murphy, E.; et al. Field Evaluation of Carbon Dioxide as a Fish Deterrent at a Water Management Structure along the Illinois River. *Manag. Boil. Invasions* **2018**, *9*, 299–308. [CrossRef]
89. Hasler, C.T.; Woodley, C.M.; Schneider, E.V.; Hixson, B.K.; Fowler, C.J.; Midway, S.R.; Suski, C.D.; Smith, D.L. Avoidance of Carbon Dioxide in Flowing Water by Bighead Carp. *Can. J. Fish. Aquat. Sci.* **2019**, *76*, 961–969. [CrossRef]
90. Sampson, S.J.; Chick, J.H.; Pegg, M.A. Diet Overlap among Two Asian Carp and Three Native Fishes in Backwater Lakes on the Illinois and Mississippi Rivers. *Boil. Invasions* **2008**, *11*, 483–496. [CrossRef]
91. Deters, J.E.; Chapman, D.C.; McElroy, B. Location and Timing of Asian Carp spawning in the Lower Missouri River. *Environ. Boil. Fishes* **2012**, *96*, 617–629. [CrossRef]
92. Réale, D.; Garant, D.; Humphries, M.M.; Bergeron, P.; Careau, V.; Montiglio, P.-O. Personality and the Emergence of the Pace-Of-Life Syndrome Concept at the Population Level. *Philos. Trans. R. Soc. B Boil. Sci.* **2010**, *365*, 4051–4063. [CrossRef]
93. Myles-Gonzalez, E.; Burness, G.; Yavno, S.; Rooke, A.C.; Fox, M.G. To Boldly Go Where No Goby Has Gone Before: Boldness, Dispersal Tendency, and Metabolism at the Invasion Front. *Behav. Ecol.* **2015**, *26*, 1083–1090. [CrossRef]
94. Cockrem, J.F. Stress, Corticosterone Responses and Avian Personalities. *J. Ornithol.* **2007**, *148*, 169–178. [CrossRef]
95. Koolhaas, J.; Korte, S.M.; De Boer, S.; Van Der Vegt, B.; Van Reenen, C.; Hopster, H.; De Jong, I.; Ruis, M.; Blokhuis, H. Coping Styles in Animals: Current Status in Behavior and Stress-Physiology. *Neurosci. Biobehav. Rev.* **1999**, *23*, 925–935. [CrossRef]
96. Réale, D.; Reader, S.M.; Sol, D.; McDougall, P.T.; Dingemanse, N.J. Integrating Animal Temperament within Ecology and Evolution. *Boil. Rev.* **2007**, *82*, 291–318. [CrossRef]
97. Tucker, E.K.; Suski, C.D.; Philipp, M.A.; Jeffrey, J.D.; Hasler, C.T. Glucocorticoid and Behavioral Variation in Relation to Carbon Dioxide Avoidance across Two Experiments in Freshwater Teleost Fishes. *Boil. Invasions* **2018**, *21*, 505–517. [CrossRef]
98. Tucker, E.K.; Suski, C.D. Presence of Conspecifics Reduces Between-Individual Variation and Increases Avoidance of Multiple Stressors in Bluegill. *Anim. Behav.* **2019**, *158*, 15–24. [CrossRef]
99. Killen, S.S.; Marras, S.; Metcalfe, N.B.; McKenzie, D.J.; Domenici, P. Environmental Stressors Alter Relationships between Physiology and Behaviour. *Trends Ecol. Evol.* **2013**, *28*, 651–658. [CrossRef] [PubMed]
100. Metcalfe, N.B.; Van Leeuwen, T.E.; Killen, S.S. Does Individual Variation in Metabolic Phenotype Predict Fish Behaviour and Performance? *J. Fish Boil.* **2015**, *88*, 298–321. [CrossRef] [PubMed]
101. Suski, C.D.; Philipp, M.A.; Hasler, C.T. Influence of Nutritional Status on Carbon Dioxide Tolerance and Avoidance Behavior in a Freshwater Teleost. *Trans. Am. Fish. Soc.* **2019**, *148*, 914–925. [CrossRef]

102. Nadler, L.E.; Killen, S.S.; McClure, E.C.; Munday, P.L.; McCormick, M.I. Shoaling Reduces Metabolic Rate in a Gregarious Coral Reef Fish Species. *J. Exp. Boil.* **2016**, *219*, 2802–2805. [CrossRef] [PubMed]
103. Dennis, C.E.; Adhikari, S.; Wright, A.W.; Suski, C.D.; Dennis, C.E. Molecular, Behavioral, and Performance Responses of Juvenile Largemouth Bass Acclimated to an Elevated Carbon Dioxide Environment. *J. Comp. Physiol. B* **2016**, *186*, 297–311. [CrossRef]
104. Cupp, A.R.; Tix, J.; Smerud, J.; Erickson, R.A.; Fredricks, K.; Amberg, J.; Suski, C.D.; Wakeman, R. Using Dissolved Carbon Dioxide to Alter the Behavior of Invasive round Goby. *Manag. Boil. Invasions* **2017**, *8*, 567–574. [CrossRef]
105. Tix, J.A.; Cupp, A.R.; Smerud, J.R.; Erickson, R.; Fredricks, K.T.; Amberg, J.J.; Suski, C.D. Temperature Dependent Effects of Carbon Dioxide on Avoidance Behaviors in Bigheaded Carps. *Boil. Invasions* **2018**, *20*, 3095–3105. [CrossRef]
106. Beitinger, T.L.; Lutterschmidt, W.I. Measures of thermal tolerance. In *Encyclopedia of Fish Physiology*, 1st ed.; Farrell, A.P., Ed.; Elsevier: Waltham, MA, USA, 2011; pp. 1695–1702, ISBN 9780080923239.
107. Somero, G. Temporal Patterning of Thermal Acclimation: From Behavior to Membrane Biophysics. *J. Exp. Boil.* **2015**, *218*, 167–169. [CrossRef]
108. Hasler, C.T.; Bouyoucos, I.A.; Suski, C.D. Tolerance to Hypercarbia Is Repeatable and Related to a Component of the Metabolic Phenotype in a Freshwater Fish. *Physiol. Biochem. Zool.* **2017**, *90*, 583–587. [CrossRef] [PubMed]
109. Killen, S.S.; Adriaenssens, B.; Marras, S.; Claireaux, G.; Cooke, S.J. Context Dependency of Trait Repeatability and Its Relevance for Management and Conservation of Fish Populations. *Conserv. Physiol.* **2016**, *4*, cow007. [CrossRef] [PubMed]
110. Gelwicks, K.R.; Zafft, D.J.; Bobbitt, J.P. Efficacy of Carbonic Acid as an Anesthetic for Rainbow Trout. *N. Am. J. Fish. Manag.* **1998**, *18*, 432–438. [CrossRef]
111. Fivelstad, S.; Waagbø, R.; Stefansson, S.; Olsen, A.B. Impacts of Elevated Water Carbon Dioxide Partial Pressure at Two Temperatures on Atlantic Salmon (*Salmo salar L.*) Parr Growth and Haematology. *Aquaculture* **2007**, *269*, 241–249. [CrossRef]
112. Neiffer, D.L.; Stamper, M.A. Fish Sedation, Anesthesia, Analgesia, and Euthanasia: Considerations, Methods, and Types of Drugs. *ILAR J.* **2009**, *50*, 343–360. [CrossRef]
113. Fredricks, K.T.; Hubert, T.D.; Amberg, J.J.; Cupp, A.R.; Dawson, V.K. Chemical Controls for an Integrated Pest Management Program. *N. Am. J. Fish. Manag.* **2019**. Available online: https://afspubs.onlinelibrary.wiley.com/doi/abs/10.1002/nafm.10339 (accessed on 2 June 2020). [CrossRef]
114. Schneider, E.V.; Hasler, C.T.; Suski, C.D. Swimming Performance of a Freshwater Fish during Exposure to High Carbon Dioxide. *Environ. Sci. Pollut. Res.* **2018**, *26*, 3447–3454. [CrossRef]
115. Ruebush, B.; Sass, G.; Chick, J.; Stafford, J. In-Situ Tests of Sound-Bubble-Strobe Light Barrier Technologies to Prevent Range Expansions of Asian Carp. *Aquat. Invasions* **2012**, *7*, 37–48. [CrossRef]
116. Hasler, C.T.; Midway, S.R.; Jeffrey, J.D.; Tix, J.A.; Sullivan, C.; Suski, C.D. Exposure to Elevated pCO_2 Alters Post-Treatment Diel Movement Patterns of Largemouth Bass over Short Time Scales. *Freshw. Boil.* **2016**, *61*, 1590–1600. [CrossRef]
117. Hasler, C.T.; Jeffrey, J.D.; Butman, D.; Suski, C. Freshwater Biota and Rising pCO_2? *Ecol. Lett.* **2016**, *19*, 98–108. [CrossRef]
118. Jeffrey, J.D.; Hannan, K.D.; Hasler, C.T.; Suski, C.D. Hot and Bothered: Effects of Elevated pCO_2 and Temperature on Juvenile Freshwater Mussels. *Am. J. Physiol. Integr. Comp. Physiol.* **2018**, *315*, R115–R127. [CrossRef] [PubMed]
119. Hannan, K.D.; Jeffrey, J.D.; Hasler, C.T.; Suski, C.D. Physiological Responses of Three Species of Unionid Mussels to Intermittent Exposure to Elevated Carbon Dioxide. *Conserv. Physiol.* **2016**, *4*, cow066. [CrossRef] [PubMed]
120. Hasler, C.T.; Jeffrey, J.D.; Schneider, E.V.; Hannan, K.D.; Tix, J.A.; Suski, C.D. Biological Consequences of Weak Acidification Caused by Elevated Carbon Dioxide in Freshwater Ecosystems. *Hydrobiologia* **2017**, *806*, 1–12. [CrossRef]
121. Midway, S.R.; Hasler, C.T.; Wagner, T.; Suski, C.D. Predation of Freshwater Fish in Elevated Carbon Dioxide Environments. *Mar. Freshwater Res.* **2017**, *68*, 1585–1592. [CrossRef]

122. Tix, J.A.; Hasler, C.T.; Sullivan, C.; Jeffrey, J.D.; Suski, C.D. Elevated Carbon Dioxide Has the Potential to Impact Alarm Cue Responses in Some Freshwater Fishes. *Aquat. Ecol.* **2016**, *51*, 59–72. [CrossRef]
123. Robertson, M.; Hernandez, M.F.; Midway, S.R.; Hasler, C.T.; Suski, C.D. Shelter-Seeking Behavior of Crayfish, Procambarus Clarkii, in Elevated Carbon Dioxide. *Aquat. Ecol.* **2018**, *52*, 225–233. [CrossRef]

Review

Biocontrol of the Common Carp (*Cyprinus carpio*) in Australia: A Review and Future Directions

Kenneth A McColl * and Agus Sunarto

CSIRO-Health and Biosecurity, Australian Animal Health Laboratory, PO Bag 24, Geelong, VIC 3220, Australia; agus.sunarto@csiro.au

* Correspondence: kenneth.mccoll@csiro.au

Received: 22 April 2020; Accepted: 29 May 2020; Published: 2 June 2020

Abstract: Invasive pest species are recognized as one of the important drivers of reduced global biodiversity. In Australia, the 267 invasive plant, animal and microbial species, established since European colonization in the 1770s, have been unequivocally declared the most important threat to species diversity in this country. One invasive pest, the common carp (*Cyprinus carpio*), has been targeted in an integrated pest management plan that might include cyprinid herpesvirus 3 (CyHV-3) as a potential biocontrol agent. The species-specificity of the released virus (and of field variants that will inevitably arise) has been assessed, and the virus judged to be safe. It has also been hypothesised that, because the virulence of the CyHV-3 will likely decline following release, the virus should be used strategically: initially, the aim would be to markedly reduce numbers of carp in naive populations, and then some other, as yet uncertain, complementary broad-scale control measure would knock-down carp numbers even further. Brief results are included from recent studies on the modelling of release and spread of the virus, the ecological and social concerns associated with virus release, and the restoration benefits that might be expected following carp control. We conclude that, while further work is required (on the virus, the target species, environmental issues, and especially the identification of a suitable broad-scale complementary control measure), optimism must prevail in order to ensure an eventual solution to this important environmental problem.

Keywords: biocontrol; Australia; common carp; *Cyprinus carpio*; cyprinid herpesvirus 3; safety; efficacy; modelling; risks

1. Introduction

In 2013, in his comprehensive book on invasive species, Simberloff [1] suggested that biological invasions are (along with climate change and habitat destruction) one of the great anthropogenic threats to global diversity. Subsequently, a United Nations-backed panel of scientists (representing over 130 nations) produced a report in 2019 that unequivocally identified invasive species as one of the important drivers of the global decline in numbers, and frequent extinction, of native animal and plant species. The report of the Intergovernmental Science-Policy Platform on Biodiversity and Ecosystem Sciences (IPBES) [2] noted that at least 680 species of vertebrates, alone, have been lost due to human actions taken since 1500.

The IPBES Report suggested that there were five direct drivers of reduced global biodiversity: (1) changes in the use of land and sea, (2) direct exploitation of the plants and animals of the world, (3) climate change, (4) pollution, and (5) invasive pest species. There has been a 70% increase in numbers of the latter since 1970 across 21 countries where detailed records were maintained. At about the same time that the IPBES Report was released, an Australian group declared [3] that invasive species were, in fact, the major threat to species diversity in Australia followed by modifications to ecosystems and agriculture. Furthermore, they found that this hierarchy of threats was consistent for almost all native plants and animals in Australia, the only exceptions being native fish where pollution replaced agriculture as the third major threat.

It is estimated that, of the thousands of exotic species that have arrived since European colonization of Australia in the 1770s, 267 have become genuine invasive pest species (207 plants, 57 animals, and three microbial pathogens) [3]. Concomitant with this influx of invasive species, at least 93 native species of plants and animals have officially become extinct with the demise of many others either recognised informally, or likely to have gone unrecorded [3]. Australian inland water communities have been forced to contend with cane toads (*Rhinella marina*) and around 43 known invasive freshwater fish species including eastern gambusia (*Gambusia holbrooki*), goldfish (*Carassius auratus*) and perhaps the most disliked of all, the common carp (*Cyprinus carpio*). Surprisingly, of about 300 species of Australian freshwater fish that are recognised in 59 families, none are known to have become extinct since European colonization although there is evidence of regional extinctions, and recovery actions have probably saved several species from extinction [4]. In addition, different federal and state bodies have listed 74 freshwater species as 'threatened' [4]. It is likely that invasive fish species are, directly or indirectly, associated with the dire status of many Australian native freshwater fish.

C. carpio (known simply as 'carp' in Australia) was probably first introduced to waters around Sydney, Australia in 1908 (earlier records possibly being confused with goldfish, *Carassius auratus*) [5,6]. However, it was not until the 1960s that they were recognized as a serious invasive pest species, particularly in the Murray-Darling Basin (MDB), a regulated river system that covers 14% of the continent on the eastern side of the nation [4]. Carp comprise up to 90% of the fish biomass in parts of the MDB, and it is recognized that they are responsible for a deleterious cascading effect on the aquatic environment: they uproot and consume aquatic vegetation which increases the turbidity of the water. This change then leads to further reductions in aquatic vegetation, invertebrate communities, aquatic birdlife, and native fish [7].

In the early years of the new millennium, the Australian Federal Government began a program to develop innovative measures for the control of several important terrestrial and aquatic invasive pest species. An integrated pest management (IPM) plan was developed for carp in Australia, and, following the recognition of cyprinid herpesvirus 3 (CyHV-3; also known as koi herpesvirus) in Israel and the USA in 1998 [8], a research program was initiated to investigate the potential of this virus as a biological control (biocontrol) agent within the IPM plan for carp. This review summarizes the work already completed, and it also identifies the outstanding requirements before CyHV-3 could be considered as a biocontrol agent in Australia.

2. The Essential Information Required for Potential Viral Biocontrol of Carp

Since the 1950s, Australia's use of two different viruses for rabbit biocontrol has demonstrated many generic lessons for future viral biocontrol programs of invasive vertebrates [9]. In broad terms, these lessons indicate the necessity for an understanding of the biology of both the targeted pest species and the putative biocontrol virus. For carp control in Australia, in particular, a vast amount of information has already accumulated on carp biology in this country, although key pieces of information were still required. However, as an exotic (or foreign animal disease) virus for Australia, specific information about CyHV-3 in Australian conditions was almost non-existent. Table 1 summarizes the essential additional information required to not only understand carp biology and ecology in this country, but also the far greater needs to provide insights into CyHV-3 activity. Much of this information on the host and the virus has now been acquired through government-sponsored research programs ("Knowns" in Table 1), but deficiencies in our knowledge still exist ("Unknowns"). Due to space considerations, the following sections briefly focus on recent additions to our knowledge about carp and CyHV-3, especially for Australia's needs.

Table 1. The essential information required for potential viral biocontrol of carp.

Information Required	Knowns	Unknowns
Carp biology in Australia	• Modelling of carp biomass across Australia in 2018 • Future estimates of carp biomass in different hydrological scenarios	• Genomic and transcriptomic study of carp in Australia
Viral epidemiology	• Global and laboratory epidemiology • Genome of the virus	• Viral epidemiology in Australian conditions • Latency
Safety of the virus(species-specificity)	• The released virus • Human safety • Asian and North American field outbreaks	
Efficacy of the virus	• Virus-host interactions ○ Transmission ○ Virulence • PCR survey of MDB carp for cyprinid herpesviruses	• Virus transmission ○ Determine R_0 • Virulence ○ Virulence of different CyHV-3 isolates ○ Virome of carp in Australia ○ Polymorphisms in immune genes of carp in Australia ○ Modelling of carp-goldfish hybrids in MDB
Epidemiological modelling of virus release and spread	• Hydrological model • Habitat suitability model • Demographic model • Epidemiological model	• Validation of models • Extension of models across entire MDB • Key epidemiological rates

Table 1. *Cont.*

Information Required	Knowns	Unknowns
Evolution of the released virus	• Rabbit biocontrol viruses • Absence of native Australian cyprinids • Insusceptibility of most closely related fish	• Amelioration of virulence of CyHV-3?
Broad- scale control measure(s) to complement the virus	• Many regional measures available • Options for broad-scale measures are available	• Ideal broad-scale measure to complement the virus • Next generation of CyHV-3 (aquaculture isolates)
Ecological concerns	• Ecological risk assessment • Environmental clean-up procedures after fish kill events • Options for utilisation of large volumes of carp	• Prey switching • Viral epidemiology in Australian conditions
Social risks	• Views of urban versus MDB populations on CyHV-3 and biocontrol	• Views of urban versus MDB populations post-release of CyHV-3
Restoration benefits from carp control	• Expert elicitation study on the ecological consequences of reduced carp numbers	• Control of other environmental stressors

2.1. Carp Biology in Australia

2.1.1. Distribution Models of Carp in Australia and Biomass Estimates

A great deal of excellent work has been conducted on carp biology since the 1960s when the species was first recognized as an important pest in Australian waterways; as examples, see [7,10–13]. However, while estimates of carp biomass in the MDB have been proffered in the past, any national carp biocontrol program would require a modern, more extensive and more accurate approximation. Working across numerous jurisdictions, Stuart et al. [14] used catch-based models to generate "heat maps" that depicted the biomass and spatial distribution of carp throughout the waterways of south-eastern Australia in 2011 and 2018. It was already known that when carp exceed a threshold density of 80–100 kg/ha, detrimental ecological impacts may occur [11]. Stuart et al. [14] found that modelled carp biomass exceeds this threshold across large areas of south-eastern Australia and therefore is consistent with the view that carp may have landscape-scale impacts manifested by the decline of water quality, native flora, fauna biodiversity, and recreational values. In short, carp represent a serious threat to freshwater ecosystems.

The 2011 and 2018 biomass estimates [14] were based on static spatial mapping. However, because carp populations can respond rapidly to changes in hydrological conditions, these estimates cannot be applied to future scenarios when a biocontrol virus might be released. Therefore, Stuart et al. [14] recommended the use of a dynamic model to provide future estimates of carp biomass, taking into account a variety of possible hydrological scenarios. Todd et al. [15] undertook dynamic modelling using an established carp population model [13] and the static biomass estimate for 2018 [14]. They then provided a range of estimates for the biomass of carp for 2023 in four regions of south-eastern Australia. Their results highlighted the variability of populations with differing hydrological and ecological conditions, and this, in turn, emphasized the advantage of dynamic modelling: it provides managers with a current estimate of carp populations in different locations, which then assists managers when considering where to release a biocontrol virus, and also where to focus clean-up operations to ameliorate the impacts of large numbers of dead carp.

2.1.2. Genomic and Transcriptomic Map of Carp in Australia

Using variability in 14 microsatellite loci, Haynes et al. [5] studied the population genetics of carp at each of 34 locations throughout the MDB. They confirmed the presence of the four recognized strains in Australia: Boolarra, Yanco, koi and Prospect, despite Prospect being originally restricted to Sydney (outside of the MDB). They also concluded that there was significant genetic structuring of carp that was associated with barriers to dispersal. In fact, they divided the MDB into 15 management units, each unit based on man-made or natural barriers to dispersal of carp. They noted that, while invasive species often show decreased levels of genetic diversity in a new location, some actually have similar, or greater, diversity due to the invasives being introduced a number of times from different sources. This apparently applies to carp in the MDB which have high levels of genetic diversity (with multiple strains in all regions). They warned that the 15 management units should be interpreted with caution because fish-ladders may increase connectivity.

While the work of Haynes et al. [5] has been valuable in providing a preliminary understanding of the population genetics of carp in Australia, there is a dire need for a genomic study of this pest species in the MDB. This would provide a level of information on the targeted pest that is commensurate with our knowledge of the Indonesian strain of CyHV-3 that has been identified as a biocontrol virus in Australia (see Section 2.2). There are at least two immediate needs that could be addressed by a genomic study of carp in Australia: information on the virome of carp in this country, and an understanding of the polymorphisms in some immune response genes that may be critical in determining the virulence of CyHV-3 (see Section 2.4).

2.2. Viral Epidemiology

Since 1998, virulent CyHV-3 has been identified in many countries, but there is no evidence for its presence in Australia [16]. Although the origin of the virus remains problematic, it appears to have arisen in recent decades, possibly from avirulent variants in Europe [17], but there is also evidence for unusual variants in New York State [18] and Oregon [19] in the USA. Other molecular studies [20] also support the idea that an avirulent variant(s) of CyHV-3 has been present in *C. carpio* for tens of thousands of years although Kopf et al. [21] highlighted two major assumptions that perhaps cast some doubt on this time-frame—firstly, that the evolutionary rate of CyHV-3 has been constant, and secondly, that this rate for an alloherpesvirus from an exothermic host is similar to an alphaherpesvirus from an endotherm.

Under permissive conditions, CyHV-3 can cause 70–100% mortality in juvenile and adult *C carpio* [22–24]. However, larvae less than 1 cm in length are completely resistant to infection due to the protective effect of skin mucus. Larvae gradually become susceptible with increasing size, culminating in complete susceptibility when they are longer than 2 cm [25–27]. Only very low doses of virus are required for infection [28], the main portal for both infection and excretion being the skin [29]. Once excreted from an infected fish, a virus survives in the aquatic environment for only about three days, regardless of water temperature [30]. While most of these epidemiological data have been acquired from overseas studies, there is little reason to expect major differences under Australian conditions. However, due to biosecurity concerns with CyHV-3, this view has not been proven because field trials are yet to be conducted in this country.

An Indonesian strain of CyHV-3 (the C07 isolate) will potentially be used as the biocontrol virus in Australia. The full genome sequence has been determined [31], revealing that the gene layout is very similar to CyHV-3-U (a US reference genome) although 310 genetic variations between the C07 strain and the reference genome were identified. Phylogenetic analysis inferred from comparisons of whole-genome sequences revealed that the Indonesian isolate is more closely related to a Japanese isolate within the Asian lineage than to isolates within the European lineage.

Being a herpesvirus, CyHV-3 is likely capable of inducing latent infections in surviving carp, but while initial studies have demonstrated that low-temperature persistent infections are possible [32,33], unequivocal latent infections are yet to be demonstrated [34].

2.3. Safety of the Virus

Two of the most important lessons from Australia's earlier work on rabbit viral biocontrol have been the necessity for assessing both the safety and efficacy of any potential biocontrol virus. 'Safety' is about species-specificity, not only of the virus isolate selected for potential release into the environment, but also of any future generations of the virus that may evolve genetic changes (mutations or recombination) following release in the field (see Section 2.6).

The most compelling evidence for the specificity of CyHV-3 is that viral-induced disease has never been reported anywhere in the world in any species other than *C. carpio* since CyHV-3 was first recognized. This includes species in polyculture systems with carp. It is an observation that has often been ignored by critics of the virus, but its importance should never be overlooked. Importantly, this broad observation includes humans whose fears of infection can also be allayed by several observations: there has been no evidence of adverse effects on humans working in CyHV-3-affected carp farms. The two closely-related viruses, CyHV-1 and -2, are not known to infect humans. There was no evidence of infection in CyHV-3-challenged mice (selected as a representative mammal in non-target species susceptibility trials) [16]. More generally, there is no evidence for any fish virus causing disease in humans [35]. These findings were corroborated by Roper and Ford [36] who, in addition, recommended that the "psychosocial effects" on human health of a mass fish kill should be investigated.

For other potential targets, numerous laboratories have suggested that many species could become infected by CyHV-3, but without causing disease. In summary, the susceptibility of 24 to 25 species of fish was tested [22,37–42], and, in all cases, there was no evidence of disease. While CyHV-3 genomic DNA was detected in 10 of 15 species of fish that were exposed to acutely- or latently-infected carp, only a small proportion of each species was supposedly infected, none showed clinical signs of disease, and only low copy numbers of CyHV-3 DNA were found [39]. Similar results were found for plankton, mussels and crustaceans [43,44]. In none of these cases, however, was there an attempt to demonstrate CyHV-3 mRNA as an indicator of virus replication, a necessary corollary of infection. It should be noted that even though one study [45] did use an RT-PCR, ostensibly to demonstrate replication of CyHV-3, their work was flawed technically in that neither primer in their RT-PCR was designed in separate viral exons nor over splice junctions. It is likely that their primers were actually detecting residual contaminating genomic DNA from virus rather than viral mRNA [16].

Yuasa et al. [46] eventually developed an RT-PCR for CyHV-3 that allowed differentiation of genomic DNA from the mRNA of replicating virus. This allowed a definitive laboratory study on the susceptibility of the following non-target species (NTS) to CyHV-3 [16]: 13 native Australian fish species, introduced rainbow trout (*Oncorhynchus mykiss*), native lamprey ammocoetes (*Mordacia mordax*), domestic chickens (*Gallus gallus domesticus*), laboratory mice (*Mus musculus*), a freshwater crustacean (*Cherax destructor*), two species of frogs (*Litoria peronii* and *Lymnodynastes tasmaniensis*), and two reptilian species (*Intellagama lesueurii* and *Emydura macquarii*). When challenging each of these NTS, CyHV-3 was given the best chance of causing disease through the use of immature, susceptible NTS that were exposed, by immersion and/or intraperitoneal inoculation to 100–1000 times the dose of virus required to infect a carp.

All challenged NTS were subjected to clinical, gross pathological and histopathological examinations, and to PCR testing (using a screening qPCR, and the specific RT-PCR [46] to re-examine any qPCR-positive samples). While low copy numbers of CyHV-3 DNA were found in occasional samples by qPCR, all such samples were negative for viral mRNA by the RT-PCR suggesting that the weakly-positive qPCR results were, in fact, due to low-level contamination events during processing of samples rather than to the presence of replicating virus. Thus, it was concluded that no evidence could be found for infection, let alone disease, in any of the NTS. Boutier et al. [47], however, offered alternative interpretations. Firstly, they suggested that "technical issues" were not addressed (although they provided no specific details on what these issues might be), and, secondly, that the deaths in NTS could have been due to a non-replicative pathogenesis such as may occur in herpesvirus latent infections of non-natural host species [48]. The latter is an interesting suggestion, but seems to overlook two important observations: (1) latency, although likely to occur in carp surviving infection with CyHV-3, has not actually been demonstrated yet in the host species, let alone a NTS [34], and (2) a productive infection, the necessary precursor to a latent infection, has not even been demonstrated in any NTS. For example, McColl et al. [16] did not find clinical signs of disease, histological lesions or any evidence of an early productive infection (in the form of viral transcripts) in any of their NTS inoculated with CyHV-3 despite examining many NTS at early and later stages following inoculation.

Kopf et al. [21], while accepting that adverse effects of the virus on native species are "highly improbable", were, however, still loathe to absolve CyHV-3 of all potential threat. They suggested that native species could be "asymptomatic carrier(s)" or transmitters of the virus. Again, this ignores the fact that to be a carrier, a non-target native species must first be infected, a claim that has never been properly demonstrated (see above). Furthermore, claims for all but very short-term transmission by non-carp species were refuted by the elegance of the simple experiments with CyHV-3 on goldfish [49]. The final argument by Kopf et al. [21], that sub-lethal infections in immunocompromised NTS be investigated, has, indirectly, already been addressed by McColl et al. [16] through the use of immature fish (with incompletely developed immune systems) in their susceptibility studies on NTS.

In summary, we believe that a robust standard protocol is required for future susceptibility testing of NTS, and we propose the following: (1) time-course sampling to demonstrate an increase or decrease of viral DNA concentration in a viral-exposed NTS, (2) using both qPCR and RT-qPCR for detecting viral DNA and mRNA, respectively, (3) attempting virus isolation from any NTS with clinical signs of disease, and (4) using histopathological examination on moribund NTS. Molecular testing, including next generation sequencing if available, should also be considered for exclusion of other known or unknown pathogens.

The results of the NTS experimental work [16] were complemented by the findings from a North American study of natural outbreaks of CyHV-3 in carp [50]. At each outbreak, no disease was observed in any co-habitating species, even in native cyprinids, thus attesting to the species-specificity of the virus. However, there are two important criticisms of other observations in the North American work: (1) there was no attempt to look for the presence of any pre-existing, potentially cross-reactive viruses that might confer protection on carp. In particular, there have been no reported serological, PCR or next-generation sequencing studies on any carp populations in North America, and (2) only one of the outbreak sites offered the opportunity for direct fish-to-fish transmission by means of dense aggregates of carp. In the Thresher et al. study [50], there appeared to be few, if any, equivalents of the limited numbers of densely populated carp breeding sites distributed throughout the MDB. Thresher et al. [50] also claimed, probably correctly, that mass die-offs would not be expected for a herpesvirus that is in equilibrium with its natural host. CyHV-3, however, is a pathogen that has only recently been recognised [8], possibly because it has only recently arisen [20]. Therefore, it has not yet had time to come into equilibrium with its host, in which case, high mortalities are probably not unexpected.

2.4. *Efficacy of the Virus*

Determining the 'efficacy' of a potential biocontrol virus is slightly more complex than determining its 'safety' because the former depends on two variables, 'transmissibility' and 'virulence'. For a biocontrol virus, 'transmissibility' is defined as the ability of the virus to establish infection in new hosts, and is often measured by the basic reproduction number, R_0, the average expected number of cases produced by a single case (in a population where all individuals are fully susceptible). 'Virulence' is a measure of the severity of the disease caused by the virus, not necessarily measured simply by mortality. For example, in the classical studies of the MYXV, Fenner and Woodroofe [51] established five grades of virulence that were based on a combination of both survival time following infection, and mortality.

Using Australia's two rabbit biocontrol viruses as examples, Di Giallonardo and Holmes [52] demonstrated that, while there are invariably strong selection pressures for transmissibility, this has been achieved for MYXV and RHDV by selection in the field of virus strains of intermediate and high virulence, respectively. These quite different paths suggest that, for any particular virus, it is not always easy to predict the outcome of the complex relationship between transmissibility and virulence. There have been no direct studies on how CyHV-3 achieves maximal transmission, but observations on viral epidemiology, particularly the virus–host interaction, may encourage the formulation of two hypotheses [53].

Firstly, the observations that CyHV-3 is excreted at low titre into an aquatic environment, and then only survives for about three days outside its host [30,54], suggest that direct transmission of virus between carp is likely to be much more important than indirect transmission via the aquatic environment. Furthermore, given that carp are highly sensitive to infection [28], and that the skin is the main portal of both infection and excretion of CyHV-3 [29], a reasonable hypothesis is that direct skin to skin contact between an infected and an uninfected fish, even if transient, is the most likely form of transmission. Such contact would likely disrupt the skin mucus layer which would enhance virus entry [26,27]. Clearly, carefully designed transmission experiments are required to test this hypothesis.

Having become infected, a viraemia develops in the carp, and the virus localizes in various tissues [55]. The adaptive immune response of the fish develops slowly [56–58], likely allowing survival of some infected hosts with potential latent infections although, as already mentioned, latency has not yet been proven unequivocally [34]. Nevertheless, assuming it does indeed occur in surviving fish (as it does in the hosts of all known herpesviruses), then recrudescence of acute infections will also occur during periods when infected fish are stressed. In Australia, massive aggregations of carp occur at annual breeding events, and such aggregations are known to induce stress and immunosuppression [59]. This, in turn, implies that annual breeding would not only allow reactivation of CyHV-3 infections, but would also favour transmission of virus by direct skin-to-skin contact of the densely aggregated fish. These observations then suggest a second hypothesis: that long-term transmission of CyHV-3 in Australian conditions may be favoured by the natural selection of low virulence strains of CyHV-3 that would allow survival of some latently-infected fish which, in turn, would lead to multiple periods of recrudescence and transmission of virus to naive fish during annual breeding events. It was postulated earlier that selection pressures may change as the density of carp declines [53], but, on reflection, this possibility may be of little importance if, indeed, most transmission occurs at densely aggregated breeding sites. The latter sites will likely form regardless of the total number of carp in the river systems because, at least in Australia, carp seem to be irresistibly attracted to these sites at certain times of the year [60]. So, while the total area of any particular breeding site may decline, the density of fish will probably remain high.

A legitimate question that arises because of the second hypothesis is that, if transmissibility drives the selection of low virulence strains of CyHV-3, can the virus be an effective biocontrol agent in Australian waters? Perhaps the answer may be found in lessons from past viral biocontrol programs involving rabbits in Australia [9]. Field experience with both MYXV and RHDV has revealed that a virus, alone, will not control the targeted invasive pest species. In fact, to be effective, biocontrol viruses must be complemented by other broad-scale control measures, a fact that has been emphasized many times from the outset of the carp biocontrol program in Australia. The use of such measures is not an admission of failure in the proposal to use CyHV-3 as a biocontrol agent; rather, it is an argument for the use of the virus in a carefully designed IPM program. The virus would markedly reduce numbers of carp in naive populations, providing the opportunity for complementary measures to then substantially knock down carp numbers even further (see Section 2.7).

The second factor affecting the efficacy of the virus, virulence, may be difficult to determine. While Fenner and Woodroofe [51] used standard inbred laboratory rabbits for their work on the virulence of MYXV, no equivalent line of carp is available in Australia to allow a standard test of the virulence of different isolates of CyHV-3, nor, indeed, to determine R_0. However, the virulence of the C07 isolate of CyHV-3 has been demonstrated in numerous studies on carp collected from all over south-eastern Australia for example, [16,61,62]. Further studies are required to test the virulence on carp collected from throughout the entire MDB.

Boutier et al. [47] expressed a number of reservations about the use of CyHV-3 as a biocontrol agent for carp in Australia. They contended that natural resistance of some carp (due to resistance-conferring polymorphisms in immune genes) and of carp-goldfish hybrids could lead to rapid proliferation of resistant phenotypes. Access to genomic and transcriptomic maps of Australian strains of carp would help to address the question of immune genes, while assessing the future importance of carp-goldfish hybrids would likely require modelling work and a better understanding of the current numbers of these hybrids in the MDB.

Boutier et al. [47] also proposed that phylogenetic studies suggested that CyHV-3 may already be present in carp in Australia, just as it may have long been present, without expressing virulence, in carp populations around the world [20]. Studies on 849 carp samples from nine sites throughout the MDB in Australia, utilizing a nested PCR (with primary and nested primers aligning perfectly with sites in the DNA polymerase gene of CyHV-1, -2, and -3), failed to reveal evidence for any known or undescribed cyprinid herpesviruses [63]. Nevertheless, this is recognized as only a preliminary study, and a more definitive virome study (using a next generation sequencing approach) from a similar sample of carp is essential to corroborate the PCR work.

Finally, in assessing the likely efficacy of CyHV-3 in Australia compared with natural overseas outbreaks [21], it is important to recognize a critical difference between the two situations: whereas most of the world is consumed with controlling outbreaks of CyHV-3 disease, Australia would aim to *enhance* the spread of the disease (and then to augment the effect of the virus with complementary control measures). To this end, it is essential that we have a deep understanding of the epidemiology of the disease under Australian conditions. Kopf et al. [21] state that "Lake Biwa (in Japan) and Blue Springs Lake (in the USA) are not good models for Australian conditions if the virus was (sic) released", but, nevertheless, most of their criticism of Australian activities is based on findings from these overseas outbreaks. Based on the known biology of the virus and of carp in Australia, we have proposed two hypotheses that account for virus pathogenesis and transmission under Australian conditions; aspects of these hypotheses have informed the work on epidemiological modelling.

2.5. Epidemiological Modelling of Virus Release and Spread

A large multi-disciplinary team developed four inter-related models, namely hydrological, habitat suitability, carp demographic, and epidemiological models, with the intention of informing any future staged release of CyHV-3 in the MDB [64].

The hydrological model focussed on the water temperature and connectivity of waterways for five diverse catchments in south-eastern Australia. It concluded that, while CyHV-3 will generally be effective from Spring through Autumn throughout south-eastern Australia, a staged release of virus would demand precise estimations of water temperature prior to release in any particular catchment to ensure conditions were permissive for virus activity. However, water temperature alone was insufficient for determining the time of virus release. It was also found that the major environmental factors influencing the distribution and abundance of carp in south-eastern Australia, and the manner in which these factors interacted with each other, were also essential in selecting a time for virus release.

This conclusion was reached through a habitat suitability workshop that utilized expert opinion within the context of a Bayesian belief network (BBN). The BBN identified river flow and water temperature as the two essential parameters determining the suitability of a habitat for adult and sub-adult carp, and both were rated as medium to high for most habitats throughout the study period. On the other hand, waterway inundation and connectivity, the essential habitat suitability factors for an abundance of larvae and young-of-year (YOY) stages, were rated poorly in most habitats during the study period. Population abundance for YOY stages, in fact, depended on a relatively small number of dense aggregations of juveniles and adults that occur in transiently flooded wetlands throughout the MDB (so-called 'recruitment hotspots'). Through the use of conversion factors guided by expert opinion, habitat suitability rankings were converted to biomass density estimates (the latter validated by recently acquired data [14]). These estimates would then allow CyHV-3 to be used in those areas where the population density of carp was approximately 80–100 kg/ha, the level at which detrimental ecological impacts may occur [11].

The biomass densities from the habitat suitability model were then used to develop a full spatio-temporal population projection (or demographic) model of carp population dynamics in which carp metapopulations were resolved into six age-stage classes (eggs, larvae, early YOY, late YOY, sub-adults, and adults). This demographic model, in turn, was integrated into a CyHV-3 epidemiological model that allowed the prediction of mortality and suppression of the subpopulations following a hypothetical release.

The epidemiological model was a variation on the standard SEIR transmission model. It included susceptible (S), exposed (E) and infectious (I) classes, but the usual recovered (R) class was replaced with classes more likely to represent a typical herpesvirus, such as CyHV-3. Thus, latent (L) and recrudescent (Z) classes were introduced, with Z representing second and subsequent infections following repeated reactivation of the virus in latently infected carp. The modelling predicted that, without recrudescence, introduction of the virus would be associated with a single mortality event in carp. However, in the more likely event of latency and recrudescence, there would be an ongoing and lasting suppression of carp populations in all catchments with reductions being to approximately 40% of the pre-release population. The impact of the virus would be sufficient to reduce carp populations in many MDB waterways to below the damage threshold of 100 kg/ha for at least 10 years. A further notable prediction was that seasonal losses would be mainly in immature carp, the mortality in the adult population being less by at least an order of magnitude (and, therefore, possibly not easily observed).

Boutier et al. [47] suggested that, at various times of the year, there could be vast tracts of water in the MDB where temperatures may be non-permissive for virus replication. These concerns have long been noted [9]. However, the modelling report [64] specifically noted that "tailoring the release of the virus to the particularities of each catchment" would be important, especially in those areas where there was a very narrow window of permissive Spring temperatures. The use of other complementary control measures (see Section 2.7) could also be important in some areas. Boutier et al. [47] also suggested the importance of 'behavioural fever' in fish as a response to infection. There is no question that this phenomenon works for individual fish, but it has not prevented mass mortalities of carp in thermally variable natural aquatic environments overseas, and it would be unlikely to do so in Australia either (particularly if a virus were to be released in relatively homogeneous shallow breeding grounds of carp).

2.6. Evolution of the Released Virus

While laboratory work and field observations strongly suggest that current strains of CyHV-3 are highly specific for *C. carpio* (and some hybrids [65,66]; see Section 2.3), the question remains about the likelihood of genetic changes in current field isolates causing a future expansion of the host range. There is no direct evidence bearing on this question for CyHV-3, but there are pertinent lessons from Australia's past experience with viral biocontrol of rabbits. Evolutionary studies [67,68] have shown that DNA viruses can, indeed, mutate (although at a much lower rate than RNA viruses). However, while this may potentially allow spill-over events or host-jumps, such events for herpesviruses occur on timescales of millions of years, and, when they occur, they are invariably into taxonomically closely related species. It is reassuring then that, while there are a number of introduced cyprinids in Australia, there are no native cyprinids. Furthermore, the most closely related native Australian species (native catfish) are insusceptible to infection with CyHV-3 [16], and, again, it should be emphasized that viral-induced disease has never been reported anywhere in the world in any species other than *C. carpio*.

Field observations on rabbit biocontrol viruses in Australia lend support to these evolutionary studies [9]. Mutations are known to have occurred in the field in both the myxoma virus (MYXV, a DNA virus present in Australia for over 60 years) and in rabbit haemorrhagic disease virus (RHDV, an RNA virus, over 20 years) [69,70], but there is no evidence that either has jumped into another species. As noted by Di Giallonardo and Holmes [71], the overall conclusion is that "host-jumps to nontarget species are not an inevitable consequence of viral evolution". As a result, all observations, whether on current isolates or potential future field variants of CyHV-3, encourage the view that the chance of cross-species transmission is very small.

2.7. Broad-Scale Control Measure(s) to Complement the Virus

Saunders et al. [72] found that there have only been three major instances where viral pathogens have been used successfully against vertebrate pest species, namely MYXV and RHDV against rabbits in Australia and feline panleukopenia virus against cats on a South African offshore island. In reviewing the lessons from these attempts at viral biocontrol of invasive vertebrates [9], it was noted that in each case complementary measures were required for sustained control or eradication of the pest species. While these measures actually included supplementary regional controls, ideally broad-scale controls would be identified and implemented. A number of regional control measures have long been implemented for carp, including commercial harvesting, electrofishing, carp traps, fishing competitions, predator stocking, poisoning, and environmental controls [7,73,74]. However, these are generally ineffectual in the long-term.

The development of broad-scale control strategies that will deliver persisting declines in carp numbers has been more problematic. Wedekind [75] reviewed possible genetic biocontrol technologies that could be used on carp populations in Australia. He broadly classified them as those that involve genetic engineering (including 'daughterless' carp and gene-drive technologies) and those that do not (including 'Trojan Y chromosome' techniques).

Of the techniques involving genetic engineering, an early major investment was made into 'daughterless' carp technology in Australia [76]. The underlying principle was that, by using an RNAi approach that suppressed expression of the female differentiating genes, natural carp populations in Australia would be biased towards all-male populations. Although initially very promising, the approach gradually fell out of favour because modelling revealed that it would take many decades to exert an impact, the corollary being that very large numbers of these genetically modified fish would need to be added to waterways annually for many years in order to force a sex bias in natural populations. Currently, a gene-drive approach that would be lethal to female offspring, or leave them infertile, is not considered a safe option for carp [75], nor indeed for any biocontrol program [77]. In the future, however, gene-drive technology may become a universal approach to controlling invasive pest species, although many modifications to current technology will be required for this approach to become acceptable.

Approaches that do not involve genetic engineering include the use of Trojan Y fish [78]. This method relies on treating young male carp with a female sex hormone, oestrogen, resulting in genetic males (with XY chromosomes) that have female sex characteristics, including the ability to produce eggs. The latter have an XY constitution, and when fertilized by a normal male, they produce a preponderance of male offspring. Extensions of this basic approach can lead to stock populations of YY individuals being produced for release into wild populations [79]. Although this strategy would require the costly regular addition of modified fish to natural populations of carp for a number of decades, it is considered the most appropriate current technique, particularly if combined with measures to increase the survival and fecundity of the manipulated carp [75].

In the immediate future, perhaps new, more virulent strains of CyHV-3 may prove to be a useful complementary measure. A situation may develop that is analogous to the commercial chicken industries where the herpesvirus, Marek's disease virus (MDV), has been an ongoing threat for many decades. Strains of MDV of increasing virulence have evolved due to the use of imperfectly immunizing vaccines. Similar vaccines have been used in carp aquaculture to protect farmed carp from outbreaks of CyHV-3 for about a decade, and it is hypothesised that they too may lead to the evolution of more virulent strains of CyHV-3 that could be used as the next-generation biocontrol viruses in Australia [53].

In summary, a number of options for a broad-scale complementary control exist (Table 2). Each has their strengths and weaknesses, but currently, we are in complete agreement with Boutier et al. [47] and Kopf et al. [21] in declaring that the ideal broad-scale complementary measure(s) has not yet been identified. However, as previously mentioned [34], new genetic options continue to appear [80], providing optimism that the ideal measure will soon be developed. Until then, it would be unwise, even wasteful, to release CyHV-3 into the Australian environment.

Table 2 Broad-scale control options to complement CyHV-3 biocontrol of carp in Australia.

Technology	Strengths	Weaknesses	Comments
Trojan Y chromosome	• Fish are not GMO • Technologies to produce Trojan Y carp are available • Very useful if current carp numbers reduced by CyHV-3	• Need regular releases of treated fish • Decades to achieve major reduction in numbers (if used alone)	• Already well-developed technology • Should combine with measures to increase the survival and fecundity of Trojan Y carp
Daughterless carp	• Very useful if current carp numbers reduced by CyHV-3	• GMO fish • Need regular releases of treated fish • Decades to achieve major reduction in numbers (if used alone)	• Already well-developed technology
Gene-drive	• Very few carp need to be released to affect the population • Very useful if current carp numbers reduced by CyHV-3	• GMO fish • Potentially difficult to contain or reverse in case of unexpected outcomes	• Need to identify germline (sex) specific promoters in carp • Other potential technical problems
Self-stocking incompatible-male system (see [78])	• System is both self-amplifying and self-limited. • Control achieved with low biomass • Reduced production overhead • Potential applications include carp elimination or prophylactic barrier.	• GMO fish • Stocking rate high compared with gene-drive, but may be lower than other options	• Technology that will likely be applicable to other invasive species • Potential technical problems
More virulent strain of CyHV-3	• Virus is not GMO • Marek's disease precedent suggests that vaccination of carp in aquaculture will generate more virulent strains of CyHV-3	• Identifying more virulent strains will likely be an ongoing process	• More virulent strains of virus as a result of long-term vaccination programmes

2.8. Ecological Concerns

Kopf et al. [21] raised the possibility of "broad ecological risks of unintended and perverse outcomes from biocontrol with CyHV-3". They suggested a number of potential problems that could arise as a result of mass mortality, and subsequent decomposition, of carp following release of virus. Australia's National Carp Control Program [81] undertook a number of studies to address such concerns.

Beckett et al. [82] conducted a comprehensive ecological risk assessment of the consequences of the proposed release of CyHV-3 in a variety of aquatic settings including transient wetlands, river systems, lakes and other water bodies. Impacts on water quality were considered most likely in locations characterised by a high carp biomass and low water flow such as occurs during carp breeding in transient wetlands. It was suggested that risks to native fish and birds, in particular, could be avoided by releasing the virus during high-flow seasons, or by the partial removal of carp from waterways prior to the release of virus. Whether this would be a practical option for the release strategy would need to be considered. If not, then mitigation strategies might include physical removal of carp carcasses from affected areas, or the use of water regulation to flush not only carcasses but also cyanobacterial blooms from affected areas. These strategies would likely also reduce the risk of outbreaks of botulism in native species, although both a literature review and field experience in south-eastern Australia (where there have been many fish kills due to blackwater events) suggest that botulism is unlikely to have much practical importance anyway. The loss of many juvenile carp from wetlands treated with CyHV-3 raised the spectre of prey-switching by piscivorous waterbirds. While the potential impact on native fish and other species must certainly be considered, insufficient research has been conducted in Australia to allow firm conclusions on the many potential interactions. It is, however, noteworthy that a number of earlier studies on prey-switching have revealed complex dynamics between predators and prey species, but little cause for long-term concern about the survival of native species [83–85].

Finally, Beckett et al. [82] noted that residual uncertainty necessarily remains because it is not possible to predict, with confidence, the epidemiology of CyHV-3 in Australia. The extent of carp mortality, more or less than predicted, is a key uncertainty. Similarly, the impacts of low dissolved oxygen levels (DO) on the many and varied native aquatic species can never be certain, although water quality modelling studies [86] suggested that dangerously low DO was only likely to be a problem where carp biomass was high and there was a concomitant severely compromised (or absent) water flow. A similar situation is predicted for widespread cyanobacterial blooms.

In broad terms, other very important ecological considerations are, firstly, the clean-up procedures for carp following a mass mortality due to CyHV-3, and, secondly, the potential waste utilisation of the subsequent large masses of dead carp. A literature review revealed very limited information about clean-up processes following a fish-kill [87], and therefore, not surprisingly, most of the documented responses were of a reactive nature. However, the Atlantic region of Canada is one of the few locations that does have a well-documented clean-up procedure. Silva et al. [87] declared that the biomass of carcasses and the location of the fish-kill should be important determinants, among others, of the extent of the clean-up operation, particularly if the affected waterway is part of a town water supply (in which case the suggested importance of a cost-benefit analysis for the operation almost seems paradoxical). Globally, most fish-kills have relied on landfills for the disposal of carcasses.

Tilley et al. [88] investigated alternative methods of disposal, and found that composting methods, that are able to use even severely degraded material, are likely to be the best option on a large commercial scale. However, flexibility and scalability of the process would also allow small scale operations in remote regions up to larger scale operations by councils or smaller commercial organizations. A large-scale rendering option at a meat rendering facility was also shown to be possible, although only fish carcasses < 24 h post-mortality would be acceptable for processing.

2.9. Social Risks

Zhang et al. [89] undertook a risk assessment to determine public perceptions about the ecological and social risks associated with the proposed use of CyHV-3 as a biocontrol agent. They conducted wide-reaching qualitative and quantitative surveys of the Australian public, focussing in the former on the general public, and in the latter on those in urban settings versus those living near major waterways. An important finding was that people who live in the MDB and who are closely connected to the river system were more likely to accept the need for carp control while still retaining some reservations about aspects of the process. Overall, the studies emphasized the importance of early, effective communication programs in order to allay the concerns of various communities.

2.10. Restoration Benefits from Carp Control

Finally, Kopf et al. [21] questioned whether Australia could expect to see any ecological restoration benefits from carp control. Casual observation of affected waterways has long suggested that carp must be exerting a profound negative ecological impact, especially in those regions of the MDB where they account for up to 90% of the biomass.

However, the very comprehensive study conducted by Nichols et al. [90] relied on more than casual observation. Assuming the proposed biocontrol program would be successful in reducing carp numbers, an expert elicitation study was conducted on the expected medium- (5–10 years) to long-term (beyond 10 years) ecological consequences of a reduced carp population in Australia. The study addressed the effects on ecosystems as a whole, along with effects on the animal and plant components, and, on water quality.

In summary, the study found that Australia's waterways are complex ecological systems, and they will almost certainly continue to degrade if nothing is done to control carp. On the other hand, if carp populations could be sustainably reduced by 70–100%, experts believe there would likely be clear long-term ecosystem benefits [90]. The same experts also emphasised that carp are not the only ecological stressor, and that other widespread environmental problems must also be addressed. However, even under ideal conditions where all stressors are identified and controlled, the experts agreed that, rather than restoration of a degraded system to its original state, an unexpected new aquatic ecosystem may be generated.

3. Final Comments and Conclusions

Australia's experience with MYXV and RHDV for rabbit biocontrol has taught us a great deal about the principles of viral biocontrol for any invasive pest vertebrate species. Perhaps the most important lessons are that it is essential to have a deep understanding of the biology of both the targeted pest species and any potential biocontrol virus. Equally important is the lesson that a viral biocontrol agent, alone, can never be expected to completely eradicate an invasive pest species; to be successful, biocontrol agents must be complemented by other broad-scale control measures in an integrated pest management program.

Is there a need for a contingency plan for any proposed biocontrol virus in the event that, following release, it fails to work as expected? Assuming all the necessary precautionary studies have been conducted, particularly on the safety, efficacy, and epidemiology of the virus prior to its release, then the most likely reason it would be judged a 'failure' is that, after an initial burst of mortality, it apparently became ineffectual. This was the experience with MYXV in Australia in the 1950s. However, in later years, it was shown that, even as rabbit mortality due to MYXV declined, the virus still held rabbit numbers below the level at which they caused environmental damage [72]. Then, with the advent of effective complementary control measures, there has been sustainable control of rabbits for approximately 60 years in Australia, and, even in the absence of eradication, biocontrol of rabbits has delivered significant economic benefits to Australia [91]. Assuming that appropriate additional controls are found to complement CyHV-3, we would anticipate a similar trajectory for carp control in Australia.

Our world currently faces myriad local and global challenges including, but not restricted to, climate change, overpopulation and loss of species biodiversity. In Australia, the control of invasive pest species, including carp, sits comfortably in this list of challenges. We need to do something about carp to improve the quality of our waterways in this country, and it is only reasonable and rational debate that will form the foundation of future decisions and actions. CyHV-3 appears to be a rare opportunity to control carp, although, as mentioned earlier, we recognise the importance of identifying a broad-scale control measure to complement the future activity of the virus in Australia. To release the virus prior to implementation of such a complementary measure would be very unwise.

However, placing a temporary embargo on the use of the virus is not to endorse inactivity. As a marine biologist recently noted in a general commentary, "unrelenting doom and gloom in the absence of solutions is not effective. Social scientists have known for decades that large problems without solutions lead to apathy, not action" [92]. We must all recognise our current progress and successes so that in 5–10 years we can all take pride in the contribution we made to carp control in Australia.

Author Contributions: K.A.M. and A.S. combined to write and edit this paper. All authors have read and agreed to the published version of the manuscript.

Funding: This research was partly funded by the Invasive Animals Co-operative Research Centre (Australia) through Freshwater Projects 4.F.7 and 3.W.1.

Acknowledgments: We thank to the Invasive Animals Co-operative Research Centre (Australia) through Freshwater Projects 4.F.7 and 3.W.1. The Australian National Carp Control Program, under the leadership of Matt Barwick at the time, provided encouragement for our work, and funded many other projects in the carp control program. Jawahar Patil (University of Tasmania, Taroona TAS 7053), Maciej Maselko (Macquarie University, North Ryde NSW 2109) and Mark Tizard (CSIRO Health and Biosecurity, Geelong VIC 3220) contributed substantially to Table 2.

Conflicts of Interest: The authors declare no conflict of interest.

References

1. Simberloff, D. *Invasive Species: What Everyone Needs to Know*; Oxford University Press: New York, NY, USA, 2013; p. 11.
2. *Nature's Dangerous Decline 'Unprecedented'; Species Extinction Rates 'Accelerating'. Media Release*; Intergovernmental Science-Policy Platform on Biodiversity and Ecosystem Services (IPBES): Bonn, Germany, 2019; Available online: https://www.ipbes.net/news/Media-Release-Global-Assessment (accessed on 28 December 2019).
3. Kearney, S.G.; Carwardine, J.; Reside, A.E.; Fisher, D.O.; Maron, M.; Doherty, T.S.; Legge, S.; Silcock, J.; Woinarski, J.C.Z.; Garnett, S.T.; et al. The threats to Australia's imperilled species and implications for a national conservation response. *Pac. Conserv. Boil.* **2019**, *25*, 231. [CrossRef]
4. Cresswell, I.D.; Murphy, H. Biodiversity: Freshwater species and ecosystems. In *Australia State of the Environment 2016*; Australian Government Department of the Environment and Energy: Canberra, Australia, 2016. Available online: https://soe.environment.gov.au/theme/biodiversity/topic/2016/freshwater-species-and-ecosystems (accessed on 28 December 2019). [CrossRef]
5. Haynes, G.D.; Gilligan, D.; Grewe, P.; Nicholas, F. Population genetics and management units of invasive common carpCyprinus carpioin the Murray-Darling Basin, Australia. *J. Fish Boil.* **2009**, *75*, 295–320. [CrossRef] [PubMed]
6. Shearer, K.; Mulley, J. The Introduction and Distribution of the Carp, Cyprinus carpio Linnaeus, in Australia. *Mar. Freshw. Res.* **1978**, *29*, 551. [CrossRef]
7. Koehn, J.D. Carp (Cyprinus carpio) as a powerful invader in Australian waterways. *Freshw. Boil.* **2004**, *49*, 882–894. [CrossRef]
8. Hedrick, R.P.; Gilad, O.; Yun, S.; Spangenberg, J.V.; Marty, G.D.; Nordhausen, R.W.; Kebus, M.J.; Bercovier, H.; Eldar, A. A Herpesvirus Associated with Mass Mortality of Juvenile and Adult Koi, a Strain of Common Carp. *J. Aquat. Anim. Health* **2000**, *12*, 44–57. [CrossRef]

9. McColl, K.A.; Cooke, B.D.; Sunarto, A. Viral biocontrol of invasive vertebrates: Lessons from the past applied to cyprinid herpesvirus-3 and carp (Cyprinus carpio) control in Australia. *Biol. Control* **2014**, *72*, 109–117. [CrossRef]
10. Vilizzi, L.; Walker, K.F. Age and growth of the common carp, Cyprinus carpio, in the River Murray, Australia: Validation, consistency of age interpretation, and growth models. *Environ. Biol. Fishes* **1999**, *54*, 77–106. [CrossRef]
11. Brown, P.; Gilligan, D. Optimising an integrated pest-management strategy for a spatially structured population of common carp (Cyprinus carpio) using meta-population modelling. *Mar. Freshw. Res.* **2014**, *65*, 538–550. [CrossRef]
12. Conallin, A.J.; Smith, B.B.; Thwaites, L.A.; Walker, K.F.; Gillanders, B.M. Exploiting the innate behaviour of common carp, Cyprinus carpio, to limit invasion and spawning in wetlands of the River Murray, Australia. *Fish. Manag. Ecol.* **2016**, *23*, 431–449. [CrossRef]
13. Koehn, J.D.; Todd, C.R.; Zampatti, B.P.; Stuart, I.G.; Conallin, A.; Thwaites, L.; Ye, Q. Using a Population Model to Inform the Management of River Flows and Invasive Carp (Cyprinus carpio). *Environ. Manag.* **2017**, *61*, 432–442. [CrossRef]
14. Stuart, I.; Fanson, B.; Lyon, J.; Stocks, J.; Brooks, S.; Norris, A.; Thwaites, L.; Beitzel, M.; Hutchison, M.; Ye, Q.; et al. *A National Estimate of Carp Biomass for Australia*; Final Report; Arthur Rylah Institute for Environmental Research, Department of Environment, Land, Water and Planning: Heidelberg, Australia; Australia for Fisheries Research and Development Corporation: Canberra, Australia, 2019.
15. Todd, C.R.; Koehn, J.D.; Brown, T.R.; Fanson, B.; Brooks, S.; Stuart, I. *Modelling Carp Biomass: Estimates for the Year 2023*; Final Report; Arthur Rylah Institute for Environmental Research, Department of Environment, Land, Water and Planning: Heidelberg, Australia; Australia for Fisheries Research and Development Corporation: Canberra, Australia, 2019.
16. McColl, K.A.; Sunarto, A.; Slater, J.; Bell, K.; Asmus, M.; Fulton, W.; Hall, K.; Brown, P.; Gilligan, D.; Hoad, J.; et al. Cyprinid herpesvirus 3 as a potential biological control agent for carp (Cyprinus carpio) in Australia: Susceptibility of non-target species. *J. Fish Dis.* **2016**, *40*, 1141–1153. [CrossRef] [PubMed]
17. Engelsma, M.; Way, K.; Dodge, M.; Voorbergen-Laarman, M.; Panzarin, V.; Abbadi, M.; El-Matbouli, M.; Skall, H.F.; Kahns, S.; Stone, D. Detection of novel strains of cyprinid herpesvirus closely related to koi herpesvirus. *Dis. Aquat. Org.* **2013**, *107*, 113–120. [CrossRef] [PubMed]
18. Grimmett, S.G.; Warg, J.V.; Getchell, R.; Johnson, D.J.; Bowser, P.R. An Unusual Koi Herpesvirus Associated with a Mortality Event of Common Carp Cyprinus carpio in New York State, USA. *J. Wildl. Dis.* **2006**, *42*, 658–662. [CrossRef] [PubMed]
19. Xu, J.-R.; Bently, J.; Beck, L.; Reed, A.; Miller-Morgan, T.; Heidel, J.R.; Kent, M.L.; Rockey, D.D.; Jin, L. Analysis of koi herpesvirus latency in wild common carp and ornamental koi in Oregon, USA. *J. Virol. Methods* **2013**, *187*, 372–379. [CrossRef] [PubMed]
20. Gao, Y.; Suárez, N.M.; Wilkie, G.; Dong, C.; Bergmann, S.M.; Lee, P.-Y.A.; Davison, A.J.; Vanderplasschen, A.; Boutier, M. Genomic and biologic comparisons of cyprinid herpesvirus 3 strains. *Vet. Res.* **2018**, *49*, 40. [CrossRef] [PubMed]
21. Kopf, R.K.; Boutier, M.; Finlayson, C.M.; Hodges, K.; Humphries, P.; King, A.; Kingsford, R.T.; Marshall, J.; McGinness, H.; Thresher, R.; et al. Biocontrol in Australia: Can a carp herpesvirus (CyHV-3) deliver safe and effective ecological restoration? *Boil. Invasions* **2019**, *21*, 1857–1870. [CrossRef]
22. Perelberg, A.; Smirnov, M.; Hutoran, M.; Diamant, A.; Bejerano, Y.; Kotler, M. Epidemiological description of a new viral disease afflicting cultured Cyprinus carpio in Israel. *Isr. J. Aquac. Bamidgeh* **2003**, *55*, 5–12.
23. Sunarto, A.; Rukyani, A.; Itami, T. Indonesian experience on the outbreak of koi herpesvirus in koi and carp (Cyprinus carpio). *Bull. Fish Res. Agen.* **2005**, *2*, 15–21.
24. Uchii, K.; Matsui, K.; Iida, T.; Kawabata, Z. Distribution of the introduced cyprinid herpesvirus 3 in a wild population of common carp, Cyprinus carpiol. *J. Fish Dis.* **2009**, *32*, 857–864. [CrossRef]
25. Ito, T.; Sano, M.; Kurita, J.; Yuasa, K.; Iida, T. Carp Larvae Are Not Susceptible to Koi Herpesvirus. *Fish Pathol.* **2007**, *42*, 107–109. [CrossRef]
26. Raj, V.S.; Fournier, G.; Rakus, K.; Ronsmans, M.; Ouyang, P.; Michel, B.; Delforges, C.; Costes, B.; Farnir, F.; Leroy, B.; et al. Skin mucus of Cyprinus carpio inhibits cyprinid herpesvirus 3 binding to epidermal cells. *Vet. Res.* **2011**, *42*, 92. [CrossRef] [PubMed]

27. Ronsmans, M.; Boutier, M.; Rakus, K.; Farnir, F.; Desmecht, D.; Ectors, F.; Vandecan, M.; Lieffrig, F.; Melard, C.; Vanderplasschen, A. Sensitivity and Permissivity of Cyprinus Carpio to Cyprinid Herpesvirus 3 during the Early Stages of Its Development: Importance of the Epidermal Mucus as an Innate Immune Barrier. *Vet. Res.* **2014**, *45*, 100. [CrossRef] [PubMed]
28. Gilad, O.; Yun, S.; Adkison, M.A.; Way, K.; Willits, N.H.; Bercovier, H.; Hedrick, R.P. Molecular comparison of isolates of an emerging fish pathogen, koi herpesvirus, and the effect of water temperature on mortality of experimentally infected koi. *J. Gen. Virol.* **2003**, *84*, 2661–2667. [CrossRef]
29. Costes, B.; Raj, V.S.; Michel, B.; Fournier, G.; Thirion, M.; Gillet, L.; Mast, J.; Lieffrig, F.; Bremont, M.; Vanderplasschen, A. The Major Portal of Entry of Koi Herpesvirus in Cyprinus carpio is the Skin. *J. Virol.* **2009**, *83*, 2819–2830. [CrossRef] [PubMed]
30. Shimizu, T.; Yoshida, N.; Kasai, H.; Yoshimizu, M. Survival of Koi Herpesvirus (KHV) in Environmental Water. *Fish Pathol.* **2006**, *41*, 153–157. [CrossRef]
31. McColl, K. *Final Report: Phase 3 of the Carp Herpesvirus Project (CyHV-3)*; PestSmart Toolkit publication; Invasive Animals Cooperative Research Centre: Canberra, Australia, 2016.
32. Eide, K.E.; Miller-Morgan, T.; Heidel, J.R.; Kent, M.L.; Bildfell, R.J.; LaPatra, S.; Watson, G.; Jin, L. Investigation of Koi Herpesvirus Latency in Koi. *J. Virol.* **2011**, *85*, 4954–4962. [CrossRef]
33. Lin, L.; Chen, S.; Russell, D.S.; Löhr, C.; Milston-Clements, R.; Song, T.; Miller-Morgan, T.; Jin, L. Analysis of stress factors associated with KHV reactivation and pathological effects from KHV reactivation. *Virus Res.* **2017**, *240*, 200–206. [CrossRef]
34. McColl, K.A.; Sunarto, A.; Neave, M.J. Biocontrol of Carp: More Than Just a Herpesvirus. *Front. Microbiol.* **2018**, *9*, 2288. [CrossRef]
35. *Assessment of Zoonotic Risk from Infectious Salmon Anaemia Virus*; Scientific Committee on Animal Health and Animal Welfare: European Commission, Health & Consumer Protection Directorate-General: Brussels, Belgium, 2000; Available online: https://ec.europa.eu/food/sites/food/files/safety/docs/sci-com_scah_out44_en.pdf (accessed on 2 January 2019).
36. Roper, K.; Ford, L. *Cyprinid Herpesvirus 3 and Its Relevance to Human Health*; Final Report; Australian National University: Canberra, Australia; Australia for Fisheries Research and Development Corporation: Canberra, Australia, 2019.
37. Bretzinger, A.; Fischer-Scherl, T.; Oumouna, M.; Hoffmann, R.; Truyen, U. Mass mortalities in koi, Cyprinus carpio, associated with gill and skin disease. *Bull. Eur. Ass. Fish Pathol.* **1999**, *19*, 182–185.
38. Kempter, J.; Sadowski, J.; Schütze, H.; Fischer, U.; Dauber, M.; Fichtner, D.; Panicz, R.; Bergmann, S.M. Koi Herpes Virus: Do Acipenserid Restitution Programs Pose a Threat to Carp Farms in the Disease-Free Zones? *Acta Ichthyol. Piscat.* **2009**, *39*, 119–126. [CrossRef]
39. Fabian, M.; Bäumer, A.; Steinhagen, D. Do wild fish species contribute to the transmission of koi herpesvirus to carp in hatchery ponds? *J. Fish Dis.* **2012**, *36*, 505–514. [CrossRef] [PubMed]
40. Cho, M.-Y.; Won, K.-M.; Kim, J.-W.; Jee, B.-Y.; Park, M.A.; Hong, S. Detection of koi herpesvirus (KHV) in healthy cyprinid seed stock. *Dis. Aquat. Org.* **2014**, *112*, 29–36. [CrossRef] [PubMed]
41. Gaede, L.; Steinbrück, J.; Bergmann, S.M.; Jäger, K.; Gräfe, H.; Schoon, H.A.; Speck, S.; Truyen, U. Koi herpesvirus infection in experimentally infected common carp Cyprinus carpio (Linnaeus, 1758) and three potential carrier fish species Carassius carassius (Linnaeus, 1758); Rutilus rutilus (Linnaeus, 1758); and Tinca tinca (Linnaeus, 1758) by quantita. *J. Appl. Ichthyol.* **2017**, *30*, 776–784. [CrossRef]
42. Pospichal, A.; Pokorova, D.; Vesely, T.; Piackova, V. Susceptibility of the topmouth gudgeon (Pseudorasbora parva) to CyHV-3 under no-stress and stress conditions. *Vet. Med.* **2018**, *63*, 229–239. [CrossRef]
43. Minamoto, T.; Honjo, M.N.; Yamanaka, H.; Tanaka, N.; Itayama, T.; Kawabata, Z. Detection of cyprinid herpesvirus-3 DNA in lake plankton. *Res. Vet. Sci.* **2011**, *90*, 530–532. [CrossRef]
44. Kielpinski, M.; Kempter, J.; Panicz, R.; Sadowski, J.; Schutze, H.; Ohlemeyer, S.; Bergmann, S.M. Detection of KHV in freshwater mussels and crustaceans from ponds with KHV history in common carp (Cyprinus carpio). *Isr. J. Aquac. Bamidgeh* **2010**, *62*, 28–37.
45. El-Matbouli, M.; Soliman, H. Transmission of Cyprinid herpesvirus-3 (CyHV-3) from goldfish to naïve common carp by cohabitation. *Res. Vet. Sci.* **2011**, *90*, 536–539. [CrossRef]
46. Yuasa, K.; Kurita, J.; Kawana, M.; Kiryu, I.; Ohseko, N.; Sano, M. Development of mRNA-specific RT-PCR for the detection of koi herpesvirus (KHV) replication stage. *Dis. Aquat. Org.* **2012**, *100*, 11–18. [CrossRef]

47. Boutier, M.; Donohoe, O.; Kopf, R.K.; Humphries, P.; Becker, J.A.; Marshall, J.; Vanderplasschen, A. Biocontrol of Carp: The Australian Plan Does Not Stand Up to a Rational Analysis of Safety and Efficacy. *Front. Microbiol.* **2019**, *10*, 882. [CrossRef]
48. Palmiera, L.; Sorel, O.; Van Campe, W.; Boudry, C.; Roels, S.; Myster, F.; Reschner, A.; Coulie, P.G.; Kerkhofs, P.; Vanderplasschen, A.; et al. An essential role for gamma-herpesvirus latency-associated nuclear antigen homolog in an acute lymphoproliferative disease of cattle. *Proc. Natl. Acad. Sci. USA* **2013**, *110*, E1933–E1942. [CrossRef]
49. Yuasa, K.; Sano, M.; Oseko, N. Goldfish is Not a Susceptible Host of Koi Herpesvirus (KHV) Disease. *Fish Pathol.* **2013**, *48*, 52–55. [CrossRef]
50. Thresher, R.E.; Allman, J.; Stremick-Thompson, L. Impacts of an invasive virus (CyHV-3) on established invasive populations of common carp (Cyprinus carpio) in North America. *Boil. Invasions* **2018**, *20*, 1703–1718. [CrossRef]
51. Fenner, F.; Woodroofe, G.M. Changes in the virulence and antigenic structure of strains of myxoma virus recovered from Australian wild rabbits between 1950 and 1964. *Aust. J. Exp. Boil. Med. Sci.* **1965**, *43*, 359–370. [CrossRef] [PubMed]
52. Di Giallonardo, F.; Holmes, E.C. Viral biocontrol: Grand experiments in disease emergence and evolution. *Trends Microbiol.* **2014**, *23*, 83–90. [CrossRef] [PubMed]
53. McColl, K.A.; Sunarto, A.; Holmes, E.C. Cyprinid herpesvirus 3 and its evolutionary future as a biological control agent for carp in Australia. *Virol. J.* **2016**, *13*, 206. [CrossRef]
54. Yuasa, K.; Ito, T.; Sano, M. Effect of Water Temperature on Mortality and Virus Shedding in Carp Experimentally Infected with Koi Herpesvirus. *Fish Pathol.* **2008**, *43*, 83–85. [CrossRef]
55. Pikarsky, E.; Ronen, A.; Abramowitz, J.; Levavi-Sivan, B.; Hutoran, M.; Shapira, Y.; Steinitz, M.; Perelberg, A.; Soffer, D.; Kotler, M. The pathogenesis of acute viral diseases in fish induced by the carp interstitial nephritis and gill necrosis virus. *J. Virol.* **2004**, *78*, 9544–9551. [CrossRef]
56. Sunarto, A.; Liongue, C.; McColl, K.A.; Adams, M.M.; Bulach, D.M.; Crane, M.; Schat, K.A.; Slobedman, B.; Barnes, A.; Ward, A.C.; et al. Koi Herpesvirus Encodes and Expresses a Functional Interleukin-10. *J. Virol.* **2012**, *86*, 11512–11520. [CrossRef]
57. Sunarto, A.; McColl, K.A. Expression of immune-related genes of common carp during cyprinid herpesvirus 3 infection. *Dis. Aquat. Org.* **2015**, *113*, 127–135. [CrossRef]
58. Neave, M.; Sunarto, A.; McColl, K.A. Transcriptomic analysis of common carp anterior kidney during Cyprinid herpesvirus 3 infection: Immunoglobulin repertoire and homologue functional divergence. *Sci. Rep.* **2017**, *7*, 41531. [CrossRef]
59. Watanuki, H.; Yamaguchi, T.; Sakai, M. Suppression in function of phagocytic cells in common carp Cyprinus carpio L. injected with estradiol, progesterone or 11-ketotestosterone. *Comp. Biochem. Physiol. Part C Toxicol. Pharmacol.* **2002**, *132*, 407–413. [CrossRef]
60. Stuart, I.G.; Jones, M. Large, regulated forest floodplain is an ideal recruitment zone for non-native common carp (Cyprinus carpio L.). *Mar. Freshw. Res.* **2006**, *57*, 333–347. [CrossRef]
61. Sunarto, A.; McColl, K.A.; Crane, M.; Sumiati, T.; Hyatt, A.D.; Barnes, A.; Walker, P.J. Isolation and characterization of koi herpesvirus (KHV) from Indonesia: Identification of a new genetic lineage. *J. Fish Dis.* **2010**, *34*, 87–101. [CrossRef]
62. Sunarto, A.; McColl, K.A.; Crane, M.S.J.; Schat, K.A.; Slobedman, B.; Barnes, A.; Walker, P.J. Characteristics of cyprinid herpesvirus 3 in different phases of infection: Implications for disease transmission and control. *Virus Res.* **2014**, *188*, 45–53. [CrossRef]
63. McColl, K.A.; Crane, M.S.J. *Cyprinid Herpesvirus 3, CyHV-3: Its Potential as a Biological Control Agent for Carp in Australia*; PestSmart Toolkit publication; Invasive Animals Cooperative Research Centre: Canberra, Australia, 2013.
64. Durr, P.A.; Davis, S.; Joehnk, K.; Graham, K.; Hopf, J.; Arakala, A.; McColl, K.A.; Taylor, S.; Chen, Y.; Sengupta, A.; et al. Development of Hydrological, Ecological and Epidemiological Modelling to Inform a CyHV3 Release Strategy for the Biocontrol of Carp in the Murray Darling Basin. Part A. In *Integrated Ecological and Epidemiological Modelling*; Final Report; CSIRO-Australian Animal Health Laboratory: Geelong, Australia; Australia for Fisheries Research and Development Corporation: Canberra, Australia, 2019.
65. Hedrick, R.P.; Waltzek, T.B.; McDowell, T.S. Susceptibility of koi carp, common carp, goldfish, and goldfish x common carp hybrids to cyprinid herpesvirus-2 and herpesvirus-3. *J. Aquat. Anim. Health* **2006**, *18*, 26–34. [CrossRef]

66. Bergmann, S.M.; Sadowski, J.; Kielpinski, M.; Bartlomiejczyk, M.; Fichtner, D.; Riebe, R.; Lenk, M.; Kempter, J. Susceptibility of koi x crucian carp and koi x goldfish hybrids to koi herpesvirus (KHV) and the development of KHV disease (KHVD). *J. Fish Dis.* **2010**, *33*, 267–272. [CrossRef]
67. Geoghegan, J.; Duchene, S.; Holmes, E.C. Comparative analysis estimates the relative frequencies of co-divergence and cross-species transmission within viral families. *PLoS Pathog.* **2017**, *13*, e1006215. [CrossRef]
68. Geoghegan, J.; Holmes, E.C. Predicting virus emergence amid evolutionary noise. *Open Boil.* **2017**, *7*, 170189. [CrossRef]
69. Kerr, P.; Liu, J.; Cattadori, I.; Ghedin, E.; Read, A.; Holmes, E.C. Myxoma Virus and the Leporipoxviruses: An Evolutionary Paradigm. *Viruses* **2015**, *7*, 1020–1061. [CrossRef]
70. Mahar, J.; Nicholson, L.; Eden, J.-S.; Duchêne, S.; Kerr, P.J.; Duckworth, J.; Ward, V.K.; Holmes, E.C.; Strive, T. Benign Rabbit Caliciviruses Exhibit Evolutionary Dynamics Similar to Those of Their Virulent Relatives. *J. Virol.* **2016**, *90*, 9317–9329. [CrossRef]
71. Di Giallonardo, F.; Holmes, E.C. Exploring Host–Pathogen Interactions through Biological Control. *PLoS Pathog.* **2015**, *11*, e1004865. [CrossRef]
72. Saunders, G.R.; Cooke, B.; McColl, K.; Shine, R.; Peacock, T. Modern approaches for the biological control of vertebrate pests: An Australian perspective. *Biol. Control* **2010**, *52*, 288–295. [CrossRef]
73. Conallin, A.J.; Smith, B.B.; Thwaites, L.A.; Walker, K.F.; Gillanders, B.M. Environmental Water Allocations in regulated lowland rivers may encourage offstream movements and spawning by common carp, Cyprinus carpio: Implications for wetland rehabilitation. *Mar. Freshw. Res.* **2012**, *63*, 865–877. [CrossRef]
74. Davidson, S. Carp crusades. *Ecos* **2002**, *112*, 8–12.
75. Wedekind, C. *Synergistic Genetic Biocontrol Options for Common Carp (Cyprinus Carpio)*; Final Report; Department of Ecology and Evolution, University of Lausanne: Lausanne, Switzerland; Fisheries Research and Development Corporation: Canberra, Australia, 2019.
76. Thresher, R.; Van De Kamp, J.; Campbell, G.; Grewe, P.; Canning, M.; Barney, M.; Bax, N.J.; Dunham, R.; Su, B.; Fulton, W. Sex-ratio-biasing constructs for the control of invasive lower vertebrates. *Nat. Biotechnol.* **2014**, *32*, 424–427. [CrossRef]
77. Akbari, O.S.; Bellen, H.J.; Bier, E.; Bullock, S.L.; Burt, A.; Church, G.M.; Cook, K.; Duchek, P.; Edwards, O.R.; Esvelt, K.M.; et al. Safeguarding gene drive experiments in the laboratory. *Science* **2015**, *349*, 927–929. [CrossRef]
78. Neville, H. Trojan Males and the Genetics of Non-Native Control. 2016. Available online: https://www.tu.org/blog-posts/trojan-males-and-the-genetics-of-non-native-control (accessed on 14 August 2017).
79. Cotton, S.; Wedekind, C. Control of introduced species using Trojan sex chromosomes. *Trends Ecol. Evol.* **2007**, *22*, 441–443. [CrossRef]
80. Maselko, M.; Heinsch, S.; Chacón, J.M.; Harcombe, W.R.; Smanski, M. Engineering species-like barriers to sexual reproduction. *Nat. Commun.* **2017**, *8*, 883. [CrossRef]
81. NCCP. *The National Carp Control Plan Strategic Research and Technology Plan 2017–2019*; Fisheries Research and Development Corporation: Canberra, Australia, 2019. Available online: http://carp.gov.au/what-we-are-doing/research/nccp-research-projects (accessed on 10 March 2020).
82. Beckett, S.; Caley, P.; Hill, M.; Nelson, S.; Henderson, B. *Biocontrol of European Carp: Ecological Risk Assessment for the Release of Cyprinid Herpesvirus 3 (CyHV-3) for Carp Biocontrol in Australia*; Final Report; CSIRO Data 61; Canberra, Australia Australia for Fisheries Research and Development Corporation: Canberra, Australia, 2019.
83. Olsen, J.; Cooke, B.D.; Trost, S.; Judge, D. Is wedge-tailed eagle, Aquila audax, survival and breeding success closely linked to the abundance of European rabbits, Oryctolagus cuniculus? *Wildl. Res.* **2014**, *41*, 95–105. [CrossRef]
84. Cliff, H.B.; Jones, M.E.; Johnson, C.N.; Pech, R.P.; Heyward, R.P.; Norbury, G.L. Short-term pain before long-term gain? Suppression of invasive primary prey temporarily increases predation on native lizards. *Boil. Invasions* **2020**, *22*, 2063–2078. [CrossRef]
85. Bureau of Resource Sciences. *Rabbit Calicivirus Disease: A Report under the Biological Control Act 1984*; Bureau of Resource Sciences: Canberra, Australia, 1996.
86. Brookes, J.D.; Hipsey, M.R. *Water Quality Risk Assessment of Carp Biocontrol for Australian Waterways*; Final Report; Environment Institute, University of Adelaide, Australia for Fisheries Research and Development Corporation: Canberra, Australia, 2019.

87. Silva, L.G.M.; Bell, K.; Baumgartner, L.J. *Clean-Up Procedures Applied for Fish Kill Events: A Review for the National Carp Control Plan*; Final Report; Australia for Fisheries Research and Development Corporation: Canberra, Australia, 2019.
88. Tilley, A.; Colquhoun, E.; O'Keefe, E.; Nash, S.; McDonald, D.; Evans, T.; Gillespie, G.; Hardwick, D.; Beavis, S.; Francina, C.; et al. *Options for Utilisation of Carp Biomass*; Final Report; Australia for Fisheries Research and Development Corporation: Canberra, Australia, 2019.
89. Zhang, A.; Carter, L.; Curnock, M.; Mankad, A. *Biocontrol of European Carp: Ecological and Social Risk Assessment for the Release of Cyprinid Herpesvirus 3 (CyHV-3) for Carp Biocontrol in Australia*; Final Report; CSIRO Land and Water: Dutton Park, Queensland, Australia; Australia for Fisheries Research and Development Corporation: Canberra, Australia, 2019.
90. Nichols, S.J.; Gawne, B.; Richards, R.; Lintermans, M.; Thompson, R. *NCCP: The Likely Medium-to Long-Term Ecological Outcomes of Major Carp Population Reductions*; Final Report; Institute for Applied Ecology, University of Canberra, Australia for Fisheries Research and Development Corporation: Canberra, Australia, 2019.
91. Cooke, B.; Chudleigh, P.; Simpson, S.; Saunders, G.R. The Economic Benefits of the Biological Control of Rabbits in Australia, 1950–2011. *Aust. Econ. Hist. Rev.* **2013**, *53*, 91–107. [CrossRef]
92. Knowlton, N. Doom and gloom won't save the world. *Nature* **2017**, *544*, 271. [CrossRef]

Article

Geographic-Scale Harvest Program to Promote Invasivorism of Bigheaded Carps

Wesley W. Bouska [1], David C. Glover [2], Jesse T. Trushenski [3], Silvia Secchi [4], James E. Garvey [5,*], Ruairi MacNamara [6], David P. Coulter [5], Alison A. Coulter [5], Kevin Irons [7] and Andrew Wieland [5]

1 U.S. Fish and Wildlife Service, 555 Lester Avenue, La Crosse, WI 54650, USA; wesley_bouska@fws.gov
2 Illinois Department of Natural Resources, 700 South 10th Street, Havana, IL 62644, USA; dave.glover@illinois.gov
3 Riverence Holdings LLC, 604 W Franklin Street, Boise, ID 83702, USA; jesse@riverence.com
4 Department of Geographical and Sustainability Sciences, 310 Jessup Hall, University of Iowa, Iowa City, IA 52242, USA; silvia-secchi@uiowa.edu
5 Center for Fisheries, Aquaculture, and Aquatic Sciences, School of Biological Sciences, 1125 Lincoln Drive, Southern Illinois University, Carbondale, IL 62901, USA; dcoulter@siu.edu (D.P.C.); acoulter@siu.edu (A.A.C.); andrew.wieland@siu.edu (A.W.)
6 Hubbs-SeaWorld Research Institute, 2595 Ingraham Street, San Diego, CA 92109, USA; rmacnamara@hswri.org
7 Illinois Department of Natural Resources, One Natural Resources Way, Springfield, IL 62702, USA; kevin.irons@illinois.gov
* Correspondence: jgarvey@siu.edu; Tel.: +1-618-536-7761

Received: 17 July 2020; Accepted: 24 August 2020; Published: 1 September 2020

Abstract: Invasive bigheaded carps, genus *Hypophthalmichthys*, are spreading throughout the Mississippi River basin. To explore the efficacy of a consumer-based market (i.e., invasivorism) to manage them, we developed a conceptual model and evaluated three harvest approaches—direct contracted removal, volume-based incentives ("fisher-side" control), and set-quota harvest ("market-side" control). We quantified the efficacy of these approaches and potential population impact in the Illinois River. Contracted removal was effective for suppressing small populations at the edge of the range but cannot support a market. "Fisher-side" removals totaled 225,372 kg in one year. However, participation was low, perhaps due to reporting requirements for fishers. The "market-side", set-quota approach removed >1.3 million kg of bigheaded carp in less than 6 months. Larger, older fish were disproportionately harvested, which may hinder the ability to suppress population growth. Total density declined in one river reach, and harvest may reduce upstream movement toward the invasion fronts. With sufficient market demand, harvest may control bigheaded carp. However, lack of processing infrastructure and supply chain bottlenecks could constrain harvest, particularly at low commodity prices. Given the geographical scale of this invasion and complicated harvest logistics, concerns about economic dependence on invasivorism that encourage stock enhancement are likely unmerited.

Keywords: invasivorism; bigheaded carp; commercial fishing; *Hypophthalmichthys*; Illinois River

1. Introduction

Invasive species threaten biodiversity worldwide [1], costing $120 billion USD annually in the United States [2]. Removal programs may control invasives [3]. For example, humans routinely overharvest fish stocks [4–6]. Thus, controlled harvest may help control invasive populations of fish and other taxa [7–10]. Yet, factors such as time, effort, and expense often limit success [11–13]. Whereas government assistance is necessary to control invasive species that have low market value such as sea lamprey, *Petromyzon marinus* (Linnaeus; [14]), fishes with commercial value could be marketed to reduce financial burden on government agencies. This consumer-based control of invasive species is popularly called invasivorism. Potential candidates for invasivorism are silver carp *Hypophthalmichthys molitrix* (Valenciennes) and bighead carp, *H. nobilis* (Richardson), collectively known as bigheaded carp. These species invaded the lower Mississippi River basin of the US in the 1970s, expanded northward, are now more abundant in the Illinois River than anywhere else globally [15–17], and may invade the Laurentian Great Lakes via Lake Michigan. Establishment of bigheaded carp in the Great Lakes may jeopardize fisheries valued at $7 billion USD per year [18,19].

Nearly a decade ago, the U.S. Army Corps of Engineers (USACE) explored options to prevent interbasin transfer of aquatic nuisance species (ANS) between the Mississippi River and Great Lakes basins, with bigheaded carp being a primary species of concern and the Chicago Area Waterway System (CAWS) the primary focus area. The CAWS contains five aquatic pathways [20], one of which, the Chicago Sanitary and Shipping Canal (CSSC), is the only permanently open connection between the basins, with Lake Michigan being the recipient Great Lake. The CSSC has previously allowed movement of ANS between the basins [21] and is near the edge of the bigheaded carp range, which is approximately 80 km south downstream in the upper Illinois River [22].

Removal of bigheaded carp is included in all eight management strategies for stopping interbasin movement of these fish [20], and non-structural control plus harvest is the only strategy that can be initiated immediately. Since 2010, contracted removal of greater than 3200 tons of bigheaded carp has occurred near the CAWS in the upper Illinois River, where commercial harvest is prohibited [23]. Removal at this range edge is expected to be agency funded for the foreseeable future [24] and has likely prevented upstream range expansion toward Lake Michigan [22]. In the lower Illinois River from where bigheaded carp in the upper Illinois River derive, commercial harvest is legal. If reliable moderate- to high-value markets can be developed for bigheaded carp in the lower river, exploitation should remain high and reduce upstream migrants via invasivorism. Although the likelihood of reducing bigheaded carp to extinction in such a large, open system is low, the capacity for population suppression and reducing further expansion may be high.

The idea of harvesting near the center of the invading population to reduce densities at range edges is supported by modeling that assesses the influence of harvest and other control measures in lower river reaches where population densities are high. The Spatially Explicit Asian Carp Population (SEAcarP) model [24] links movement probabilities among river pools or reaches with demographic responses to harvest removal to predict the likelihood of population density declines at the edge of the species' range (see Erickson et al. [25] for similar approach with grass carp). The model has been applied to the Illinois River using movement data from Coulter et al. [22], predicting that increasing mortality of bigheaded carp in the lower river will effectively reduce densities at the upstream invasion front.

Stimulating market demand to accomplish control via invasivorism may seem like a simple task. Bigheaded carp are valued food fish in much of the world. With native, wild bigheaded carp stocks threatened or extirpated [26], global demand is now primarily met by aquaculture, with these species being among the most cultured fish in the world [27]. Globally, over 5.3 million tons of silver carp are cultured annually, primarily in China, India, Bangladesh, Iran, the Russian Federation, and Cuba [27]. There may be a high demand for bigheaded carp from the U.S., since consumers in countries such as China are willing to pay a premium price for wild-caught fish [28], and may perceive U.S.-sourced fish as being of a higher quality than cultured products.

Although recent surveys have suggested that there is potential domestic consumer demand for bigheaded carp as food [29], most US markets are for rendered carp products (meals and oils), as ingredients in livestock and aquaculture feeds [30–32] and as hydrolyzed fertilizers. The current supply of bigheaded carp in US rivers is not a limiting factor in the growth of the industry, but rather the lack of processing plants and reliable domestic markets plus access to existing international markets to monetarily compensate commercial fishers. A well-developed fishery infrastructure does not exist in the central area of the invasion in the US. Thus, a fishery must be built to implement control via invasivorism.

To determine how a long-term, self-maintained fishery may be developed to control the bigheaded carp invasion, we first developed a broad conceptual supply chain model for economic development of a controlled harvest fishery to promote invasivorism in the Illinois River system, throughout the invaded US, and more broadly for any invaded ecosystem where market-driven consumption is an option. We then evaluated three harvest strategies against invasive bigheaded carp in the upper and lower Illinois River along the conceptual supply chain continuum, quantified the resulting exploitation, and examined economic factors affecting removal by harvest.

The harvest strategies evaluated during 2010 through 2012 were (1) an ongoing contracted harvest program in the upper Illinois River, (2) a "fisher-side" incentives program that offered select commercial fishers progressive economic rewards for participating (i.e., sharing harvest data) and harvesting increasing amounts of bigheaded carp for direct-consumption markets, and (3) a "market-side" incentives program that set a quota-based harvest of bigheaded carp for indirect-consumption markets of fish meal. Harvest removal of fish from populations is often biased toward larger, older individuals, which affects population responses [10]. We hypothesized that a removal effort of this large, geographic-scale magnitude should impact bigheaded carp population demographics including abundance, size structure, and age structure within the study area in the lower Illinois River, and stimulate direct and indirect consumption markets for bigheaded carp, with implications for basin-wide population dynamics and reduced risk to the Great Lakes.

2. Conceptual Model

The commercial fishing industry in the Illinois and Mississippi River basins is significantly reduced from its former past [16], where harvest of mussels and fish peaked in the early 1900s. Much of the fish harvest in this region was for local consumption and as consumer demand waned, river water quality declined, and sedimentation increased, fish harvest as a source of protein and income declined by the 1930s. For a fishery to develop to effectively reduce bigheaded carp from both free flowing and pooled reaches of these and other river systems, industry infrastructure must be reestablished. We propose a simplified conceptual model of bigheaded carp supply, transport, processing, and demand in flowing and non-flowing waters of the US Midwest (Figure 1).

The model identifies major components of a developing fishery for bigheaded carp. The first primary bottleneck to commercial-level invasivorism is the fishers and fishing fleet (Figure 1). The size of the fishing crews, training, time investment, and many other fixed and variable costs influence the reliability and effectiveness of fishing bigheaded carps. At this juncture, fishing is largely conducted by part-time crews with limited gears, because of a lack of funds for purchasing and maintaining boats and equipment. These crews fish for natives and carp, mostly for local businesses or contracted fishing in the upper Illinois River, maintaining local expertise to drive potential harvest expansion. However, with no reliable market for bigheaded carps, capital investment in large, sustained fishing operations will not occur.

The supply of bigheaded carps in the Illinois River and other systems is driven by myriad factors that vary with water body type and location (Figure 1). Lohmeyer and Garvey [33] conducted a recruitment assessment of bigheaded carps in pooled and unpooled reaches of the Upper Mississippi River, finding that recruitment was low but consistent in unpooled reaches and higher and variable in pools (also see Chick et al. [34]). Recruitment variability will affect the reliability of harvest and supply. Harvest will vary with location because of the logistics of reaching fishing locations and transporting fish back to boat landings. Unpooled rivers typically have fewer access points and longer travel times due to their longitudinal geomorphology. Ensuring that harvested bigheaded carps enter a market that provides an economic benefit is necessary to create a system where invasivorism is feasible. Currently, most bigheaded carps removed from the Illinois River system are sold directly to ethnic or local fish markets from the fishers or collected at no cost by fish processors following agency-contracted removal efforts in the upper Illinois River (see [32]). Harvested bigheaded carps also may be transported by fishers to larger processors, although the number of processors in the region is limited and often geographically distant from harvest areas, making this a costly and often prohibitive option. A second bottleneck to developing market-based invasivorism is ensuring that a steady supply of fresh bigheaded carp is procured by fishers and transported to processors in a way that is safe for the consumer, economical, and efficient.

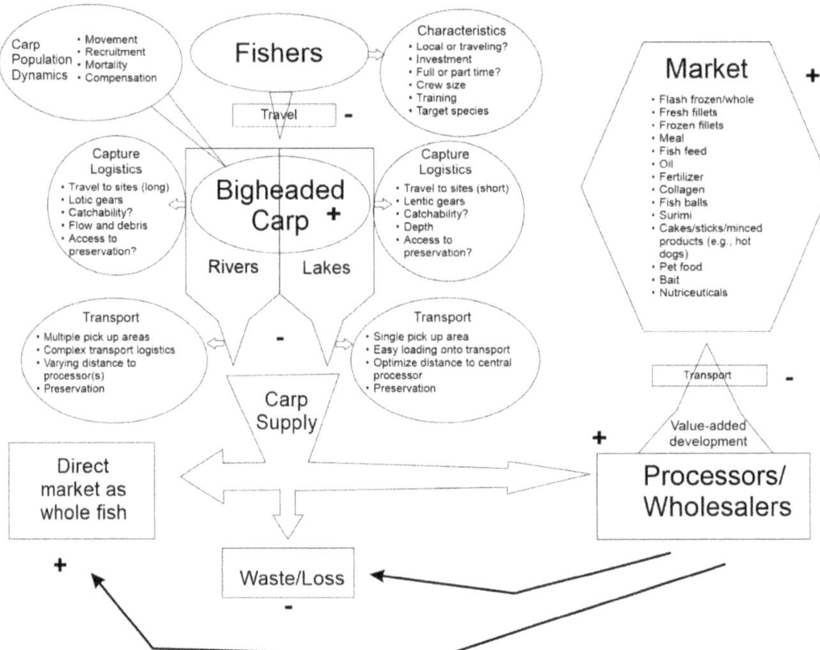

Figure 1. Conceptual model of a fishery for bigheaded carps in the rivers and lakes of the invaded south and north central United States. Economic factors affecting fishers will influence the species composition and biomass of bigheaded carp available to the supply chain. Transportation and markets will drive the level of harvest. Each component along the supply chain has costs (− sign) and economic gains (+) that determine how much fish are removed from the environment.

Providing bigheaded carp-derived products in ways that maximize their value is the goal of wholesalers and processors, where there is clearly global demand for bigheaded carp products. However, costs of setting up local processing facilities is high and potentially risky, while transporting whole fish to existing processors on the coasts or overseas is logistically difficult, involves federal regulation, and is ultimately costly (Figure 1). Creating local demand while developing export and processing facilities is likely the most economically feasible model for establishing harvest as control. However, most investors are wary of supporting such facilities without assurances of dependable supplies of high-quality bigheaded carp from the rivers, which is currently limited by undeveloped fishing capacity, uncertainties about regional fish production, government red tape, and a virtually non-existent transport network.

This conceptual model is not intended to be exhaustive, but it does provide several areas where investments or support may help develop a market for removing bigheaded carp at areas of high density. In the following sections, we describe an effort to stimulate fishing at the "fishing end" and at the "market demand" end of the bigheaded carp supply chain (Figure 1), and the collection of bigheaded carp population demographics through subsampling of commercially harvested carps at the processing plant, and fishery-independent sampling in the field before and after the harvest programs were implemented.

3. Materials and Methods

The two programs we created to stimulate harvest in the lower Illinois River were compared to the ongoing contracted fishing program in the upper Illinois River [35]. Commercial fishing does not occur in the upper Illinois River at the edge of invaded range, so market-driven removal is not an option in this region (Figure 2). In the lower Illinois River, commercial harvest has occurred for more than a century and has the potential to develop into a control method via invasivorism (Figure 2).

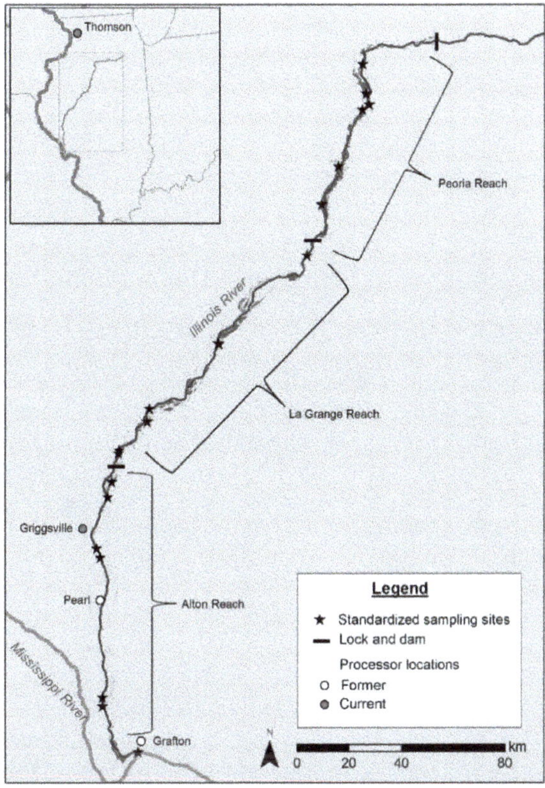

Figure 2. Map of the Illinois River study areas and processing plants during the 2011–2012 study period. Note that Griggsville became non-operational at an unknown time since the study concluded.

3.1. *Training, Certification, and Incentives-Based Approach*

To accomplish the goals of the "fisher-side" commercial fishing strategy in the lower Illinois River, a stakeholders' meeting was held in Grafton, IL, in the period 20–21 September 2010, during which academic researchers, regulatory authorities, commercial fishers, fish processors, marketers, and distributors discussed using commercial harvest of bigheaded carp in the lower Illinois River as a means of augmenting contracted fishing occurring in the upper river (IISG 2010). Key findings that shaped the design of our removal experiment included the need to (1) improve safety and quality of harvested bigheaded carp as food, (2) properly brand and find markets for the product [29,32], (3) provide financial support for commercial fishers, (4) form a public–private partnership to stimulate harvest, and (5) use associated data to inform an adaptive management framework [36].

We initiated the pilot-scale training, certification, and incentives-based approach to support harvest of bigheaded carp from the lower three reaches of the Illinois River in 2011 to augment the contracted fishing in the upper river (Figure 2). Names of licensed commercial fishers were obtained from the Illinois Commercial Fishing Association (ICFA), and participants were selected by lottery. The training related to (1) safe handling of bigheaded carp for consumption in foreign and domestic markets, (2) licensing and safe operation of commercial fishing vessels, (3) biosecurity practices to prevent transmission of aquatic nuisance species and pathogens, and (4) coordination and sharing of bigheaded carp harvest data with stakeholders.

In addition to receiving $0.42 USD/kg for their catch, incentives for participating fishers included reimbursement for the cost of two ICFA memberships (i.e., fisherperson and deckhand; $100 USD total), the annual Illinois commercial fishing license fee ($35 USD), and net tags required for commercial fishing in Illinois waters ($250 USD). Furthermore, participating fishers received $1000 USD to offset fuel costs after harvesting 22,680 kg of bigheaded carp and $3000 USD to offset gear purchase/repair/replacement costs if they harvested a total of 45,359 kg. For their catch to be eligible for incentives, participating fishers had to report the date/time, location (using a provided handheld GPS), and species composition of their catch. Data had to match information recorded on processor receipts. Only bigheaded carp caught from the Illinois River were considered, and all fish had to be sold to processors for human consumption.

3.2. Set-Quota Harvest Approach

The quota commercial fishing strategy ("market-side") program was conducted in spring 2012, whereby a set-quota fishing effort was implemented to explore the biological and ecological effects of increased bigheaded carp harvest. Following a competitive bidding process, a third-party logistics company (Select Logistics Network, Inc., Clinton, IL, USA) and a local fish processing facility (Big River Fish Company, Pearl, IL, USA) were selected to coordinate the harvest and processing of bigheaded carp at a price of approximately $0.42 USD/kg to fishers, from the lower three reaches of the Illinois River in order to yield up to 453,600 kg of dried fish meal. The bigheaded carp removed through this approach were not eligible for incentives and were processed into fish meal by Protein Products, Gainesville, FL, USA.

3.3. Field and Processor Subsampling

Fisheries are typically highly selective for sizes, ages, and species and thus may have unique impacts on the populations. We visited processing plants approximately every two weeks during the period 1 February 2012 through 8 May 2012 while contracted fishing was occurring. In order to characterize bigheaded carp population demographics within the commercial catch during each biweekly sample, we randomly selected up to 100 silver carp and 100 bighead carp that were harvested from each of the lower three reaches of the Illinois River (depending on availability) and recorded total length (TL) and weight data. Post-cleithra were removed from up to five fish per species per 50 mm length group per river reach for age determination.

We sampled the Illinois River using standardized, fishery-independent sampling to compare bigheaded carp population metrics. We recognized that the sampling was insufficient to detect a fishing impact in such a short time with limited effort. However, we were able to assess annual variability in stock and potential harvest selectivity. Sampling occurred in August 2011 before contracted harvest began and August 2012 after harvest concluded. Sampling was conducted along the main channel of the Illinois River at four fixed locations within each of the three lower pool reaches, as well as nearby backwater lakes and side channels (Figure 2). Pulsed-DC electrofishing transects (Smith-Root GPP 5.0 electrofisher; 15 min each), with two netters, were conducted along each main channel and backwater site during the day. Captured fish were euthanized by immersion in 300 ppm tricaine methanesulfonate (MS-222) until opercular movement ceased. All fish were weighed and measured (total length), and post-cleithra were collected from a subsample of ten silver carp per 10 mm length group per reach; due to smaller sample sizes, post-cleithra were removed from all collected bighead carp.

To determine age, post-cleithra were sectioned transversely across the center with a 1.5 amp diamond-blade low-speed isomet saw (Buehler, Lake Bluff, IL, USA) following Johal et al. [37]. Two independent readers used side illumination from an MI-150 fiber optic light (Dolan-Jenner Industries, Boxborough, MA, USA); if disagreement between readers could not be resolved, the sample was omitted. Age distributions were developed for the entire dataset using an age–length key [38].

Electrofishing catch per unit effort was compared between 2011 and 2012 for silver and bighead carp by reach and for all reaches combined using paired *t*-tests with each site being treated as the unit of replication. All statistical analyses were conducted using SAS 9.2. (SAS Institute, Cary, NC, USA). An alpha level of 0.05 was used to judge statistical significance.

4. Results

4.1. Training, Certification, and Incentives Approach

Although the fishers harvested bigheaded carp from the Illinois River (Figure 2), the "fisher-side" data collection and fish removal goals of the 2011 incentives approach were not fully achieved. Despite the promise of incentive payments, commercial fishers infrequently reported data such as fishing location (GPS coordinates). Of the ten commercial fishers enrolled in the incentives program, only three fully participated, providing GPS locations and dates related to their fishing efforts. These individuals harvested a combined total of 225,372 kg of bigheaded carp from the Illinois River, and received a total of $8000 USD in incentive payments. Although they technically fulfilled the obligations of the incentives program, review of the data revealed the GPS coordinates provided were typically locations of boat access ramps from which they launched, not the precise locations of harvest. This provided little detailed information other than from which river reach fish were harvested. Harvest effect in the Illinois River was likely minimal due to the low harvest quantities.

4.2. Set-Quota Harvest and Processor Subsampling Approach

The "market-side" fishing approach, with a goal of harvesting enough bigheaded carp to yield an open order of 453,600 kg of dried fish meal, was successful in meeting removal and data goals. Between 25 January 2012 and 11 June 2012, commercial fishers harvested 805,878 kg of bigheaded carp from the Alton reach, 223,910 kg from the La Grange reach, and 276,559 kg from the Peoria reach, for a total of 1,306,346 kg, which yielded enough dried meal to meet the contracted goal.

4.3. Bigheaded Carp Standardized Sampling Catch Rates Pre- and Post-Harvest

Mean silver carp electrofishing CPUE for the three lower reaches of the Illinois River combined had a 2011 rate of 100.4 fish/h (SE = 14.6) and 2012 rate of 81.0 fish/h (SE = 17.4). This difference was not significant ($t_{17} = 1.60$; $p = 0.128$). Silver carp mean CPUE declined from 2011 to 2012 in the Alton reach by nearly half ($t_5 = -3.77$; $p = 0.01$; Table 1), but was not significantly different for the La Grange reach ($t_6 = -1.53$; $p = 0.18$) or the Peoria reach ($t_4 = 0.33$; $p = 0.76$; Table 1). Mean bighead carp CPUE was not different from 2011 to 2012 for the lower three reaches of the Illinois River combined ($t_{19} = 1.12$; $p = 0.28$) or among reaches ($t_6 \leq 1.05$; $p \geq 0.34$). Overall bighead carp CPUE was 2.9 fish/h in 2011 (SE = 2.3) and 0.3 fish/h in 2012 (SE = 0.2) among all reaches.

Table 1. Standardized densities (mean fish electrofished per hour; catch per unit effort, CPUE) of silver carp in each reach pool of the Illinois River before (2011) and after (2012) harvesting occurred.

Year	Reach	N Sites	Silver Carp Mean CPUE	SE
2011	Alton	7	61.6	10.7
	LaGrange	7	107.3	27.3
	Peoria	6	126.7	23.2
2012	Alton	6	33.3	10.8
	LaGrange	7	75.5	15.6
	Peoria	5	147.1	46.1

4.4. Size and Length at Age

Comparisons of length frequency distributions and length at age from commercially harvested and electrofished silver carp in 2011 and 2012 demonstrated the size selectivity of commercial gears. The length frequency histogram comparing commercial harvest to standardized sampling showed a bimodal distribution of commercially caught silver and bighead carp (Figure 3). Examination of the age frequency of silver carp showed over 25% of commercially harvested silver carp were age 5 and older, compared to just over 7% of silver carp collected during standardized sampling (Figure 4), with harvested fish having larger length at age than fish in the standard samples.

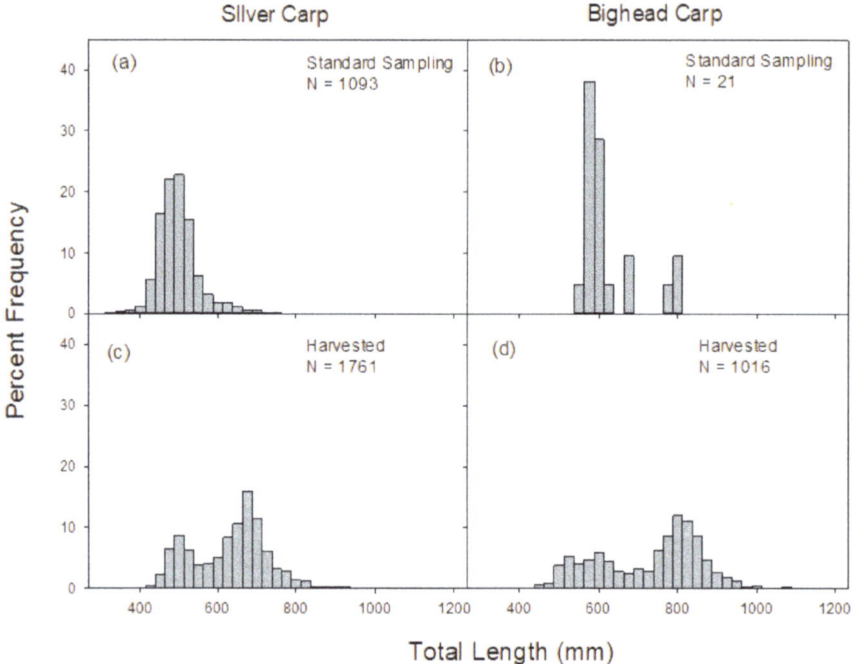

Figure 3. Percent frequency of total lengths of (**a**) silver carp and (**b**) bighead carp sampled with standardized gear (upper panels) and harvested by fishers (**c**,**d**) in the lower Illinois River during 2012.

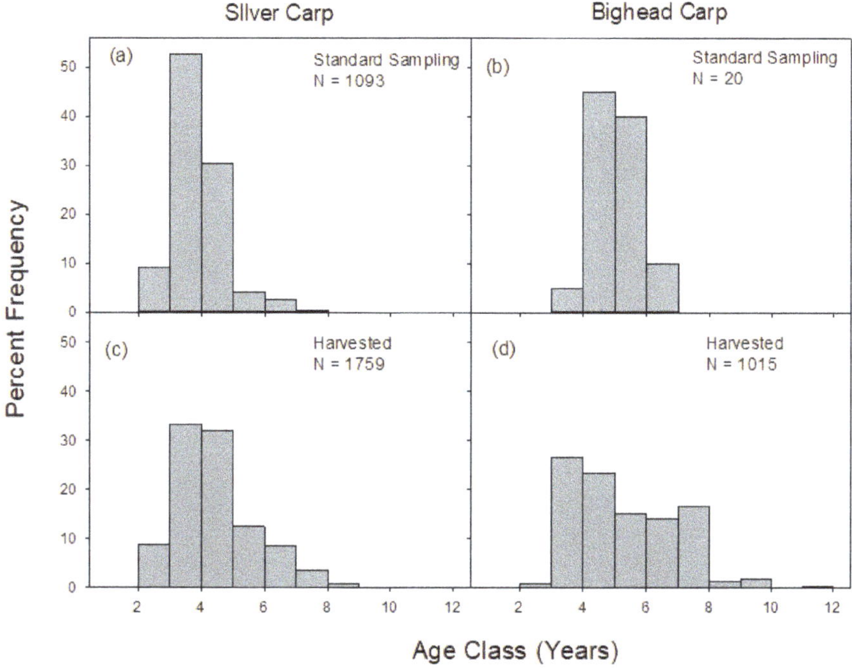

Figure 4. Percent frequency of ages of (**a**) silver carp and (**b**) bighead carp sampled with standardized gear and harvested by fishers (**c**,**d**) in the lower Illinois River during 2012.

5. Discussion

Our "experimental" approach toward stimulating invasivorism in the lower Illinois River yielded important information about how to effectively and economically remove large quantities of bigheaded carp and potentially other invasive species. There are three examples for comparison.

First, the completely subsidized, decade-long contracted removal program in the upper Illinois River is an example of one extreme where market demand for fish does not drive harvest, although harvested bigheaded carps are acquired for free by processors. As we noted earlier, this approach has likely prevented density increases in the upper river, although societal, taxpayer costs are high and bigheaded carp densities continually rebound as fish immigrate from downstream [17]. Along the supply chain model we developed, this approach is completely decoupled from fish availability, market needs, or economic costs and benefits of fishing. Because processors obtain fish at no cost, there is no need to develop a consistent demand for fish. Without contracted removal at the range edge, populations will build in the upper river, greatly increasing the probability of movement toward the CAWS and potentially into Lake Michigan and the other Great Lakes.

The second approach we conducted at the "fisher side" of the model was considered unsuccessful. In the lower Illinois River, when it came to enlisting fishers to assist in the collection of data, the incentive levels did not substantially increase fishing, with our participants not complying. Non-compliant fishers may have placed a greater value on their proprietary information (i.e., fishing locations and methods) than what they could gain from the incentives program. Providing incentives without confirming information about fishing location, effort, and species composition may lead to potential misinformation about the source of the fish, compromising the efficacy of control programs where targeted harvest is necessary. Any successful fisheries management program requires significant buy in and compliance by fishers [39], and this was not accomplished with this approach. In summary, for this incentives program to work, stakeholders need to work together to reduce impediments to compensation and information sharing, which will require considerable effort, time, and resources.

The third approach we conducted at the "market side" of the conceptual model was considered a success, with our removal goal achieved within 6 months. Creating a fish meal demand using a logistics company to arrange transport and processing allowed fishers to quickly deploy and meet the market price we set. The processor was required by contract to ensure that the source of catch by fishers was recorded accurately, allowing us to match catch data with field-derived sampling. Presumably, this scenario shows that, if demand increased with prices similar to those we set ($0.42 USD/kg), consistent fishing would occur. At the time, prices for bigheaded carp commodities were approximately half of what we set ($0.26 USD/kg) for fish intended for rendering or other industrial purposes (personal communication, Lisa McKee, Big River Fish Co.; personal communication, Gray Magee Jr., CEO, American Heartland Fish Product LLC, Grafton, IL, USA). Fishers also likely preferred the increased income (i.e., they were paid directly by the processor) over the paperwork and documentation necessary to meet incentive benchmarks set in the "fisher-side" approach (i.e., funding required application).

The experimental large-scale removal approaches we instituted during the first half of 2012 were at that time the primary market for bigheaded carp on the Illinois River. The State of Illinois had contracted with China to export over 13.6 million kg annually for direct consumption [40]; however, this volume was not exported due to the lack of fish processing infrastructure and logistic difficulties in transportation and distribution. At the time of the project, "market side"-driven harvest appeared to stimulate the expansion of the bigheaded carp fisheries in the region, at least temporarily. Big River Fish Company subsequently relocated to a larger facility in Griggsville, IL. Upon reopening in 2013, they purchased significant amounts of bigheaded carp for direct consumption markets in China (personal communication, Lisa McKee, Big River Fish Company). Despite this promising start, this processor appeared to have closed permanently by 2020. Schaefer Fisheries in northern Illinois (Thomson, IL, USA, Figure 2) purchased bigheaded carp for domestic direct consumption markets, and for use in liquid organic fertilizer [41]. Direct consumption purchases by this processor have declined since a fire burned the original building in 2015 and reduced capacity for manufacture of bigheaded carp food products such as hotdogs and extruded meats [41,42]. A rendering facility operated by American Heartland Fish Products LLC opened in Grafton, IL near the confluence of the Illinois and Mississippi Rivers in May 2014 (Figure 2). This facility was processing up to 27,000 kg of bigheaded carp per day into fish oil and fish meal until noxious odor violations led to its closure (personal communication, Gray Magee Jr., CEO, American Heartland Fish Product LLC). Other processors and distributors have opened in the region since the early 2010s, with varied success and capacity [32]. Despite continued interest among stakeholders in elevating market demand for bigheaded carp from the Illinois River and other inland waters, harvest has fluctuated rather than increased in the lower Illinois River since 2010 (Figure 5). Likely contributing to the lack of increased harvest, market prices have remained relatively unchanged since the inception of this project [32].

Population modeling suggests that, to deplete bigheaded carp populations in the Illinois River, all age classes must be targeted for removal [10]. Although a large amount of bigheaded carp biomass was removed in 2012, no large-scale removal of bigheaded carp under 500 mm total length occurred. Based on processor subsampling and informal surveys, we determined that a limited number of commercial fishers appeared to use seines or other gears which would harvest all sizes of bigheaded carp. The majority of fishers used gill, trammel or hoop nets that target larger fish. Not only were commercial fishers harvesting larger, older fish, compared to standardized sampling, but they were also harvesting the largest fish within younger age classes. This is supported by the lack of a bimodal distribution in the age-frequency histogram of commercially harvested fish. As an alternative to the collapse approach, the SEAcarP model suggests that market-based fishing in the lower river will effectively reduce densities of carp near the edge of the invaded range in the upper river [24]. This allows for more effective contracted removal efforts at the range edge and reduces the probability of a breach through barrier systems.

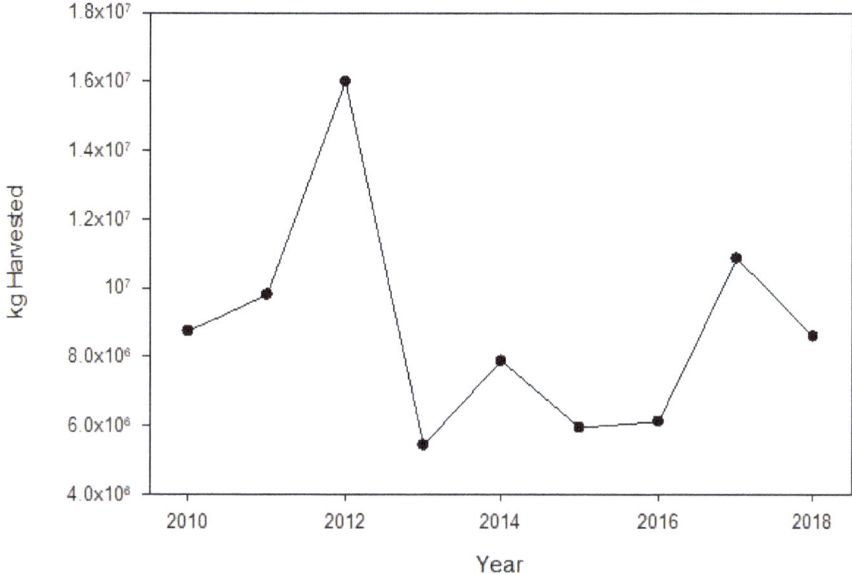

Figure 5. Total biomass in kg of bigheaded carp harvested from the lower Illinois River during the 2010–2018 period, derived from Illinois Department of Natural Resources records [43].

6. Conclusions

Bigheaded carp standing stock and biomass will vary among river reaches and years as a function of variable recruitment, emigration, immigration, harvest impacts, and environmental conditions. In our study, silver carp densities in the Alton Pool did decline, whereas they did not in the other reaches, perhaps because Alton Pool was in proximity to the contracted processor and received greater than 60% of the total harvest. However, without control reaches, more intensive sampling, and multiple years, it is impossible to infer a direct, causal relationship. With sufficient market demand, commercial harvest may control bigheaded carp. However, lack of processing infrastructure and supply chain bottlenecks could constrain harvest, particularly at low commodity prices. Whether a commercial harvest approach to fighting bigheaded carp will be successful can only be assessed over time. Any such evaluation must consider that once a nuisance species becomes an economic resource or a part of local culture, it may no longer be considered a nuisance, but an asset. This could result in pressure to maintain the species or even expand its range to uninvaded regions [9,44], causing a paradox for managers trying to restrict or prevent the spread of certain invasives or mitigate their effects on native species and ecosystems (see Settle et al. [45], and a related discussion on socioeconomic feedbacks of invasive lake trout control in Yellowstone Lake [3]). Given the large, geographical scale of this invasion and complicated harvest logistics, concerns about economic dependence on invasivorism that encourage stock enhancement are likely unmerited.

While harvest-driven extirpation is unlikely, more realistic reduction and control benchmarks might be achieved, although the nature of such benchmarks has yet to be fully articulated. Identifying bigheaded carp density thresholds that would lessen their impacts on native ecosystems or reduce the risk of density-dependent upstream movement toward the Great Lakes or other uninvaded regions are options. Once population goals are determined, management agencies must monitor market prices relative to population densities. If harvest can achieve significant decreases in bigheaded carp density, flexibility in the fishery to move to other river reaches where bigheaded carps are abundant or switch to alternative stocks would be necessary to maintain stability. It is likely that higher (possibly subsidized) prices for bigheaded carp would be needed to compensate for the greater effort necessary to maintain harvest levels in a declining abundance scenario. Nonetheless, partially subsidized fishing may still be a more cost-effective, efficient, and publicly acceptable means of bigheaded carp control. Agencies must be prepared to provide the economic flexibility and stability necessary in a widespread, complex region such as the Mississippi River basin to ensure that removal by harvest remains robust to maintain control through time without creating unintended dependencies.

Author Contributions: Conceptualization, W.W.B., D.C.G., J.T.T., S.S., J.E.G., R.M., and K.I.; methodology, W.W.B., D.C.G., J.T.T., S.S., and J.E.G.; software, W.W.B., D.C.G., and J.E.G.; validation, W.W.B., D.V.G., J.T.T., S.S., J.E.G., R.M., D.P.C., A.A.C., K.I., and A.W.; formal analysis, W.W.B. and J.E.G.; investigation, W.W.B.; resources, W.W.B.; data curation, W.W.B. and A.W.; writing—original draft preparation, W.W.B.; writing—review and editing, W.W.B., D.C.G., J.T.T., S.S., J.E.G., R.M., D.P.C., A.A.C., K.I., and A.W.; visualization, W.W.B.; supervision, J.E.G.; project administration, J.E.G.; funding acquisition, J.E.G. All authors have read and agreed to the published version of the manuscript.

Funding: This research was funded by the Illinois Department of Natural Resources.

Acknowledgments: We would like to thank the staff and graduate students that assisted with data collection and processing, Paul Hitchens for his assistance with logistics and harvest data compilation, and Kristen Bouska for creating the map figure and helpful manuscript edits. The findings and conclusions in this article are those of the author(s) and do not necessarily represent the views of the U.S. Fish and Wildlife Service.

Conflicts of Interest: The authors declare no conflict of interest.

References

1. Rahel, F.J. Homogenization of freshwater faunas. *Annu. Rev. Ecol. Syst.* **2002**, *33*, 291–315. [CrossRef]
2. Pimentel, D.; Zuniga, R.; Morrison, D. Update on the environmental and economic costs associated with alien-invasive species in the United States. *Ecol. Econ.* **2005**, *52*, 273–288. [CrossRef]
3. Koel, T.M.; Arnold, J.L.; Bigelow, P.E.; Brenden, T.O.; Davis, J.D.; Detjens, C.R.; Doepke, P.D.; Ertel, B.D.; Glassic, H.C.; Gresswell, R.E.; et al. Yellowstone Lake Ecosystem Restoration: A Case Study for Invasive Fish Management. *Fishes* **2020**, *5*, 18. Available online: https://www.mdpi.com/2410-3888/5/2/18 (accessed on 1 July 2020). [CrossRef]
4. Allan, J.D.; Abell, R.; Hogan, Z.; Revenga, C.; Taylor, B.W.; Welcomme, R.L.; Winemiller, K. Overfishing of inland waters. *BioScience* **2005**, *55* (Suppl. 12), 1041–1051. [CrossRef]
5. Quinn, J.W. Harvest of Paddlefish in North America. In *Paddlefish Management, Propagation, and Conservation in the 21st Century: Building from 20 Years of Research and Management*; Paukert, C.P., Scholten, G.D., Eds.; American Fisheries Society: Bethesda, MD, USA, 2009; pp. 203–221.
6. Worm, B.; Hilborn, R.; Baum, J.K.; Branch, T.A.; Collie, J.S.; Costello, C.; Fogarty, M.J.; Fulton, E.A.; Hutchings, J.A.; Jennings, S.; et al. Rebuilding global fisheries. *Science* **2009**, *325*, 578–585. [CrossRef]
7. Roman, J. Eat the Invaders. *Audubon Mag.* **2004**, *0410*, 1.
8. Weber, M.J.; Hennen, M.J.; Brown, M.L. Simulated population responses of common carp to commercial exploitation. *N. Am. J. Fish. Manag.* **2011**, *31*, 269–279. [CrossRef]
9. Nunez, M.A.; Kuebbing, S.; Dimarco, R.D.; Simberlof, D. Invasive species: To eat or not to eat, that is the question. *Conserv. Lett.* **2012**, *5*, 334–341. [CrossRef]
10. Tsehaye, I.; Catalano, M.; Sass, G.; Glover, D.; Roth, B. Prospects for fishery-induced collapse of invasive Asian carp in the Illinois River. *Fisheries* **2013**, *38*, 445–454. [CrossRef]
11. Meronek, T.G.; Bouchard, P.M.; Buckner, E.R.; Burri, T.M.; Demmerly, K.K.; Hatleli, D.C.; Klumb, R.A.; Schmidt, S.H.; Coble, D.W. A review of fish control projects. *N. Am. J. Fish. Manag.* **1996**, *16*, 63–74. [CrossRef]
12. Wydoski, R.S.; Wiley, R.W. Management of undesirable fish species. In *Inland Fisheries Management in North America*, 2nd ed.; Kohler, C.C., Hubert, W.A., Eds.; American Fisheries Society: Bethesda, MD, USA, 1999; pp. 403–430.
13. Colvin, M.E.; Pierce, C.L.; Stewart, T.W.; Grummer, S.E. Strategies to control a common carp population by pulsed commercial harvest. *N. Am. J. Fish. Manag.* **2012**, *32*, 1251–1264. [CrossRef]
14. Great Lakes Fisheries Commission. Budget. 2019. Available online: http://sealamprey.org/budget.php (accessed on 30 June 2020).

15. Sass, G.G.; Cook, T.R.; Irons, K.S.; McClelland, M.A.; Michaels, N.N.; O'Hara, T.M.; Stroub, M.R. A mark-recapture estimate for invasive Silver Carp (Hypopthalmichthys molitrix) in the La Grange Reach, Illinois River. *Biol. Invasions* **2010**, *12*, 433–436. [CrossRef]
16. Garvey, J.; Ickes, B.; Zigler, S. Challenges in merging fisheries research and management: The Upper Mississippi River experience. *Hydrobiologia* **2010**, *640*, 125–144. [CrossRef]
17. MacNamara, R.; Glover, D.; Garvey, J.; Bouska, W.; Irons, K. Bigheaded carps (*Hypophthalmichthys* spp.) at the edge of their invaded range: Using hydroacoustics to assess population parameters and the efficacy of harvest as a control strategy in a large North American river. *Biol. Invasions* **2016**, *18*, 3293–3307. [CrossRef]
18. Buck, E.H.; Upton, H.F.; Stern, C.V.; Nicols, J.E. Asian Carp and the Great Lakes Region. *Congr. Res. Rep.* **2010**, *28*, 1–28.
19. Lauber, T.B.; Stedman, R.C.; Connelly, N.A.; Ready, R.C.; Rudstam, L.G.; Poe, G.L. The effects of aquatic invasive species on recreational fishing participation and value in the Great Lakes: Possible future scenarios. *J. Great Lakes Res.* **2020**, *46*, 656–665. [CrossRef]
20. USA. Army Corps of Engineers. Great Lakes and Mississippi River Interbasin Study Report. 2014. Available online: http://glmris.anl.gov/glmris-report/ (accessed on 30 June 2020).
21. Kolar, C.S.; Lodge, D.M. Ecological predictions and risk assessment for alien fishes in North America. *Science* **2002**, *298*, 1233–1236. [CrossRef]
22. Coulter, A.A.; Brey, M.K.; Lubejko, M.; Kallis, J.L.; Coulter, D.P.; Glover, D.C.; Whitledge, G.W.; Garvey, J.E. Multistate models of bigheaded carps in the Illinois River reveal spatial dynamics of invasive species. *Biol. Invasions* **2018**, *20*, 3255–3270. [CrossRef]
23. Asian Carp Regional Coordinating Committee. Asian Carp Monitoring and Response Plan. 2018. Available online: http://asiancarp.us/Documents/MRP2018.pdf (accessed on 30 June 2020).
24. Asian Carp Regional Coordinating Committee. Monitoring and Response Plan for Asian Carp in the Upper Illinois River and Chicago Area Waterway System. 2020. Available online: http://asiancarp.us/PlansReports.html (accessed on 30 June 2020).
25. Erickson, R.A.; Eager, E.E.; Kocovsky, P.; Glover, D.C.; Kallis, J.L.; Long, K.R. A spatially discrete, integral projection model and its application to invasive carp. *Ecol. Model.* **2018**, *387*, 163–171. [CrossRef]
26. International Union for Conservation of Nature and Natural Resources. The IUCN Red List of Threatened Species. 2020. Available online: https://www.iucnredlist.org/species/166081/6168056 (accessed on 30 June 2020).
27. Food and Agriculture Organization. Cultured Aquatic Species Information Programme, Food and Agriculture Organization of the United Nations. 2020. Available online: http://www.fao.org/fishery/culturedspecies/Hypophthalmichthys_molitrix/en (accessed on 30 June 2020).
28. Wang, F.; Zhang, J.; Mu, W.; Fu, Z.; Zhang, X. Consumers' perception toward quality and safety of fishery products, Beijing, China. *Food Control* **2009**, *20*, 918–922. [CrossRef]
29. Varble, S.; Secchi, S. Human consumption as an invasive species management strategy: A preliminary assessment of the marketing potential of invasive Asian carp in the US. *Appetite* **2013**, *65*, 58–67. [CrossRef] [PubMed]
30. Bowzer, J.; Trushenski, J.; Glover, D.C. Potential of Asian carp from the Illinois River as a source of raw materials for fish meal production. *N. Am. J. Aquac.* **2013**, *75*, 404–415. [CrossRef]
31. Bowzer, J.; Trushenski, J. Growth performance of hybrid Striped Bass, Rainbow Trout, and Cobia utilizing Asian carp meal-based aquafeeds. *N. Am. J. Aquac.* **2015**, *77*, 59–67. [CrossRef]
32. Tetra Tech. *Illinois Department of Natural Resources Asian Carp Business Process Analysis: Final Report and Action Plan*; Tetra Tech: Chicago, IL, USA, 2018.
33. Lohmeyer, A.M.; Garvey, J.E. Placing the North American invasion of Asian carp in a spatially explicit context. *Biol. Invasions* **2009**, *11*, 905–916. [CrossRef]
34. Chick, J.H.; Colaninno, C.; Beyer, A.; Brown, K.; Dopson, C.; Enzerink, A.; Goesmann, S.; Higgins, T.; Knutzen, N.; Laute, E.; et al. Following the edge of the flood: Use of shallow-water habitat by larval silver carp Hypophthalmichthys molitrix in the upper Mississippi river system. *J. Freshw. Ecol.* **2020**, *35*, 95–104. [CrossRef]
35. Coulter, D.P.; MacNamara, R.; Glover, D.C.; Garvey, J.E. Possible unintended effects of management at an invasion front: Reduced prevalence corresponds with high condition of invasive bigheaded carps. *Biol. Conserv.* **2018**, *221*, 118–126. [CrossRef]

36. Illinois-Indiana Sea Grant. *Asian Carp Marketing Summit*; Sea Grant Publication: Springfield, IL, USA, 2010; Volume IISG-11–04, pp. 1–27.
37. Johal, M.S.; Esmaeili, H.R.; Tandon, K.K. Postcleithrum of Silver Carp, Hypophthalmichthys molitrix (Van. 1844), an authentic indicator for age determination. *Curr. Sci.* **2000**, *79*, 945–946.
38. Isley, J.J.; Grabowski, T.B. Age and growth. In *Analysis and Interpretation of Freshwater Fisheries Data*; Guy, C.S., Brown, M.L., Eds.; American Fisheries Society: Bethesda, MD, USA, 2007; pp. 187–228.
39. Boenish, R.; Willard, D.; Kritzer, J.P.; Reardon, K. Fisheries monitoring: Perspectives from the United States. *Aquac. Fish.* **2020**, *5*, 131–138. [CrossRef]
40. Illinois Government News Network. Governor's Office Press Release. 2010. Available online: http://www3.illinois.gov/PressReleases/ShowPressRelease.cfm?SubjectID=3&RecNum=8624 (accessed on 30 June 2020).
41. Schaefer Fisheries. Available online: http://schaferfish.com (accessed on 30 June 2020).
42. Kent, A. Fire Officials: Schaefer Fisheries Deemed a Total Loss. *Clint. Her.* **2015**, 1. Available online: https://www.clintonherald.com/news/fire-officials-schafer-fisheries-deemed-a-total-loss/article_576f8e96-eabb-11e4-9d98-1799509de8c3.html (accessed on 30 June 2020).
43. Maher, R. *Commercial Catch Report Exclusive of Lake Michigan*; Illinois Department of Natural Resources: Brighton, IL, USA, 2019; pp. 1–27.
44. Pasko, S.; Goldberg, J. Review of harvest incentives to control invasive species. *Manag. Biol. Invasions* **2014**, *5*, 263–277. [CrossRef]
45. Settle, C.; Crocker, T.D.; Shogren, J.F. On the joint determination of biological and economic systems. *Ecol. Econ.* **2002**, *42*, 301–311. [CrossRef]

Article

Numeric Simulation Demonstrates That the Upstream Movement of Invasive Bigheaded Carp Can Be Blocked at Sets of Mississippi River Locks-and-Dams Using a Combination of Optimized Spillway Gate Operations, Lock Deterrents, and Carp Removal

Daniel Patrick Zielinski [1,*] and Peter W. Sorensen [2]

1 Great Lakes Fishery Commission, 2200 Commonwealth Blvd., Suite 100, Ann Arbor, MI 48105, USA
2 Department of Fisheries, Wildlife and Conservation Biology, University of Minnesota, St. Paul, MI 55018, USA; soren003@umn.edu
* Correspondence: dzielinski@glfc.org

Citation: Zielinski, D.P.; Sorensen, P.W. Numeric Simulation Demonstrates That the Upstream Movement of Invasive Bigheaded Carp Can Be Blocked at Sets of Mississippi River Locks-and-Dams Using a Combination of Optimized Spillway Gate Operations, Lock Deterrents, and Carp Removal. *Fishes* **2021**, *6*, 10. https://doi.org/10.3390/fishes6020010

Academic Editor: Maria Angeles Esteban

Received: 13 March 2021
Accepted: 24 March 2021
Published: 26 March 2021

Abstract: Invasive bigheaded carp are advancing up the Upper Mississippi River by passing through its locks-and-dams (LDs). Although these structures already impede fish passage, this role could be greatly enhanced by modifying how their spillway gates operate, adding deterrent systems to their locks, and removing carp. This study examined this possibility using numeric modeling and empirical data, which evaluated all three options on an annual basis in both single LDs and pairs under different river flow conditions. Over 100 scenarios were modeled. While all three approaches showed promise, ranging from 8% to 73% reductions in how many carp pass a single LD, when employed together at pairs of LDs, upstream movement rates of invasive carp could be reduced 98–99% from current levels. Although modifying spillway gate operation is the least expensive option, its efficacy drops at high flows, so lock deterrents and/or removal using fishing/trapping are required to move towards complete blockage. Improved deterrent efficacy could also offset the need for more efficient removal. This model could help prioritize research and management actions for containing carp.

Keywords: integrated pest management; model; hydraulic; acoustic deterrent; invasive fish; conservation

1. Introduction and Mini-Review

The spread of invasive fish has contributed to the extirpation of many species of fish as well as a loss of biodiversity and ecosystem integrity across the globe [1–3]. When eradication is not possible, as is almost always the case [4,5], containment is the only option [2,3]. In rivers, containment can be complicated by the presence of migratory native fishes and flooding. Developing ways to selectively control the upstream movement of invasive fish has challenged North American fisheries managers since the turn of the 19th century, when the common carp, *Cyprinus carpio*, and sea lamprey, *Petromyzon marinus*, [6–8] became abundant. Only a few solutions have been identified, and none for large rivers where testing options are expensive and difficult. These complexities make numerical simulations of control options a valuable tool. Here, we use numerical models to evaluate three control options for invasive bighead carp, *Hypophthalmichthys nobilis*, and silver carp, *H. molitrix*, (collectively known as bigheaded carp) at the locks-and-dams (LD) they must pass to move upstream in a large river. Our findings describe several promising ways that a targeted and integrated approach can effectively control an important invasive fish. In this introduction, we review the bigheaded carp problem, Mississippi River LDs, and three ways to control bigheaded carp at these choke points; we then outline our study objectives and approach before proceeding to the methods.

1.1. The Bigheaded Carp Problem

Recently, bighead carp and silver carp from Asia have become a serious problem in the Mississippi River Basin of North America [9]. Bigheaded carp were introduced to Arkansas from Asia in the 1960s, escaped into the Mississippi River [10] and continue to invade the upper reaches of the Mississippi River Basin. These species are large (>20 kg), microphagous filter-feeding fish that compete with native planktivorous fish for food, driving reductions in their abundance, size and condition, while altering food webs [11–13]. Additionally, silver carp can jump up to 3 m out of the water, interfering with recreational boating [14]. Bigheaded carps reproduce in areas of flowing water and have semi-buoyant eggs that require long stretches of flowing water to hatch and recruit, making the pools between LDs a good place to control and remove adults because LDs restrict fish movement [15–17]. Carp are also sensitive to sound, making them susceptible to being blocked with acoustic (non-physical) deterrent systems [18–20]. Finally, bigheaded carps are not particularly strong swimmers [21], so their movement through LDs is open to manipulation, especially in systems with multiple LDs that create impassible water velocities.

Bigheaded carp presently comprise the majority of the fish biomass in many areas of the Mississippi River Basin, although they have yet to establish themselves in either the headwaters of the Mississippi River or the Laurentian Great Lakes. While carp passage into the Great Lakes is currently protected by an electrical barrier in the Illinois River [17], the headwaters of the Mississippi River remain unprotected because they are wide and prone to flooding and thus cannot support a simple electrical barrier, so new approaches at LDs are sought.

1.2. Mississippi River Locks-and-Dams

The Upper Mississippi River (UMR) is regulated by a series of 29 LDs operated by the US Army Corps of Engineers (USACE) and are named (and numbered) in a sequential fashion from north to south (Figure 1). Nearly all LDs have both a navigational lock and a gated spillway system. The USACE operates these structures in a manner that permits navigation while protecting the structures from erosion/scour by limiting water velocities. Spillway gates are seated at the bottom of the river and progressively raised to pass water and regulate water depth, but in so doing, create water velocities underneath them that fish may struggle to overcome. As LD spillway gates are lifted, the velocity of water passing underneath them is reduced, dropping to a minimal value when/as they come out of the water entirely (a condition known as "open-river"). In contrast, flow in navigational locks is very low to allow boats (and fish) to pass, but access is regulated by miter gate opening and the locks are a relatively small (~10%) part of most dams. Together, spillway gates and locks inhibit upstream fish passage. However, their effects on fish vary: some LDs exert large effects on fish passage and some very little—depending on their design, local river conditions, spillway gate operations (e.g., the number, location, and opening height of each gate), and the fish species (fish swimming ability varies greatly) and their size.

Several LDs whose spillways rarely open fully are known to greatly reduce upstream fish passage of native migratory species including lake sturgeon, *Acipenser fulvescens*, and paddlefish, *Polyodon spathula*, [22–24] as well as invasive species including both bigheaded carps [25] and common carp [26]. Notably, the swimming abilities of carps are very similar when size is considered [21,26]. While some migratory fishes have disappeared from the Upper Mississippi River (UMR) since LDs were installed, analyses of the current fish population structure suggest that LDs likely have little effect on the remaining populations of native fishes [27], although their effects on newly arriving invasive carp appear quite substantial. The abilities of fishes to surmount spillway gates varies with environmental conditions that include water velocity, water temperature, fish species, fish size, and physiological condition, LDs that experience open-river conditions less frequently are more likely, on average, to impede upstream fish movement [28]. Many LDs in upper regions of the UMR experience open-river conditions far less often than those in the lower portion of the river (Table 1, [29]). While some LDs have overflow systems that operate during high flow (Table 1), many do not, or they could be screened, and thus these LDs can be used in carp control.

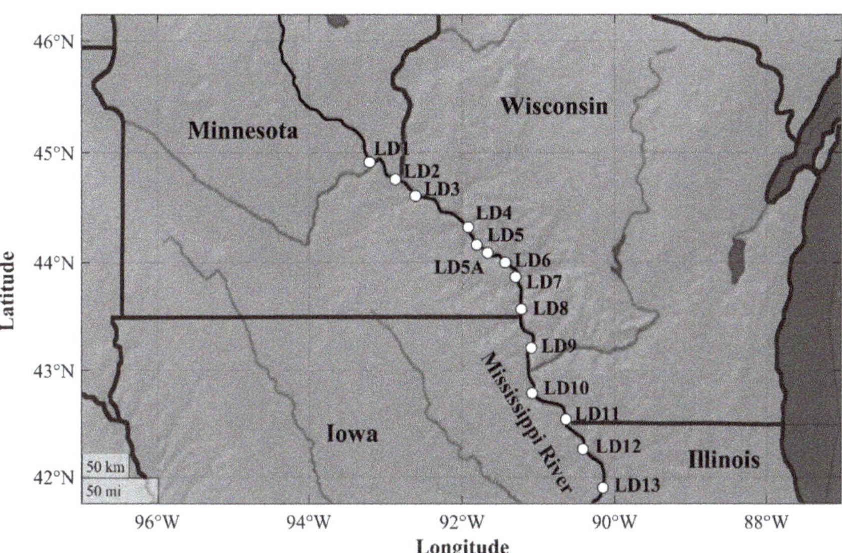

Figure 1. Location of locks-and-dams (LDs) in the upper portion of the Upper Mississippi River (UMR). See Table 1 for details on LDs.

Importantly, LDs influence each other and the fish that pass through them, synergizing the ability of each to impede overall fish movement upstream; although this has not been well studied. Further, it is likely that adjacent (consecutive and proximate) LDs could have greater influence on bigheaded carp populations more than other LDs separated by great distances because bigheaded carp require 50–100 km of turbulent open river to reproduce successfully [30]. Of course, short pools (50–100 km) also create excellent opportunities for fisheries managers to sample, catch, and remove carp that might pass the LD immediately below them.

1.3. Options to Control Carp Passage at Locks-and-Dams

Three good options exist to control carp at LDs: the spillway gates, the lock, and the pools above LDs into which fish must pass and where capture is possible. Of these, the spillway gates are of singular importance because they typically comprise 90–95% of the structure size and are at least partially open most of the time. Adjusting spillway gate openings is a good option to reduce carp passage. Its potential has been shown by both modeling [28] and descriptions of fish passage from fish tracking studies [26,31–33], the latter showing a strong correlation between spillway gate opening, water velocity, and passage. Numerical modeling at two relatively typical Mississippi River LDs, LD 2 and LD 8, has shown that fish passage through their gated spillways is dependent on hydraulic conditions that include velocities that exceed 5 m/s at lower gate openings through which very few fish can pass [26,28,34]. Further, we have developed a numeric fish passage model (FPM) that uses three-dimensional water velocities found around LD spillways gates to determine whether and/or how fish with known swimming abilities can (and do) swim through gates with different settings and river flow [28]. FPM simulations have also shown that the spillway gate operations presently used by the USACE can result in slightly unbalanced flow regimes at LDs, and thus create regions of low velocity that fish (carp) can swim through. Remarkably, this validated FPM describes ways ("optimized operating conditions") that spillway gate settings can be re-balanced to reduce carp passage, sometimes by as much as 50–75% [28,34]. As these modifications reduce scour, they have proven to be acceptable to the USACE [28]. Thus, modifying/optimizing spillway gate operations to balance water velocities at LDs when they are not in an open-river condition has great potential to restrict upstream carp passage at little to no cost.

Table 1. Summary of locks-and-dams (LDs) in the Upper Mississippi River (UMR). The percent time spent in open-river was calculated from historical records between 1970–2000 [29]. Δ River km is the distance (pool size) between that LD and the next one upstream [29]. The Upper and Lower Saint Anthony Falls Dam (upstream of LD 1 and lacking an operational lock) and Chain of Rocks Lock (downstream of LD 26) differ structurally from LDs 1–26 and are not included. The final column indicates whether the lock-and-dam has an additional uncontrolled overflow spillway that functions during high flow conditions. Consecutive LDs that experience open-river conditions less than 5% of the time are shaded. Two LDs that do not go into open-river (0%), because they do not have spillway gates are also shown (LD 1, LD 19).

Lock-and-Dam	% Open-River	River km	Δ River km	# of Gates	Other Spillway?
1	0.0%	1365	8	0	Yes
2	1.3%	1312	53	19	No
3	15.6%	1282	29	4	Yes
4	3.9%	1212	71	28	No
5	1.7%	1188	24	34	No
5A	13.9%	1172	15	10	Yes
6	9.7%	1149	23	15	Yes
7	4.7%	1131	19	16	Yes
8	3.9%	1093	37	15	Yes
9	18.4%	1043	50	13	Yes
10	19.7%	990	53	12	Yes
11	1.8%	938	51	16	No
12	13.9%	896	42	10	Yes
13	5.5%	841	55	13	Yes
14	0.5%	794	47	17	No
15	1.3%	777	17	11	No
16	16.8%	736	41	19	Yes
17	31.9%	703	33	11	Yes
18	12.1%	661	43	17	Yes
19	0.0%	586	75	119	No
20	33.9%	552	34	43	No
21	21.3%	523	29	13	Yes
22	16.5%	485	38	13	Yes
24	17.6%	440	45	15	Yes
25	20.5%	388	51	17	Yes
26	19.7%	323	65	9	Yes

A second option to control carp passage at LDs is to add deterrent systems to the lock chambers. LD lock chambers are designed to support barge navigation and thus have little measurable flow, making them well suited to these systems. Upstream fish passage through open lock chambers has been observed in the summer months for a number of fishes, including bigheaded carps [25,26,31]. Non-physical deterrent systems that use sound, or sound paired with other stimuli (i.e., air bubbles, strobe lights, carbon dioxide), are presently being developed for use in these systems [18,35–42]. Sound is favored because it is safe and, similarly to all ostariophysians, bigheaded carp have a wider hearing range and lower hearing threshold than many native fish. Laboratory tests using a variety of sound signals [37–41] and sound coupled with air-bubble curtains [36,38] have documented deterrent efficiencies between 75–97%. A test of a cyclic sound coupled with an air curtain and light (a bio-acoustic fish fence or "BAFF") blocked 95% of all carps in a creek, but further testing is required [42]. The effects of sound could be taxon-specific.

A third option to control bigheaded carp is fish removal in pools upstream of LDs. Removal is especially feasible in short pools where sampling to gauge effectiveness is reasonable and bigheaded carp may also be unable to reproduce. Carp removal could be achieved through subsidized targeted removal or possibly commercial ventures [43,44]. In the Illinois River, contracted harvest of bigheaded carp has been used successfully since 2010 to help reduce propagule pressure on the USACE electric barrier at Chicago [45]. Bigheaded carp are typically removed using short-set large-mesh gill and trammel nets. The gear used in the Illinois River selects for larger fish, and removal has been effective at decreasing the density of bigheaded carp populations restricted by a downstream LD [43,44,46].

1.4. Introduction to This Study

In the present study, a stochastic size-structured fish passage model (S-FPM) was developed to examine the potential for controlling bigheaded carp passage by blocking upstream passage using different combinations of modified gate operations, acoustic deterrence at navigational locks, and carp removal across pairs of consecutive UMR LDs. This model simulated passage of carp to examine ways it could be reduced. It examined many options at both single and consecutive LDs using known carp passage rates, monthly river discharge, several levels of lock deterrent systems, fish size, and different levels of removal. Our overarching goal was to determine whether and how an integrated approach to control bigheaded carp might be reasonable in the UMR and if so, what factors might best contribute to its efficacy. To address this, we asked several related questions: (1) What gains might be realized by managing bigheaded carp passage at two adjacent LDs versus just one?; (2) What benefits might be realized by modifying spillway gate operations at one or two LDs to reduce carp passage at different river flows?; (3) What are the benefits of adding a non-physical deterrent(s) to either individual or pairs of LDs and how do they compensate for increased fish passage at high flow?; (4) What additional benefits might carp removal schemes have on carp control?; and (5) How might these three options be employed together as part of an integrated pest management scheme? We focused on silver carp as it is the species of greatest concern and worked in sequential fashion, combining factors as we went to examine synergistic effects at varying river flows, the effects of which on spillway gate passage are complex but important. Lessons from silver carp should nevertheless apply to bighead and common carp as they have similar biologies. Possible effects of carp population size-structure and behavioral drive to attempt spillway gate passage were also examined. We use changes in fish passage rate as our metric, given the absence of data on silver carp population size in the upper reaches of the UMR.

2. Methods

The S-FPM was created to simulate and estimate annual upstream passage rates of bigheaded carp through either one or two LDs in the UMR. This model included 6 categories of variables including: (1) whether a single or a paired set of LDs is being managed; (2) local environmental variables (e.g., river flow); (3) carp population size-structure; (4) carp behavior/passage route; (5) carp passage rates at spillway gates and locks (and effects of deterrents on them); and (6) estimated effects of carp removal on overall passage rate (Table 2, see below). Over 100 scenarios were modeled using empirical data from LD 8, a relatively typical UMR LD (Table 1) which has 15 spillways gates and 1 operational lock. First, we describe the model, then the variables it uses, and then how it was deployed.

2.1. Model Framework

The S-FPM evaluates fish passage as a consequence of a series of junctions at either a single or two consecutive LDs (i.e., pairs of LDs with one located immediately upstream of the other so they synergize each's actions). While doing so it uses fish movement rates and route selection (i.e., the path fish pursue while swimming upstream) informed by both field data and fish spillway passage indices for LDs calculated using our fish passage model (FPM). Specific variables used in the S-FPM model include (Table 2): environmental conditions; fish population and size-structure; fish behavior—upstream movement and route selection; fish behavior—passage indices and deterrence; and carp removal. Fish passage at LD spillway gates is considered using our FPM which considers fish swimming performance with respect to species and size, as well as water velocities at specific LD spillway gates as informed by LD structure and river flow using computational fluid dynamics (CFD) [28]. When possible, silver carp data (e.g., swimming performance) were used but when not available, data were used from the closely related common carp (e.g., passage rates through spillways gates of LD).

Table 2. List of stochastic size-structured fish passage model (S-PFM) variables and values. Where available, variable mean +/− standard deviation is provided. Variables are categorized by italicized section headings and further described in the methods. Data derived from common carp are noted with an (*), otherwise data come from silver carp. Q is river discharge. P_A_{US}, P_B_{US} are the proportions of fish that move upstream at LD A and LD B. P_A_L, P_B_L and P_A_S, P_B_S are the proportions of upstream swimming fish to atempt passage through the lock and spillway gates at each LD, respectively. PI_{lock}, PI_{spill} are the passage indices at the lock and spillway gates. At is the number of passage attmpts made per month at the spillway gates. D is the efficiency of a deterrent system inhibiting passage through the lock chamber. R is the proportion of fish removed from the intermediate pool.

Variable	Notation	Value	Source(s)
2.1.1. Environment			
River discharge	Q	50%, 20%, 5%, 1% exceedance flows	[47]
2.1.2. Fish population and size structure			
Population	-	50,000 per size class	N/A
Size classes	-	≤600, 700, 800, & 900 mm total length	[48]
2.1.3. Fish behavior—Upstream movement and route selection			
Proportion of upstream movement	P_A_{US}, P_B_{US}	87, 72, 52, 52, 45, 48, 13, 11 (+/−25%)	[49]
Lock route *	P_A_L, P_B_L	7.3+/−7.1%	[33]
Spillway routes *	P_A_S, P_B_S	27+/−16%	[33]
2.1.4. Fish behavior—Passage indices and deterrence			
Lock passage index *	PI_{lock}	5+/−5%	[31,33]
Spillway passage index	PI_{spill}	f(size, operation, discharge)	[28]
Attempts	At	1, 2, 5	[25]
Deterrence from lock	D	0, 25, 50, 75, 100%	[38,40,42]
2.1.5. Carp removal			
Removal	R	0, 5, 10, 40%	[17]

The S-FPM model employs two LDs (LD A and LD B) and they have the same spillway gate operations, a realistic scenario because most UMR LDs have nearly identical structural components (Figure 2). LD 8 is used as the base conditions for each, which is reasonable because its design is typical of most LDs and it is also well studied [28]. Both LD A and LD B are associated with pools: Pool A is downstream of LD A, Pool B is located between LD A and LD B, and Pool C is located upstream of LD B. In the model, carp start in Pool A. and the S-FPM calculates passage rates of carp moving from Pool A to Pool C each month for a year (which thus includes seasonal effects). Each month a proportion of fish moves upstream (P_A_{US}, P_B_{US}) and then attempt to pass through one, or both LDs. While doing so, each upstream swimming carp is assigned to one of three routes: the spillway gate (P_A_S, P_B_S), the navigational lock (P_A_L, P_B_L), or both spillway and lock (P_A_{S+L}, P_B_{S+L}). The combined route of spillway and lock gives fish the opportunity to pass through either the lock or spillway (a scenario observed at LD8 [33]), while the other routes limit to just one route. The likelihood of passage through the lock chamber is modelled using mean passage rates of common carp observed at LD 8, while passage through the spillway gates has been determined using the fish passage index (FPI) previously calculated by Zielinski et al. [28]. Individual carp that pass either route (P_A_{pass}, P_B_{pass}) then move into the upstream pool and those in Pool B are subjected to the passage model again whereas those in Pool C remain upstream of LD B. Fish that do not either move upstream or attempt to do so and are blocked by either LD A or LD B's spillway gates (B_A_S, B_B_S) or lock chamber (B_A_L, B_B_L) return to their pool of origin and undergo the passage model the subsequent month (if/when the model simulation allows for future attempts- we tested 1–5 attempts). Those carp that pass LD A and are found in Pool B are also then subject to possible removal (R). River flow (i.e., discharge), proportion of upstream movement, route selection, and passage indices were updated monthly in the model. The total number of fish from each size class within each pool was recorded monthly and divided by the initial population size to determine the proportion of fish passing each LD (the percent). Finally, the number/proportion of carp eventually found in Pool C represents the proportion that passed both LDs while the combined proportion of fish in Pool B and Pool C reflect the proportion passing a single LD (LD A). The model was coded in Matlab (Mathworks, MA, USA) (Figure 2).

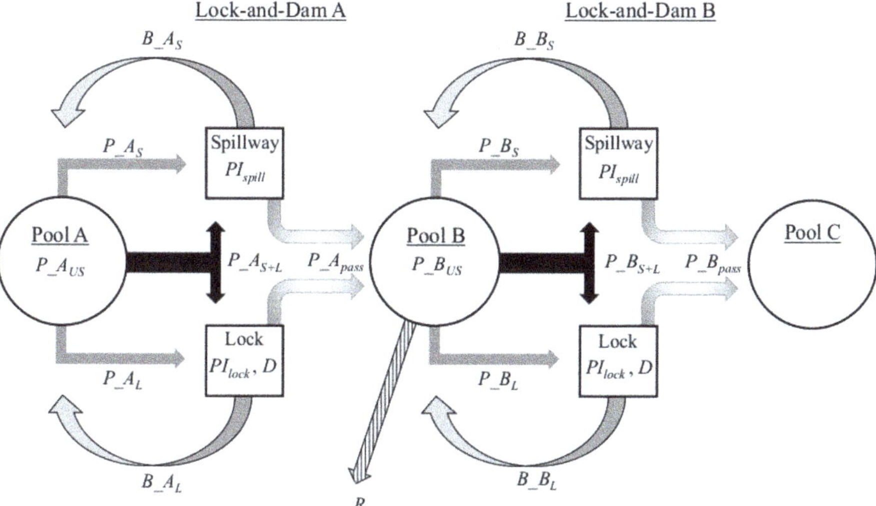

Figure 2. Schematic representation of the stochastic size-structured fish passage model (S-FPM). The model uses a silver carp population with five size classes that are released in Pool A (downstream).

2.1.1. Environment

Opportunities for fish passage at LDs vary with spillway gate openings and these were determined by river discharge. The S-FPM model examined 4 hydrologic scenarios in the UMR and which we describe as monthly exceedance values. Monthly exceedance discharge is equal to the median monthly discharge that is exceeded for some percentage of the time. Exceedance was calculated from 30 years of river discharge at LD 8; we identified 50%, 25%, 5% and 1% exceedance discharge values [47]. In our case (LD 8), a 50% and a 25% exceedance discharge condition does not require the spillway gates to be fully open anytime during the year, while the 1% exceedance discharge condition requires LD 8 to operate in open-river conditions 7 months of the year (Figure 3).

2.1.2. Fish Population and Size Structure

Each simulation used a population of 200,000 numeric silver carp from 4 size classes (50,000 carp per size class). This number was selected to minimize variance between model runs (the variance was calculated to be less than 0.5% for each size class at 50,000 fish). A size-structured approach was used because swimming performance, is influenced by fish size [21]. Each run of the model was initialized with a population of carp being placed into Pool A, which was assigned a body length from one of four 100 mm size classes based on data from either the UMR or Wabash River where carp have been established longer and are larger [48] (Table 3). Because most silver carp in the UMR have a total length of less than 600 mm, a size class whose swimming abilities are not known, the size distribution of carp used in the model was adjusted so that the smallest carp was 600 mm. This likely led to conservative (artificially high) estimates of passage rate as small fish cannot swim as fast as larger fish. For model simulations, the proportion of each size class of carp found within each pool was multiplied by the length-frequency percentage of a given population distribution (Table 3) to produce relevant size-specific results. The UMR population size-structure was used as the default in the S-FPM reported in the results although the impact of fish size-structure on the model was calculated for reference (see Supplemental data, Figure S1).

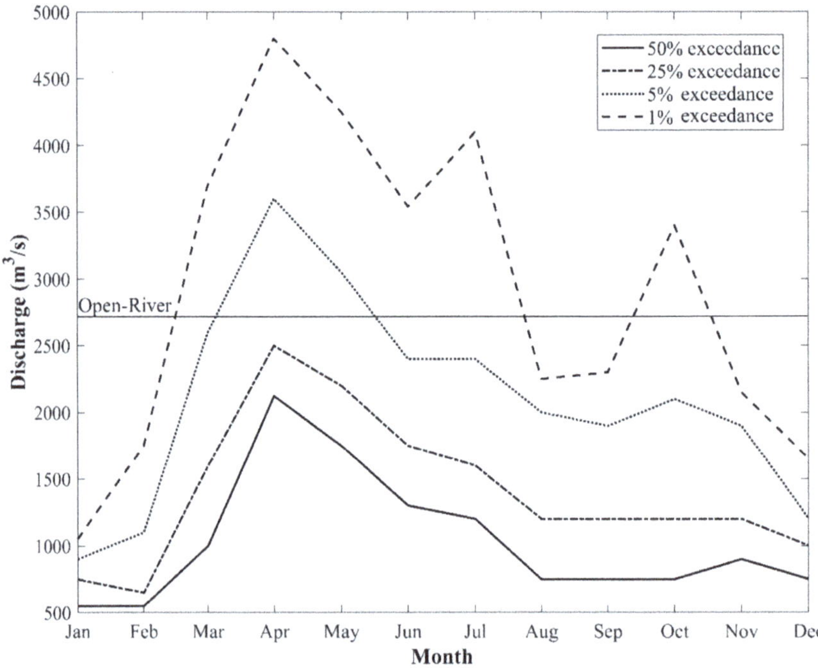

Figure 3. Monthly discharge at LD 8 based on 50%, 25%, 5%, and 1% exceedance durations between 1972–2000 [47]. Open-river conditions start when discharge is greater than 2718 m^3/s.

Table 3. Length-frequency distributions of silver carp (by 100 mm length increment) from the Upper Mississippi River (UMR) and Wabash River [48].

Total Body Length (mm)	% Frequency in the UMR *	% Frequency in the Wabash River
≤600	90	8
700	6	38
800	3	51
900	1	3

* 73% of the UMR silver carp population has a total length ≤ 500 mm.

2.1.3. Fish Behavior—Upstream Movement and Route Selection

Telemetry studies have shown that upstream movement rates of carp vary seasonally [49,50] and that carp take different paths through LDs [26,31,33] with carp moving upstream more vigorously in the spring than in summer and fall. Our S-FPM used seasonal upstream movement rates, and assumed fish did not move between November and February (Table 2). The proportion of each size class within each pool that was selected to move upstream (P_A_{US}, P_B_{US}) was randomly assigned from a normal distribution with a mean equal to the mean upstream movement measured by Coulter et al. [49] with a standard deviation of 25% of the mean. All individuals were then assigned a movement indicator (R_1) from a uniform random distribution (0–100) each month. In Pool A, individuals with $R_1 \leq P\_A_{US}$ moved upstream to challenge LD A. Any individuals that passed LD A were then assigned a new movement indicator once they entered Pool B and the selection process repeated itself.

Just as different numbers of bigheaded carp could move upstream (or not) in the river and our model, they could also choose different paths or routes, with some carp following the river's edge to encounter a lock, others moving to the center of the channel and encountering a spillway gate, and others demonstrating a mixed approach that included both options. Our model considered these three possibilities using available data. An ongoing study using acoustic telemetry is assessing the movement and passage of common carp at LD 8 [33] and we used its findings. Briefly, data collected in 2019 from over 100 transplanted, tagged common carp downstream of LD 8 found that 7.3% of all adult common carp approached only the lock chamber, 27% approached only the spillway gates, and the remainder explored both options. These values were employed and the proportions of upstream moving carp selected to move towards the lock chamber (P_A_L, P_B_L) or the spillway gates (P_A_S, P_B_S) were randomly assigned from a normal distribution with a mean and standard deviation derived from the common carp data collected by Whitty et al. [33]. All individuals moving upstream were assigned a route indicator (R_2) from a uniform random distribution (0–100) each month. In Pool A, individuals with $R_2 \leq P\_A_L$ attempted to pass through the lock chamber and individuals with $R_2 \geq 1 - P\_A_S$ attempted passage through the spillway gates. All remaining, unassigned fish attempted passage through both the lock chamber and spillway gates. Individuals passing LD A are assigned a new route indicator and the route selection process repeated for LD B.

2.1.4. Fish Behavior—Passage Indices and Deterrence

The likelihood of any fish (carp) passing through a lock chamber is dependent on a combination of opportunity and behavior. In contrast, the likelihood of them making through spillway gates is driven by opportunity, behavior, and swimming performance. Both were modeled for a single LD and consecutive LDs. First, we discuss passage rate at the spillway gates, then locks.

Spillway Gate Passage

Fish (carp) passage through spillway gates is dependent on several variables including fish species, size, behavior, and gate opening/water velocity (i.e., gate operations). To estimate the likelihood of carp passing through LD spillway gates, we used the fish passage indices (FPI) we developed earlier [28] for silver carp at LD 8. Briefly, the FPI was calculated using a FPM which pairs high resolution water velocity data at specific gate settings with known fish swimming performance data to predict if, when, and where fish could pass through a hydraulic structure assuming the fish follows the path of least resistance (a conservative assumption) [28]. This FPM and its resultant FPIs have been validated in tracking studies of common carp at LD 2 [26] and LD 8 [33].

To create estimates of spillway passage, the S-FPM used FPI values [28] to assign a spillway passage index (PI_{spill}) at both LD A and LD B that was based on river discharge and fish length (we used data from the UMR and another location, see below). We calculated FPI for silver carp assuming both base (current/historical) gate operations and gate operations modified and optimized to restrict carp passage for five river discharges including open-river conditions [28]. We used linear interpolation to calculate the spillway passage index at intermediate discharges and the nearest value for discharges outside of the range [28] (Figure 4). Individuals assigned to the spillway route were then assigned a spillway passage indicator (R_3) from a uniform random distribution (0–100) each month. In Pool A, individuals with $R_3 \leq PI_{spill}$ successfully passed through LD A spillway gates while all remaining fish were blocked. Any individuals that passed LD A and were assigned to the spillway route were assigned a new spillway passage indicator and the spillway passage process repeated. The spillway passage index calculated for the S-FPM included an attempt variable (At) that allowed fish to challenge the spillway gates multiple times per month. Based on the average number of attempts observed by silver carp at Starved Rock Lock-and-Dam [25], the model assumed each fish following the spillway gate route was assigned 2 passage attempts per month. Over the 8-month period of our model, any given fish could attempt to pass through the spillway gates up to 16 times. Simulations using 1 and 5 attempts per month were also run to evaluate how attempt rate impacts passage estimates (Supplemental data, Figure S1).

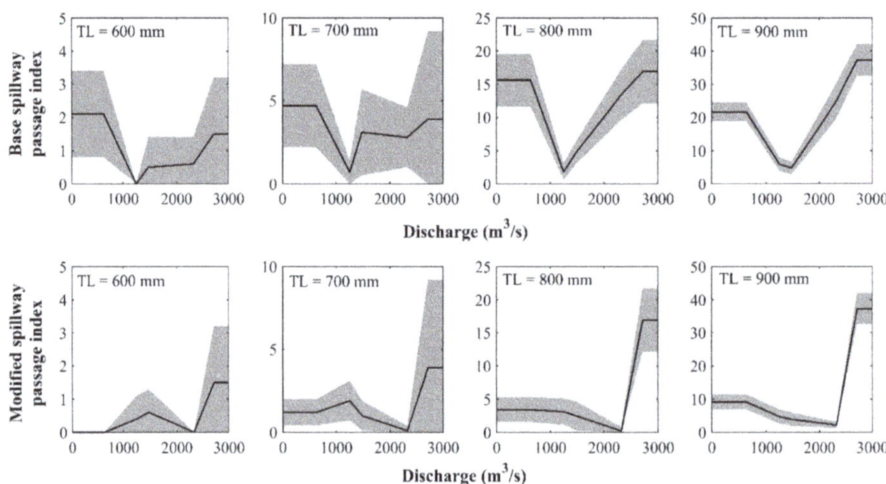

Figure 4. Spillway passage index for silver carp with a total length of 600–900 mm at LD 8 under base and modified gate operations [28]. The solid black line indicates the mean passage index and shaded area is the standard deviation. The passage index is calculated at 635, 1250, 1475, 2325, and 2720 m^3/s (open-river). Note, the different y-scales for each total length.

Lock Passage and Deterrents

Fish can only swim upstream through a lock chamber when a boat is locking through it and its miter gates are open. To do so, fish must enter the lock chamber, an area of high noise and turbulence, and their success in passing appears to be low. For instance, the rate of passages relative to the number of passage attempts was found to be 7% for silver carp at LD 26 [31] and 5% for common carp at LD 8 [33]. In our model, the lock passage index (PI_{lock}) at each LD was randomly assigned from a normal distribution with a mean and standard deviation from empirical data collected by Whitty et al. [33] and Tripp et al. [31]. As reported, passage rates [31,33] were measured relative to the number of passage attempts, so the lock passage index does not need to explicitly simulate multiple passage attempts through the lock chamber (i.e., passage rate is expected to be ~5% regardless of the number of attempts). Individuals assigned to the lock chamber route were assigned a lock chamber passage indicator (R_4) from a uniform random distribution (0–100) each month. In Pool A, individuals with $R_4 \leq PI_{lock}$ successfully passed through the lock chamber while all remaining carp were blocked. Individuals passing LD A were then assigned a new lock chamber passage indicator and the lock passage process repeated at LD B.

Of course, base passage rates through locks can, in theory, be reduced by adding deterrent systems to them. We included the possibility that a deterrent will be developed and successfully implemented for use in LD(s) in our model. Due to uncertainty in the specifics of the deterrent type and efficacy, we examined the impact of adding deterrents at one or both locks with several efficiencies: 0%, 25%, 50%, 75%, and 100%. Deterrent values were based on those already measured in the field and laboratory for acoustically based systems [36–42].

Lock and/or Spillway Passage

Finally, our model allowed for the possibility that some carp will attempt to move upstream using a combination of both locks and/or spillway gates (e.g., fish assigned to the spillway + lock chamber route). Each month these individuals were assigned both a lock chamber and spillway passage indicator. Similar to fish assigned to just the spillway route, fish were allowed multiple attempts to pass the spillway gate per month (if appropriate). If passage criteria were satisfied for either the lock chamber or spillway gates, that carp was deemed to have passed that LD.

2.1.5. Carp Removal

Physical removal of fish is commonly used to control populations of invasive species [43,51]. This approach is already being successfully employed in the Illinois River to control bigheaded carp using contracted commercial fishers [43,45,46]. Simulations using the Spatially Explicit Asian Carp (SeaCarP) model estimate 40% of the population needs to be harvested to reduce the risk of introduction into the Great Lakes, and it is possible this is presently being achieved in some areas [17] where the population seems to be constant. Several fishing techniques have been developed for this purpose and are still being improved including the "Modified Unified Method" from China [17]. We included the possibility of removal in our model as R (removal) and assign it values 0%, 5%, 10% and 40%. Each month, all individuals that move into Pool B were assigned a removal indicator (R_5) from a uniform random distribution (0–100). Individuals with $R_5 \leq R$ were then removed from the population. The likelihood of removal was the same for all sizes of fish in Pool B.

2.2. Model Simulation

Over 100 simulations were run to assess the individual and combined impacts of modified spillway gate operation, lock deterrence, and removal on silver carp passage rates through single and consecutive LDs (Table 4). For each simulation, we tracked the number of fish passing both LD A and LD B individually and the annual proportion of fish passing each structure was calculated by dividing the total passed by the initial population size. The proportion of carp passing LD A was the total passage rate expected at one LD and the proportion passing LD B is the total passage at two consecutive LDs. Modeling proceeded in 4 steps so we could systematically evaluate the role of different variables in a step-wise fashion with each variable (management action) being added to the previous case. We started by exploring the roles of the simplest management option, modified gate operation. First, passage rates during either base (current as determined from USACE historical records) or modified spillway gate operations to block silver carp were calculated and then compared at different flow (exceedance) scenarios. Second, the impact of adding non-physical (acoustic) deterrent(s) with several efficiencies to LD lock(s) were examined using modified spillway gate operations. Third, the impact of employing carp removal in the intermediate pool (Pool B) on overall annual passage was examined in combination with varying levels of lock deterrence, including none and assuming modified spillway gate operations. All cases used the carp size structure measured in the UMR distribution [48] while carp were allowed to attempt to pass twice a month, per expected values. After completing these runs at different flow (exceedance) conditions, we examined the average annual effects of several combinations of variables across all exceedance values expected in a year. We did this to evaluate the overall effects of individual variables. Finally, we assessed the impact of population size structure and spillway gate passage attempt rate assuming modified spillway gate operations, no lock deterrence, and no removal (Supplementary data, Table S3). A total of 104 simulations were run to accommodate all iterations over four hydrologic scenarios (Table 4), the results of which are presented in Supplemental data (Tables S1 and S2).

Table 4. List of unique model simulations. Brackets indicate range of values used in each simulation. Each simulation provides the annual proportion passing either a single or pair of two consecutive LDs.

No. of Simulations	Exceedance Discharge (%)	Spillway Operation	Deterrence (%)	Targeted Removal (%)	Attempts	Size Distribution
4	(1, 5, 25, 50)	Current	0	0	2	UMR
80	(1, 5, 25, 50)	Optimized	(0, 25, 50, 75, 100)	(0, 5, 10, 40)	2	UMR
8	(1, 5, 25, 50)	Optimized	0	0	(1, 5)	UMR
12	(1, 5, 25, 50)	Optimized	0	0	(1, 2, 5)	Wabasha

3. Results

3.1. Effects of Managing Carp Using Consecutive LDs

Our model suggested that approximately 18.1% of silver carp of the size presently found in the UMR can be expected to pass a single typical LD under base (historical) spillway gate operating conditions during the course of a typical year with this rate increasing to 22.4% at high flows (Figure 5, Table S1). When two LDs were considered instead of a single LD, this rate dropped by 85% across all simulated flows to approximately 2.7% (Table S1). The effects of managing carp at two adjacent LDs locations were multiplicative.

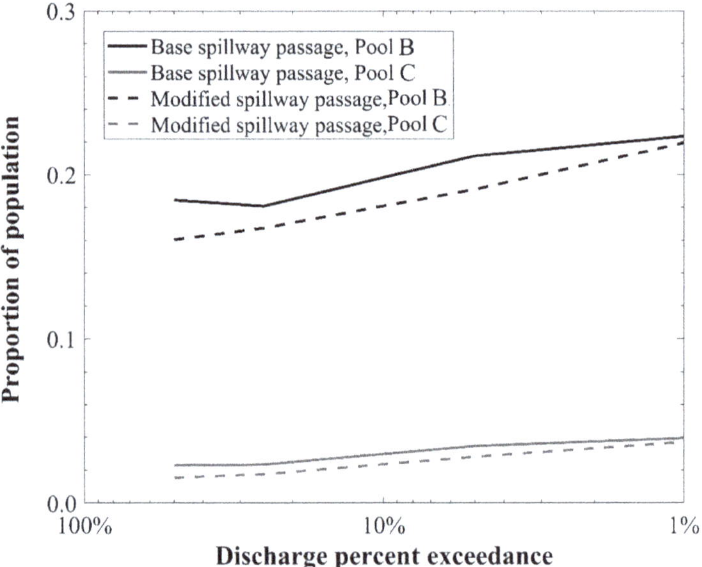

Figure 5. The proportion of silver carp passing a single LD (black lines) or two consecutive LDs (grey lines) under base spillway passage conditions (solid lines) or modified spillway gate conditions as calculated by our FPM (dashed lines).

3.2. Effects of Managing Carp by Modifying LD Spillway Gate Operations

Modifying spillway gate function at a single LD had notable effects, reducing the proportion passed by approximately 11% at an exceedance of 50% for one LD but dropping to only 2% at an exceedance of 1% when the river is mostly in open-river conditions (Figure 5, Table S2). When the effects of modifying spillway gate operations on passage through consecutive LDs was considered, the overall proportion passing two LDs decreased by about 88% to an overall value of only 1.5–3.7%. Notably, while consecutive LDs may be expected to go into open-river at similar times, they were unlikely to be identical and if the distance between them small, reproduction may be unlikely. Modifying spillway gate operations was thus especially beneficial at pairs of LDs that rarely go into open-river, but other options probably should be considered for the later scenario at higher flows (exceedances).

3.3. *Effects of Adding a Non-Physical Deterrent to One or Both Locks*

Adding a deterrent to the lock chamber of one or both LDs operating their spillway gates in a modified manner was very effective, especially when pairs of LDs were considered (Table 5, Figure 6 and Table S2). At a single LD, lock deterrence systems that were more than 50% effective reduced the number of silver carp that could pass to less than 10%. If a deterrent with 100% efficacy was used, the value dropped to 2% at the 50% exceedance flow, and to less than 10% at the 1% exceedance flow when gate operations were modified (Table S2). When two LDs were considered, each with a deterrent in the lock, the annual proportion of silver carp passing was less than 2% for a deterrent only 50% effective overall under all deterrence levels and modified gate operations. Notably, the relative impact of a lock deterrent on fish passage was relatively unaffected by flow conditions.

Table 5. Summary of the estimated effects of pairing LDs, modifying their gate operation, adding a deterrent to the lock chamber and removing carp in the intermediate pool between them on the overall annual passage rates of the silver carp population with the size structure presently found in the UMR [48]. The annual proportion of carp passed is averaged across all four flow scenarios. Percent reduction is calculated relative to the proportion passed at a single LD (1 LD) under base gate operations conditions.

Case	Proportion Passed	% Reduction
Base gate operations		
1 LD	0.200	NA
2 LD	0.030	85
Modified gate operations		
1 LD	0.185	8
2 LD	0.025	88
Modified gate operations + deterrents		
1 LD + 25% Det	0.151	24
1 LD + 50% Det	0.121	40
1 LD + 75% Det	0.088	56
1 LD + 100% Det	0.054	73
2 LD + 25% Det	0.018	91
2 LD + 50% Det	0.013	94
2 LD + 75% Det	0.009	96
2 LD + 100% Det	0.006	97
Modified gate operations + removal at intermediate pool		
2 LD + 5% removal	0.022	89
2 LD + 10% removal	0.021	90
2 LD + 40% removal	0.014	93
Modified gate operations + deterrents + removal at intermediate pool		
2 LD + 25% Det + 5% removal	0.017	92
2 LD + 50% Det + 5% removal	0.012	94
2 LD + 75% Det + 5% removal	0.008	96
2 LD + 100% Det + 5% removal	0.006	97
2 LD + 25% Det + 10% removal	0.015	92
2 LD + 50% Det + 10% removal	0.011	95
2 LD + 75% Det + 10% removal	0.008	96
2 LD + 100% Det + 10% removal	0.006	97
2 LD + 25% Det + 40% removal	0.010	95
2 LD + 50% Det + 40% removal	0.007	96
2 LD + 75% Det + 40% removal	0.006	97
2 LD + 100% Det + 40% removal	0.004	98

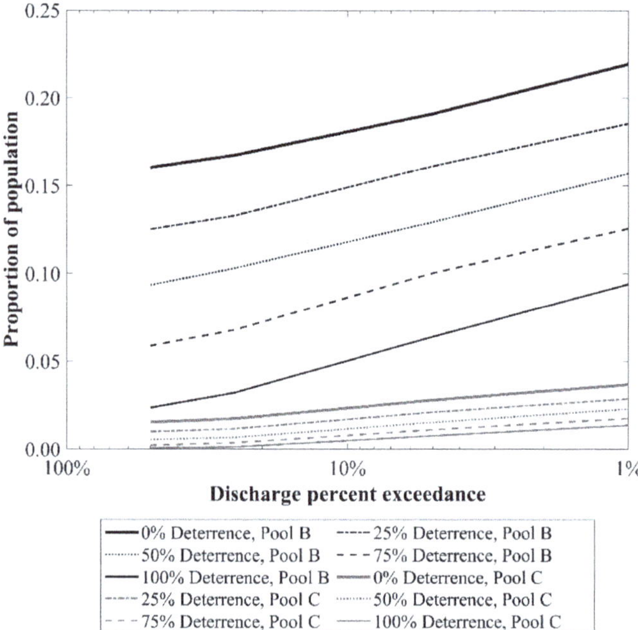

Figure 6. The proportion of upstream swimming silver carp passing through either a single LD (black lines) or two consecutive LDs (grey lines) equipped with nonphysical deterrents of different efficacies and using modified spillway gate operations.

3.4. Effects of Carp Removal in the Intermediate Pool

Adding carp removal to a control scheme while utilizing modified gate operations and a deterrent had additional effects on reducing passage. Effects were multiplicative with a removal rate of 40% without lock deterrence reducing overall annual passage by 93% compared to a single LD with base spillway gate operations (Figure 7, Table 5 and Table S2).

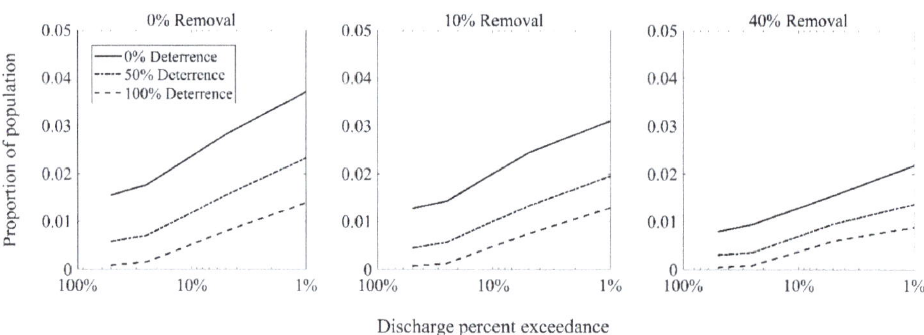

Figure 7. Proportion of silver carp passing two consecutive LDs equipped with non-physical deterrent systems with different efficacies and whose intermediate pool was subjected to carp removal.

3.5. Overview of the Averaged Combined Effects of Multiple Management Options

Lastly, we calculated average annual carp passage rates when all exceedance values were considered. These showed that when pairs of adjacent LDs were considered, only 3% of all carp attempting to pass can be expected to do so with 2 attempts, versus 20%

at 1 LD (Table 5). Modifying gate operations drops this value to 2.5% (88% drop from one LD with base spillway operations). If a 50% effective deterrent is added to two LDs the average value decreased to 1.3% and if the deterrent increases to 100% effective, the proportion passed drops to 0.6% (a 97% decrease vs. nothing occurring at one LD, the current situation). The addition of carp removal together with lock deterrents had the greatest impact on reducing silver carp passage rates. The best-case scenario reduced silver carp passage to only 0.4% and required 100% lock deterrence paired with 40% removal in the pool (Figure 7). Notably, several levels and types of carp removal and lock deterrence achieved the same level of passage reduction. For example, the annual passage rate at consecutive LDs could be reduced to less than 1% by pairing 10% removal rate with a lock deterrent with as little as 50% efficacy, even when exceedance values approached 1%. If the deterrent was close to 100% effective, values decreased by about half again (Table 5).

4. Discussion

Our simulations demonstrate that upstream passage of invasive silver carp in the UMR can be reduced to only 1–2% of current rates through an integrated approach that uses consecutive LDs and some combination of three tractable control techniques. These include reducing passage using spillway gate adjustment, adding non-physical deterrents to lock chambers, and removing carp from the intermediate pool. While modification of the spillway gate operation could occur with no modification to infrastructure, both lock deterrents and carp removal are likely to be costly, although they do not need to be highly efficient (i.e., 50% efficacy might suffice) to drive over 90% reductions in carp passage. Remarkably, carp control appears possible even during high river flows with an approach that employs pairs of strategically selected LDs. All of the control measures we describe can be implemented.

We believe that our simulations are reasonable because they are based on empirical data (ex. exceedance values, known gate settings, velocities, fish passage routes and swimming abilities) and a validated fish passage model that was designed to provide conservative overestimates of actual passage [28]. It is also promising that silver carp telemetry data suggest this species does not challenge LDs repeatedly [31,32]. The recent documented movement of significant numbers of adult bigheaded carp through both LD 19 [52] and LD 8 [53] attests to an urgent need to reduce bigheaded carp passage rates below the conditions currently existing at LD 8. A 50% reduction in passage rates seems possible using a single control option, while a 90% or greater reduction to an overall rate of just 2% appears attainable if both a deterrent and carp removal is used, even during times of high flow and need only be moderately effective (25%). Previous suppositions that carp can only be stopped at systems that lack operating gates [32] appear overly simplistic, which is important because only 2 of the 29 LDs in the UMR do not have bottom mounted spillway gates.

The most significant finding of our study is likely that bigheaded carp should not be managed at single LD, as has been the practice, but at pairs of LDs close to each other that rarely experience open-river conditions. Fortunately, three such locations exist in the UMR: LD 14–15, LD 7–8, and LD 4–5 (see management section below) (Table 1). Across all hydrologic scenarios, the cumulative impact of adding a second LD resulted in an average decrease in carp upstream passage of 85% compared to passage at a single LD. These LDs need to be located close to each other (50–100 km) to be effective, prevent spawning, and facilitate monitoring as well as possible removal.

Likely our next most important finding is that modifying LD spillway gate operations to reduce passage can be highly effective on an annual basis and would come at little cost because the predictive models have been developed and validated [26,28]. Simply modifying gate operations at a single LD decreases carp passage by about 8% overall. Multiplicative effects are expected if operations are optimized at two locations. Importantly, the modifications to gate operations we propose are safe for navigation and LD structural integrity as they do not induce additional scour [28]. While promising for both carp control and LD operations, the benefits of modified gate operations are restricted to the period when LD(s) are operating under controlled conditions (e.g., non open-river), so additional control options such as adding a lock deterrent and removing carp must be considered. Notably, the S-FPM model results we describe are conservative and estimate the upper limit of passage rates owing to our conservative assumption of fish behavioral drive and our assumption that carp can find the most efficient way upstream [28]. The hydrologic scenarios we considered were also conservative because the possibility of average monthly discharge surpassing the 1% exceedance flow for 12 consecutive months is low. For example, the 1% exceedance discharge conditions that would require LD 8 to operate in open-river conditions for 10 straight months, or 83% of the year, actually occur less than 5% of the time [29].

The third most important finding of this study is that a single approach to controlling carp is unlikely to suffice: an integrated approach is needed. Together the three options we described synergize with each other's activities, especially at times of high flow when passage through the spillway gates is high. By using all three options, none of them needed to be singularly effective. Ideally, three options would be implemented but two might suffice if used strategically.

The addition of non-physical deterrent systems to LDs had a notable effect on overall system efficacy that persisted during high river flows even if not highly effective. Typically, non-physical deterrents can be expected to reduce overall annual silver carp passage by about 5% even if the deterrents are only 25% efficient, and close to 20% overall if 100% efficient, the efficiency presently suggested for a BAFF [38,42]. If deterrents were used in two locks, the effects would be multiplicative at all flows. Notably, a BAFF guides fish away from the lock openings so it could be paired with a trap to remove carp as well as capture native species below the LD for possible movement upstream (although see [27]). A BAFF operating at 100% efficiency could thus drive a removal rate of about 20%, compensating for the cost and effort of running a removal program and supplement native fish conservation. Some level of species-specificity which might permit native fish passage may also be possible with acoustic deterrents, such as the BAFF, because carp are especially sensitive to sound [19,37,38,40–42]. Other types of deterrent systems that use CO_2 [36] could be considered, but the would not be species-specific. A BAFF is presently being tested at a LD on the Kentucky River and shows promise [54]. Deterrents appear likely to be a necessary component of an invasive carp control system and their continued development is encouraged.

Even modestly effective carp removal efforts would also be helpful in an integrated carp control program, especially if implemented in pools between paired, managed LDs. Removal would amplify the effects of modified gate operation and deterrents. Further, if a deterrent is not implemented, removal would be necessary, especially at times of high river flow when carp passage will be high. While the actual efficacy of carp removal is presently unknown, and numerous reports suggest it is low, it has adequately prevented the spread of adult silver carp further up the Illinois River [43,44]. Several techniques have been developed and improvements are being made to the "modified unified method" [17]. Notably, carp removal is likely to be especially effective in small pools where it would also limit possible spawning success, the ultimate objective of most fish control strategies [3]. The choice of LDs and the pool between them will be very important for removal strategies, and even modestly effective removal strategies, as low as 5%, would be beneficial. Admittedly, removing carp when there are low densities is difficult and may require use of radio-tagged Judas fish or perhaps eDNA [55,56]. Removal year-round is exceedingly labor-intensive, difficult [17], and expensive (Illinois spends more than a million dollars on this annually [53]). If less than 5% efficiency is realized in a UMR pool then a deterrent will be needed. More work on quantifiable removal options is needed. In any case, it is clear that an integrated approach using multiple control options at multiple LDs is highly desirable.

Our model also evaluated the importance of fish size on passage rates. Large silver carp, such as those found in the Wabash River appear nearly twice as likely to pass (Figure S1, Tables S3 and S4). The behavior of these fish is also important; an increased number of passage attempts significantly increased passage across all hydrologic conditions. For example, fish that attempted spillway passage 5 times per month had nearly a 2-fold increase in passage (Figure S1, Tables S3 and S4). This result is consistent with findings of others [57]. Fortunately, there is good reason to consider that the average attempt rate of bigheaded carp may not be higher than 5 attempts, although this requires study.

Our model has some notable strengths and weaknesses. Most important, as described above, our model assumptions are conservative and likely produce overestimates of passage. Indeed, they are based on empirical data and consistent with the slow upstream spread of bigheaded carp—over 10 years to pass LD 19 [52]. River flows are unlikely to be as consistently high as we modeled. Further, bighead carp are less likely to pass than silver carp based on their swimming performance [21]. Nevertheless, our model does have some uncertainties. First, we do not know the efficiencies of non-physical deterrents at LDs [38,42]. Second, the efficacy and size-selective nature of removal in rivers is unknown. Our model also does not account for fish population demographics.

5. Summary

This study clearly demonstrates that silver carp and likely other carps can be effectively (98%+) blocked at select pairs of LDs if they are operated in tandem and employ multiple approaches including modified gate operation, lock deterrents, and carp removal. These options could be used in multiple ways and need not be 100% efficient. Further information and improvement can come once an integrated control scheme is put into place.

6. Management Recommendations and Future Directions

It is reasonable to consider controlling invasive bigheaded carps at LDs in the UMR. Control strategies should employ pairs of LDs that are close to each other and rarely experience open-river conditions and at least two of the three options we have described. This could be extremely effective and economical. As the likelihood of carp passage increases with fish size, so does the chance of their reproducing, efforts should be timely. Three pairs of UMR LDs meet the criteria for successful control, but LD 4–LD 5 seem to have special promise because silver carp have not moved beyond them yet and bred, they are very rarely in open-river, Pool 5 is small, and they resemble LD 8 so their hydraulics are understood [28,34]. Ideally gate operations will be modeled and optimized. As with common carp control, developing ways to monitor carp abundance will be critical to success [4,5]. Detailed studies of carp movement around and through LDs will be extremely helpful as would further modeling efforts to improve model precision to guide carp control in UMR and elsewhere in the basin [58].

Supplementary Materials: The following are available online at https://www.mdpi.com/article/10.3390/fishes6020010/s1, Figure S1: Impacts of population size on passage, Table S1: Carp passage rates with and without gate modifications at different exceedances; Table S2: Carp passage rates with gate modifications at different control options at different exceedances; Table S3: Carp passage rates with gate modifications and different numbers of attempts using fish the size of those in the UMR; Table S4: Carp passage rates with gate modifications and different numbers of attempts using fish the size of those in the Wabash River.

Author Contributions: Conceptualization, P.W.S. and D.P.Z.; methodology, D.P.Z.; validation, P.W.S. and colleagues; formal analysis, D.P.Z.; resources: P.W.S.; data curation, D.P.Z.; writing—original draft preparation, D.P.Z.; writing—review and editing, P.W.S.; funding acquisition, P.W.S. Both authors have read and agreed to the published version of the manuscript.

Funding: The data described in this research was funded by the Legislative Citizens Commission for Minnesota Resources.

Institutional Review Board Statement: Not applicable.

51. Rytwinski, T.; Taylor, J.J.; Donaldson, L.A.; Britton, J.R.; Browne, D.R.; Gresswell, R.E.; Lintermans, M.; Prior, K.A.; Pellatt, M.G.; Vis, C.; et al. The effectiveness of non-native fish removal techniques in freshwater ecosystems: A systematic review. *Environ. Rev.* **2019**, *27*, 71–94. [CrossRef]
52. Larson, J.H.; Knights, B.C.; McCalla, S.G.; Monroe, E.; Tuttle-Lau, M.; Chapman, D.C.; Amberg, J. Evidence of Asian Carp spawning upstream of a key choke point in the Mississippi River. *N. Am. J. Fish. Manag.* **2017**, *37*, 903–919. [CrossRef]
53. Stanley, G. Record 51 Carp Caught. Minneapolic Star Tribune. 13 March 2020. Available online: https://www.startribune.com/record-51-asian-carp-caught-in-minnesota-a-sign-the-fish-may-have-established-permanent-populations/568775572/ (accessed on 13 March 2020).
54. U.S. Fish and Wildlife Service. Bio-Acoustic Fish Fence Now Operational at Lake Barkley. 2020. Available online: https://www.fws.gov/southeast/news/2019/11/bio-acoustic-fish-fence-now-operational-at-lake-barkley/ (accessed on 15 July 2020).
55. Bajer, P.G.; Chizinski, C.J.; Sorensen, P.W. Using the Judas technique to locate and remove wintertime aggregations of invasive common carp. *Fish. Manag. Ecol.* **2011**, *18*, 497–505. [CrossRef]
56. Coulter, D.P.; Wang, P.; Coulter, A.A.; Van Susteren, G.E.; Eichmiller, J.J.; Garvey, J.E.; Sorensen, P.W. Nonlinear relationship between Silver Carp density and their eDNA concentration in a large river. *PLoS ONE* **2019**, *14*, e0218823. [CrossRef] [PubMed]
57. Castro-Santos, T. Quantifying the combined effects of attempt rate and swimming capacity on passage through velocity barriers. *Can. J. Fish. Aqua. Sci.* **2004**, *61*, 1602–1615. [CrossRef]
58. Erickson, R.A.; Eager, E.A.; Kocovsky, P.M.; Glover, D.C.; Kallis, J.L.; Long, K.R. A spatially discrete, integral projection model and its application to invasive carp. *Ecol. Model.* **2018**, *387*, 163–171. [CrossRef]

fishes

MDPI

Review

Achieving Sea Lamprey Control in Lake Champlain

Bradley Young *, BJ Allaire and Stephen Smith

U.S. Fish and Wildlife Service, Essex Junction, VT 05452, USA; bj_allaire@fws.gov (BJA.); stephen_j_smith@fws.gov (S.S.)
* Correspondence: bradley_young@fws.gov; Tel.: +1-802-662-5304

Abstract: The control of parasitic sea lamprey in Lake Champlain has been a necessary component of its fishery restoration and recovery goals for 30 years. While adopting the approach of the larger and established sea lamprey control program of the Laurentian Great Lakes, local differences emerged that shifted management focus and effort as the program evolved. Increased investment in lamprey assessment and monitoring revealed under-estimations of population density and distribution in the basin, where insufficient control efforts went unnoticed. As control efforts improved in response to a better understanding of the population, the effects of lamprey on the fishery lessened. A long-term evaluation of fishery responses when lamprey control was started, interrupted, delayed, and enhanced provided evidence of a recurring relationship between the level of control effort applied and the measured suppression of the parasitic sea lamprey population. Changes in levels of control efforts over time showed repeatedly that measurable suppression of the parasitic population required effective control of 80% of the known larval population. Understanding the importance of assessment and monitoring and the relationship between control effort and population suppression has led to recognition that a comprehensive, not incremental, approach is needed to achieve effective control of sea lamprey in Lake Champlain.

Keywords: population suppression; lampricide; fishery restoration; sea lamprey; Lake Champlain

Citation: Young, B.; Allaire, BJ; Smith, S. Achieving Sea Lamprey Control in Lake Champlain. *Fishes* **2021**, *6*, 2. https://doi.org/10.3390/fishes6010002

Received: 6 December 2020
Accepted: 28 December 2020
Published: 26 January 2021

Publisher's Note: MDPI stays neutral with regard to jurisdictional claims in published maps and institutional affiliations.

1. Introduction

Sea lamprey (*Petromyzon marinus*) parasitism is a limiting factor to both the restoration [1] and recovery [2] of fish populations in Lake Champlain. The preferred host species of the lake include lake trout (*Salvelinus namaycush*), land-locked Atlantic salmon (*Salmo salar*), and lake sturgeon (*Acipensir fulvescens*). While sea lamprey do parasitize other species in the lake, the parasitic load on these species and the level of induced mortality place sea lamprey more in the role of a predator than parasite. The origin of sea lamprey in Lake Champlain has been the subject of debate. A series of genetic studies [3–5] concluded that they were endemic to the lake and likely remnants of the Champlain Sea, when the basin was contiguous with the Atlantic Ocean in following the last glacial event approximately 10,000 years ago. Eshenroder [6,7] challenged the assumptions of the genetics models using historical collections and canal construction timelines to propose that sea lamprey entered Lake Champlain through the New York State canal system when it joined the Hudson River to Lake Champlain through a series of connections during the end of the 19th century. Regardless of origin, Atlantic-native sea lamprey have proven to be a nuisance species in Lake Champlain and incompatible with its freshwater hosts.

Lake Champlain was historically home to lake trout and landlocked Atlantic salmon populations [8–10]. During 19th-century industrialization, the damming of tributaries and deforestation degraded riverine habitat [9]. This loss of habitat in concert with over-exploited fisheries led to the extirpation of native stocks of both species between approximately 1850 and 1900 [10]. A programmatic effort to restore these native species and introduce other salmonids began in 1973 with the formation of the Lake Champlain Fish and Wildlife Management Cooperative (Cooperative); the Cooperative comprises the

New York Department of Environmental Conservation, the Vermont Fish and Wildlife Department, and the U.S. Fish and Wildlife Service. In 1977, the Cooperative set goals to reestablish a lake trout and Atlantic salmon fishery and establish a rainbow trout (steelhead) fishery by 1985 [11]. As efforts to improve the fishery moved forward, parasitic sea lamprey populations surged. It became clear to the Cooperative that meeting fishery restoration and recovery goals would require efforts to suppress sea lamprey population.

In developing a program to control Lake Champlain sea lamprey, the Cooperative followed the existing Laurentian Great Lakes (Great Lakes) model [12,13] in establishing three fundamental management components. First, basin-wide assessments determine densities and distributions of larval sea lamprey and direct selection and implementation of control efforts. Second, as part of an integrated pest management approach, both chemical and physical control methods target larval and adult life history stages. Because the larvae of Lake Champlain consistently spend four years maturing in tributaries before emigrating to the lake as parasites, four year classes can be eliminated effectively once every four years using lampricides (selective piscicides) applied to tributaries and their associated deltas [14]. The active ingredient of the liquid and bar formulations of lampricide applied to rivers is 4-nitro-3-(trifluoromethyl)phenol (IUPAC nomenclature) and commonly referred to as TFM. The active ingredient of the lampricide applied to deltas in a granular formulation and occasionally applied to rivers in a liquid formulation as a synergist with TFM is 5-chloro-*N*-(2-chloro-4-nitrophenyl)-2-hydroxybenzamide (IUPAC nomenclature) and commonly referred to as niclosamide. All formulations of these lampricides are restricted-use pesticides and manufactured solely for application by designated federal and state government agencies. While manufacturers have refined product formulations at times, the two active ingredients used for controlling sea lamprey have remained the same for the entirety of the control program.

Second, physical control methods such as dams, temporary barriers, and screens serve to block and trap migrating adults before they reach habitat suitable for spawning [15–17]. The program benefits from dams on ten tributaries (labeled 1, 2, 4, 15, 16, 17, 19, 21, 24, 25; Figure 1) built for purposes other than lamprey control where they serve to limit the length of river accessible to adult sea lamprey migrating upstream to spawn. The program uses temporary, seasonally-installed barriers on seven tributaries that block adult sea lamprey during their spring spawning season (April–June), but are removed for the other nine months of the year (labeled 10, 11, 19, 20(2), 22, 26; Figure 1). These temporary barriers include traps which allow adult sea lamprey to be removed and killed and other aquatic species to be removed and passed above the barrier. The effectiveness of physical control methods in Lake Champlain varies from 100% with large hydropower dams to occasionally 0% with small temporary barriers subject to failure when overcome by high water events. When feasible, lampricides are a more effective, reliable, and consistent method of control. However, especially where lampricide use is restricted, physical control methods have a role in the program.

Third, we monitor and evaluate control efforts using a wounding rate index [18,19] to track changes in the frequency of lamprey parasitism on host species of interest. The wounding rate index is not a direct measure of parasitic lamprey abundance. Characteristics of host species and their population dynamics affect it in ways that are difficult to quantify [20,21]. Despite the limitations of the wounding rate index to provide direct point estimates of abundance, its consistent and standard usage over time [18,22] in Lake Champlain and the Great Lakes provides opportunities to compare general trends in the relationship between sea lamprey and host abundances.

The population dynamics of sea lamprey and the effort necessary to suppress their population has been studied and modeled by Great Lakes researchers to develop and refine their control program [23]. However, understanding the stock recruitment relationship of sea lamprey has proven challenging because of the uncertainty associated with density independent recruitment and compensation from density dependent survival [24–26]. Defining and establishing consistent or standardized levels of control effort required to achieve desired levels of lamprey population suppression have therefore also proven difficult. While long-term successful suppression of populations has been satisfactory for decades, Jones and Adams [27] propose that population eradication remains possible. We share these interests and seek to add experience from the Lake Champlain sea lamprey control program to further the understanding of how lamprey populations respond to increasing levels of control effort. The smaller scale of Lake Champlain and availability of a 30-year data set present an opportunity to consider these dynamics in ways that may lead to new insights as lamprey control efforts continue to evolve.

2. Management Phases

When the Cooperative evaluated progress toward its fishery restoration goals in 1985, approximately half of the lake trout and Atlantic salmon collected were found to be the target of sea lamprey parasitism as measured using the standardized wounding rate index [18,22]. Experience from the Great Lakes and Lake Champlain fishery data showed that efforts to restore these salmonid species would not be successful without suppression of the sea lamprey population. In 1990, the Cooperative began an 8-year experimental control program (ECP) under the guidance and in coordination with the Great Lakes program. At that time, assessments documented larval lamprey in 19 tributaries [28]. The ECP used lampricide to control populations in 13 tributaries (labeled 1, 4–8, 10, 13–15, 17, 22, 23; Figure 1) while trapping migrating adults on three others [29]. The ECP was designed as a pilot program to determine whether the model of sea lamprey control used in the Great Lakes could be applied to Lake Champlain to suppress the lamprey population. After eight years, the evaluation of both sea lamprey suppression and fishery responses led the Cooperative to pursue further and continuing sea lamprey control to support its fishery goals [29].

To transition from the ECP to a long-term control program (LTCP), a federal Environmental Impact Statement (EIS) was required. The process of writing and approval of this document took three years. Several groups opposed the use of lampricide and filed lawsuits challenging the EIS. The Cooperative ultimately received approval of the EIS in 2001 [30] and made plans to resume the control of sea lamprey in 2002 as the LTCP began. The period (1998–2001) between the ECP and LTCP has been termed the partial control program (PCP). During that time, lampricide treatments remained on schedule in New York where available state funds effectively extended the ECP there. Continued treatment of Vermont tributaries required federal funds that remained unavailable until approval of the EIS. During the PCP, the nine New York tributaries treated during the ECP remained controlled while of the four Vermont tributaries treated during the ECP, two were trapped (labeled 22, 23; Figure 1), and two were left uncontrolled (Table 1). At the time, the Cooperative believed that a reduction in control efforts during PCP would sustain some lesser level of population suppression, but would avoid surrendering all progress made during the ECP.

With the EIS in place to begin the LTCP, lampricide treatments resumed in Vermont in 2002 and continued in New York. Sea lamprey wounding rates of 25 per 100 lake trout and 15 per 100 Atlantic salmon were set as goals that the Cooperative believed could support fishery restoration goals [30] based on experience from the Great Lakes and the ECP [29]. Although the EIS enabled the LTCP to proceed, issues on individual rivers resulted in further permitting challenges. As the LTCP resumed in 2002, assessments documented larval lamprey populations in 20 tributaries in need of control [30]. Of those, nine were treated with lampricide and five were trapped (Table 1) [30]. As work progressed toward meeting the requirements for the inclusion of new and existing lamprey-producing tributaries, sea lamprey control efforts languished and wounding rates climbed higher until implementation of a more comprehensive approach. The LTCP authorized by the EIS [30] has continued to the present day. As the program progressed and incorporated experience to affect changes and improvements, the LTCP of 2020 has grown and now documents larval lamprey in 26 tributaries, controlled presently using 19 lampricide treatments and five barriers with traps.

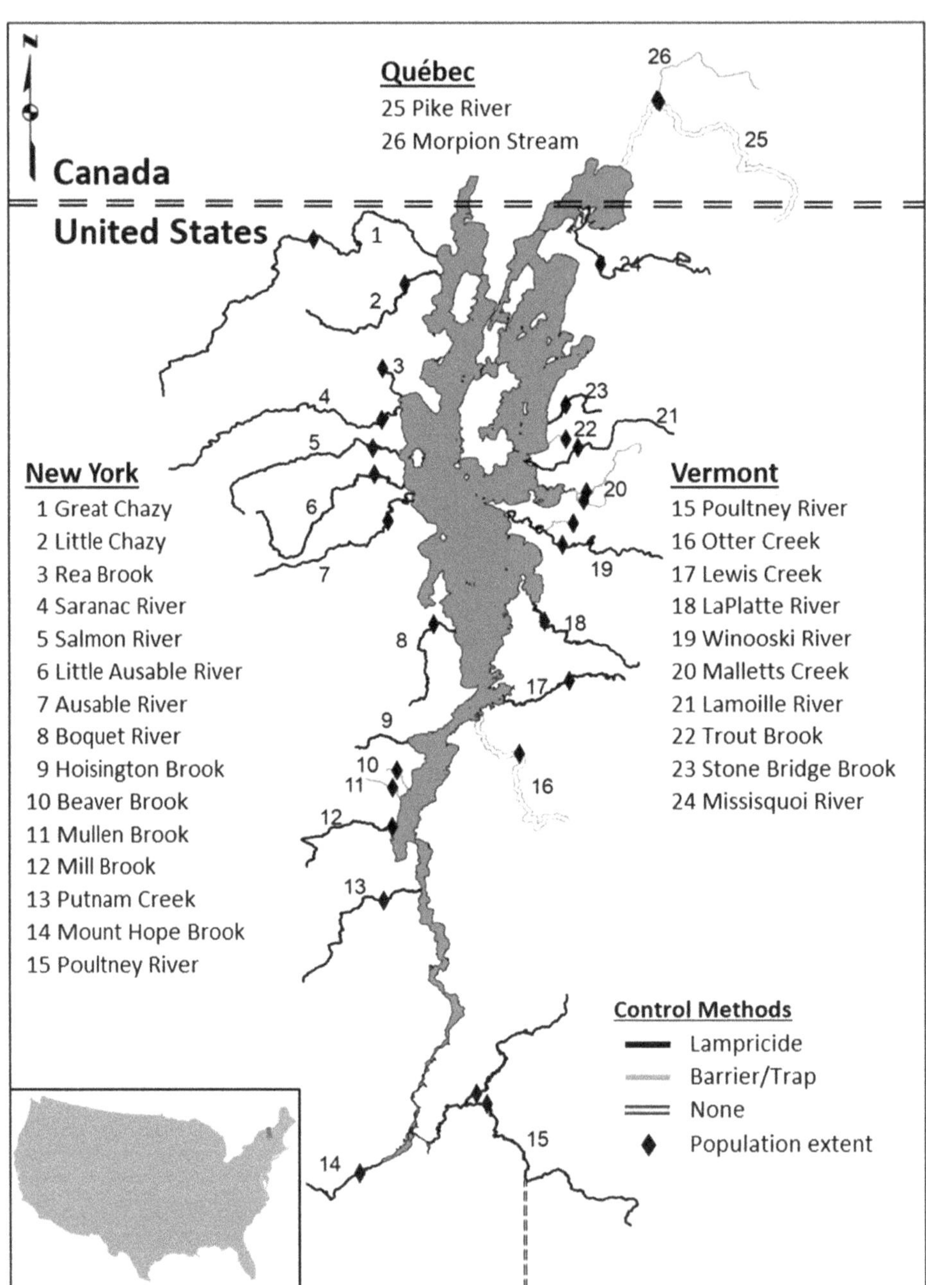

Figure 1. Lake Champlain and its 26 tributaries with currently known larval sea lamprey populations, controlled as indicated. The lake map represents the red-colored region on the inset United States map. Lamprey-producing subordinate tributaries controlled concurrently with mainstem tributaries are not included in counts.

Table 1. Historical levels of chemical (lampricide) and physical (barriers with traps) sea lamprey control efforts used on Lake Champlain. Lamprey-producing subordinate tributaries controlled concurrently with mainstem tributaries are not included in counts. ECP = experimental control program; PCP = partial control program; LTCP = long-term control program.

Year	Management Phase	Tributaries with Lamprey	Lampricide Control	Trapping Control	% of Tributaries Controlled
1992	ECP	19	13	3	84
2000	PCP	19	9	5	74
2002	LTCP	20	9	5	70
2009	LTCP	20	14	5	95
2020	LTCP	26	19	5	92

These three differing periods of Lake Champlain lamprey control unintentionally provided insight into the relationship between the lamprey population of Lake Champlain and the effort required to suppress it. Over 30 years, lamprey densities have fluctuated in individual tributaries, populations have expanded to new tributaries, control efforts have adjusted and sometimes been delayed, and technological advancements have enabled new and improved approaches to control. When viewing these programmatic adaptations over the long term, patterns emerged that lead to a better understanding of how to successfully control sea lamprey in Lake Champlain.

3. Management Review

The Lake Champlain lamprey control program presents opportunities for reviewing both short- and long-term population responses to control efforts. The lamprey control program set forth as a management initiative, not an ecological experiment [11,28,30]. Variables were not controlled, measures were coarse, and replication exists only in the form of a time series. Despite the inability to apply statistical models or tests, some general trends and patterns emerged over time that demonstrate relationships and emphasize aspects of the program in ways that will inform managers when making decisions in years to come.

By 2005, the Cooperative began questioning why the level of control effort was not showing the same type of anticipated response in lamprey reduction seen during the experimental program. With the implementation of the LTCP in 2002, fishery managers anticipated similar positive results based on the responses seen during the experimental program. The Cooperative distilled explanations for why wounding rates reached record highs in the period 2004–2006 into three general categories that led to further investigations into the need for: (1) control of additional known sources of larval production, (2) locating and controlling unknown sources of larval production, and (3) improved lampricide treatments that reduce the number of residual (surviving) larvae. While most agreed that the reason for rising wounding rates was some combination of these three, determining where to focus efforts required more data.

At the inception of the experimental program, it was logical and appropriate to believe that because Lake Champlain had a parasitic sea lamprey problem consistent with what the Great Lakes program manages, that if the Cooperative implemented the same control techniques and methodologies as the Great Lakes successfully employed, Lake Champlain would experience the same positive results. The immediate responses seen during the experimental program did appear to validate that approach as application of standard lamprey control techniques showed an expected reduction in sea lamprey wounds (Figure 2) [29]. As time passed, it became increasingly clear that while the general approach to sea lamprey control was capable of working in Lake Champlain as it does in the Great Lakes, there were nuanced differences that had not been recognized or accounted for and undermined existing assumptions.

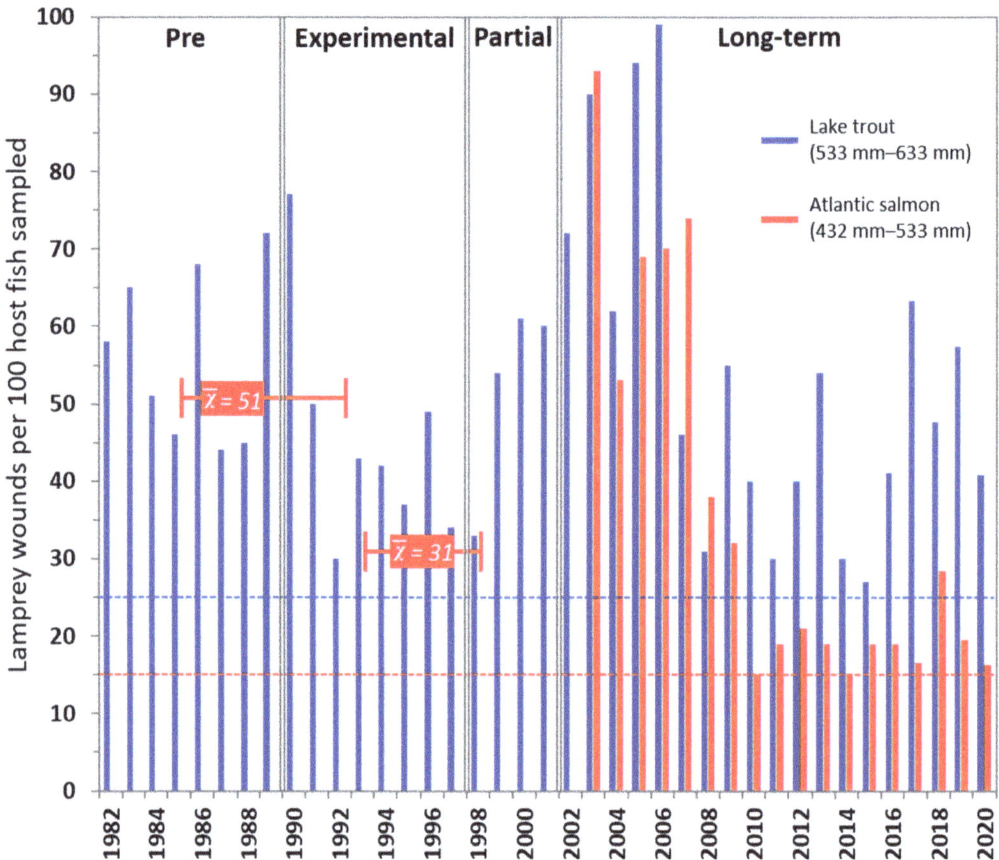

Figure 2. Sea lamprey wounding rates on lake trout and Atlantic salmon measured as number of wounds per 100 fish [18]. The vertical double lines separate the periods before lamprey control (Pre), the 8 year experimental control program (Experimental), the period of partial control (Partial), and the long-term control program (Long-term). Non-standardized collection effort among years for Atlantic salmon wounding data, prior to the Long-term program, led to grouping and averaging available data across the years 1985–1992 and 1993–1998 to approximate and reflect the time-lag responses to lamprey parasitism during the Pre and Experimental control periods, respectively. Horizontal dashed lines indicate the management goals for lake trout (25) and Atlantic salmon (15).

The three areas of concern shared a common need for more assessment and monitoring data. Enhancement of existing control actions was a simpler, more direct, a more convenient solution, and might ultimately prove necessary. However, such determinations required a more detailed understanding of the density and distribution of the larval population in the basin and site-specific measures of control efficacy. The Great Lakes program affirmed this need for enhanced assessment and its critical importance as they also placed attention on assessment in developing more effective control strategies [31,32]. Any broadly applied attempts focused on increasing existing control efforts were unlikely to address all remaining sources of lamprey production that contributed to the parasitic population. With increased attention paid, Lake Champlain assessment and monitoring developed into a more systematic approach, where quadrennial surveys provided comprehensive coverage of all tributaries in the basin for the detection of new and emerging larval populations. Implementation of standardized surveys that both preceded and followed every lampricide treatment became a permanent method for determining effectiveness. Regular surveys on tributaries with barriers and traps verified the effectiveness of the method used at each site. The data gained from these increased assessment and monitoring efforts provided new insights and helped to isolate the reasons that the LTCP was not matching the success of the ECP.

3.1. Discontinuity

3.1.1. Partial Control

During the four years of the PCP, lake trout wounding rates rose sharply from 33 in 1998 to 77 in 2002 (Figure 2). The Cooperative expected that a reduction in control effort during the PCP would result in higher wounding rates, but the resurgence of lamprey during this period to even higher wounding levels on lake trout than seen prior to the ECP was unexpected (Figure 2). A rebound effect appeared underway that partial control efforts failed to slow or lessen. While treatments continued on the nine New York tributaries treated during the ECP, the PCP did not include delta treatments previously associated with four of those tributaries during the ECP. We cannot quantify the contribution of those untreated deltas, but it amounted to further reduction in the cumulative control effort expended during the PCP. The four Vermont tributaries controlled during the ECP were not disproportionately large lamprey producers, based on larval population survey data. In fact, larval survey data indicate the nine treated New York tributaries accounted for more than a commensurate 69% of the total larval production among the 13 tributaries treated during the ECP. In light of the success of the recent ECP, expectations were that partial control efforts would produce partial population suppression and, at the very least, keep the lamprey population from returning to previous levels. Yet despite the treatments conducted in New York and attempts to trap two (labeled 22, 23; Figure 1) of the four Vermont tributaries treated during the ECP, the lake trout wounding rate incline that began during the PCP in 1998 continued to rise through the early years of the LTCP (Figure 2). The Cooperative did not track Atlantic salmon wounding data during the PCP, but once those measures resumed in 2003, they showed the same sharp increase in wounding rate as seen for lake trout (Figure 2).

3.1.2. Delayed Control

After the approval of the EIS, some local citizens continued to express concern and objection to the use of lampricide to control sea lamprey on the Poultney River (Figure 1). Through engaged conversation, the Cooperative chose to negotiate an agreement to delay chemical control for five years on that river. The agreement led to the creation of Federal Advisory Committee Act (FACA) group whose charter was to work toward alternative methods to control sea lamprey that circumvented the use of lampricides on the Poultney River. The 5 year delay ended in 2007 at which time no feasible alternatives had emerged that could effectively control the larval population estimated at over 163,000 in 2006. The wounding rates for both lake trout and Atlantic salmon in 2006 had reached a record high point (Figure 2) and led the Cooperative to proceed with application of lampricide to the Poultney in 2007. Following that treatment, wounding rates that had remained elevated even after the start of the long-term program in 2002, began to decline (Figure 2). The decline could not be attributed solely to the treatment of the Poultney River because other program improvements were also underway. The program treated the Winooski River with lampricide for the first time in 2004, making it the largest treated Vermont tributary at that time. The level of control effort applied across the basin was rising which included bringing the Poultney River back into the program. The end of the Poultney 5 year delay and other initiatives begun in 2006 marked a turning point, as seen in Figure 2.

3.2. Enhanced Assessment and Monitoring

From 1990 through 2005, we evaluated lampricide treatment effectiveness primarily by counting visible lamprey mortality the day following a treatment. Those observations provided evidence of dead lamprey that validated reasonable assumptions of treatment effectiveness. However, when wounding rates remained higher than expected and without a clear cause, we questioned whether qualitative observations of dead larvae following treatments missed quantitative measures of actual treatment effectiveness. To evaluate that, we added a new regular aspect to the assessment program in 2006. The summer following each fall lampricide treatment, assessment crews began performing post-treatment assessments for comparison to pre-treatment assessments.

Post-treatment surveys provided a new way to evaluate and understand the effectiveness of treatments. We found that measurements of treatment effectiveness based on the comparison of pre- and post-treatment assessments were a more nuanced and river-specific metric than previously understood. Perhaps the most surprising finding was that when we observed relatively large numbers of dead lamprey following some treatments, we occasionally found substantial numbers of larvae that simultaneously survived those same treatments. This led to further investigations and refinements in lampricide application approach and methodology.

When looking into the reasons that some treatments were highly successful and others were not, we discovered multiple factors that contributed to varying levels of larval lamprey survival during some treatments. We understood and addressed variables affecting dose, alkalinity, pH, seasonality, and stream-specific requirements based on toxicity testing. We were also aware of and accounted for variables affecting exposure, discharge, attenuation, dilution, channel morphology, and others. We found that the ineffective treatments were not the result of program-related miscalculations or technical errors. Instead, river-specific characteristics had been missed which required applying a more nuanced control approach to each river.

After evaluating all lampricide-controlled tributaries, post-treatment assessments revealed that most treatments had indeed been successful. However, some did show a consistent presence of residual larvae following treatments. The Ausable River and Putnam Creek (labeled 7, 13; Figure 1) provide two examples of how we identified and corrected ineffective treatments. The Ausable River has a mean annual discharge of 715 cubic feet per second (CFS), making it the second-largest New York tributary to Lake Champlain. Larval lamprey population estimates have averaged more than 600,000 over the past 15 years, not including its associated delta population, thereby ranking the Ausable as the largest producer of sea lamprey in Lake Champlain. Assuming that recruitment of larvae to the parasitic population of the lake is density independent [26] and similar to that of other tributaries in the basin, ineffective treatments there yield more considerably more net lamprey production than would ineffective treatments in smaller and lower populated tributaries. This recognition reemphasized that the importance and consequences of successful lampricide treatments increased as the size of the larval lamprey population increased.

We found that two factors in the Ausable were responsible for its insufficient treatment effectiveness. One was river morphology on the day of treatment. The Ausable splits into two mouths near its terminus. Large portions of the larval population reside in each mouth. Depending on the discharge of the river on the day of treatment, or changes in channel morphology from year to year, we found that disproportionate volumes of the mainstem followed one mouth or the other. Under ideal conditions, lampricide reaches both mouths in volumes proportional to their channel volumes. When conditions are not ideal, one mouth becomes a disproportionate route for lampricide-treated water traveling downstream and leads to sub-lethal lampricide exposure for the population of the mouth receiving lower flow.

To address this in the short term, selected portions of river that received sub-lethal doses during the 2006 and 2014 fall treatments received supplemental retreatments in the following springs (2007 and 2015). Increased secondary applications of backwaters and the addition of a supplemental downstream lampricide application point, contingent on discharge at the time of treatment, also improved delivery of lethal doses of lampricide to lamprey infested habitat. These additional steps used during lampricide applications are common, but the need to place additional application points along the river are usually obvious and arranged when first designing a treatment. The new development here was using the post-treatment assessment as a tool to identify a problem that was not otherwise recognized. For over a decade prior, presence of dead larvae following treatments served as sufficient indication of successful treatments of the Ausable. It was not until resolute post-assessment surveys identified the presence of residual survivors and their locations that we were able to isolate and address the issue. When first performed on the Ausable in 2011, that post-treatment assessment revealed the 2010 treatment had been 47% effective. Since that time, following improvements to application methodology, post-treatment assessments showed that the 2014 treatment and 2015 supplemental retreatment were cumulatively responsible for raising treatment effectiveness to 94%. The 2019 post-assessment survey of 2018 treatment found that it successfully eliminated 72% of the larval population.

Putnam Creek presented a much different set of circumstances. Despite being smaller with a mean annual discharge of 80 CFS, the abundant preferred habitat of this tributary provides conditions that support a larval population consistently estimated at over 150,000 during the last 15 years. Treatment monitoring data consistently showed that lampricide concentration and other water chemistry parameters fell within the bounds of successful treatments. However, once we started post-assessment surveys, we discovered despite seeing numerous dead larvae following treatments, there were often still large numbers of residual larvae the following year. Because larvae distribute themselves and drift over time [33], identifying any specific point sources leading to treatment survival proved difficult. Following several investigations, we discovered groundwater influence was the likely source of residual larvae in Putnam Creek. A portion of the river is in an area where groundwater routinely seeps from the banks. Through an additional series of spatial measurements in the channel using a temperature probe, we found a groundwater sublayer present within the channel sediment as well. Though this groundwater was not a substantial contributor to the overall discharge of Putnam Creek and did not affect measured treatment concentrations, we believe it provided microrefugia to sediment-dwelling larval lamprey. Fresh water recharge from below the sediment water interface countered the lethal treatment concentration present above that interface during treatments resulting in a net sub-lethal and survivable exposure in that section of the tributary. We have not yet developed a way to fully negate groundwater influence that leads to treatment residuals, but detecting its presence and extent have allowed treatments to be fine-tuned to better address the specific areas now presumed to provide lamprey with refuge during treatments.

Before post-treatment assessments began in 2006, numerous dead lamprey led managers to believe treatments were successful. The examples in the Ausable River and Putnam Creek show how that assumption led to incorrect expectations that control effort equated to control effectiveness. Until we quantified and monitored control effectiveness with additional assessment effort, remaining sources of production like these two tributaries went unnoticed. There are additional tributaries in the basin found to need river-specific adjustments to treatment strategy as well. The importance of post-treatment assessments and the way they inform control effectiveness has led us to make them a standard part of the control program.

3.3. Aggressive Programmatic Expansion

As post-treatment assessments revealed sources of residual larval lamprey production in need of additional attention, the Cooperative sought to eliminate lamprey populations in additional tributaries where assessments had detected their presence. The EIS from 2001 had prescribed plans for how and where to control the population as part of the LTCP. Unfortunately, it did not include provisions for the addition of newly colonized tributaries in the future. As efforts continued to control all known sources of production, the detection of new larval populations led to the need for a federal Environmental Assessment (EA) in 2008 to authorize the inclusion of the Lamoille River and Pond Brook in Vermont and Mill Brook in New York into the LTCP [34]. Continued annual larval assessments later found new populations of lamprey in the Little Chazy River and Rea Brook in New York (Figure 1). To keep pace with larval lamprey colonization, these two tributaries warranted production of another EA in 2018 that added them to the LTCP [35]. Following yet another detected new colonization, a third EA added Hoisington Brook in New York (Figure 1) to the LTCP in 2019 [36]. Experience from the PCP, when partial control allowed the population to expand, factored heavily into the decision to maintain an aggressive approach to addressing all sources of production. While EA's were required to expand the LTCP and deliver control at those locations, another tributary identified in the original EIS [30] showed new evidence of an emerging population. The LaPlatte River (Figure 1) did not warrant control throughout the ECP and LTCP, but when surveys showed an emerging population, the Cooperative chose to initiate lampricide treatments there in 2016.

Morpion stream and the Pike River in Québec (Figure 1) were both known sources of production, but as Canadian tributaries to Lake Champlain, they are not subject to jurisdiction under the EIS issued by the United States. Requests made to Québec provincial officials to treat both tributaries with lampricide were not successful, leaving both as uncontrolled sources of lamprey production. Through a long process of evaluating potential alternatives to using lampricide, an innovative seasonally-removable, modular screen barrier structure was designed and installed in Morpion Stream 2014. Morpion stream is approximately 10 m wide and up to 1.5 m deep at the barrier site. Each spring, prior to lamprey spawning season, seven flow-through screen modules are set into place on a concrete base laid into the sediment. Each module is composed of 5 m height aluminum frame containing a bottom-hinged screen. Each screen locks in place upright using a float barrel mechanism that lifts during flood conditions to release the top of the screen to pivot on its bottom hinge and fall flat and flush to the sediment. This feature prevents debris buildup or extreme flows from turning the flow-through screen barrier into a dam that would flood surrounding lands. When locked in place and operating, the 13-mm spaced grates on each screen block lamprey from migrating upstream, but allow the river to flow through with minimal impoundment upstream of the barrier. The barrier is also angled between banks which naturally directs sea lamprey searching for passage into a trap where they are collected. This design is a unique solution to blocking sea lamprey in a river where discharge is too high to use small-scale (channel width < 5 m) barrier solutions and where lampricide usage is prohibited. While being a smaller tributary to the larger Pike River, the larval population of Morpion Stream has been estimated as high as 135,000 and warrants control. Following installation of the barrier, larval population estimates have averaged under 50,000 with recent technical improvements expected to result in additional declines. The Pike River remains uncontrolled and is a known producer of sea lamprey. At this time, we have no options available to control lamprey there. Its size and migratory non-target species concerns preclude consideration of a barrier or lampricide. As new technologies are developed, we hope to find an agreeable form of control to use in the Pike River in the future.

4. Discussion

After 30 years of perspective since the start of the ECP, factors that influenced the long-term success of sea lamprey control in Lake Champlain have been recognized, addressed, and used to steer decisions on where to focus resources efficiently. Periods of discontinuance resulted in a disproportionate population resurgence. Insufficient attention to assessment and monitoring led to misinformed assumptions. Control techniques executed soundly and according to plan suffered from cryptic sources of unaccounted variation. The examples presented do not provide particularly novel or noteworthy management actions. The fine-tuning of sea lamprey control has been ongoing in the Great Lakes for over 70 years [12,13]. However, with 26 lamprey-producing tributaries among a watershed with 226, Lake Champlain may offer a scale where the dynamics between sea lamprey parasitism and its effects on the fishery produce detectable effects among a relatively few sources of lamprey production. With lamprey found in 450 of the 5400 tributaries of the

Great Lakes [13], changes in individual streams become less pronounced and detectable in a control program of nearly 20 times the scale. When considering the many changes to control efforts over the length of the Lake Champlain program, when they occurred, and their various effects on the lamprey population, we have formed two conclusions that we believe offer insight into sea lamprey control efforts into the future.

First, we assert that the relationship between control effort and population reduction is non-linear based on the measurements of wounding rates following changes in the control program. The wounding rate index is not a direct measure of the parasitic population and cannot be used to make empiric estimations of that relationship. Attempts to understand the relationship are therefore limited to general qualitative observations of this relationship rather than quantitative descriptive models. Even with that limitation, we believe the wounding rate data can reflect changes in trends and serve in part as a lesser surrogate measure of relative abundance to indicate when substantial changes in the lamprey population occur or are sustained over time. The relationship between control effort and population suppression appears to follow an inverse sigmoidal relationship depicted in the conceptual diagram in Figure 3. The long-term wounding rate data (Figure 2) show instances during the PCP and during the start of the LTCP when wounding rates failed to decline until additional and more effective control was administered to sources of lamprey production. If the relationship was linear, then some fractional decline should have been detected during the PCP when 69% of the tributaries controlled during the ECP continued to be treated. As the LTCP began and added tributaries to the original 13 treated during the ECP, there was an expectation of decline that failed to materialize for the first five years of the LTCP. When additional assessments and monitoring began in 2006 and led to improvements in control effectiveness, along we the resumption of delayed treatments and inclusion of new ones, the benefits of control efforts began to exceed costs as represented by point 1 on Figure 3.

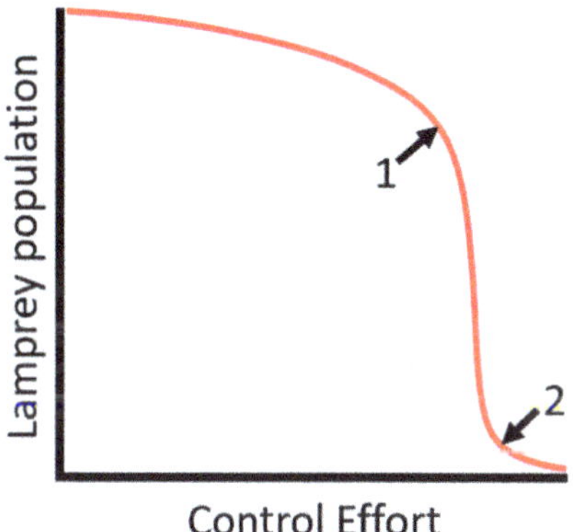

Figure 3. Conceptual representation of the relationship between the Lake Champlain sea lamprey population and efforts to control it. Point 1 indicates where benefits of population suppression begin to exceed the costs of control efforts. Point 2 indicates the beginning of diminishing returns where the costs of additional control efforts yield limited additional population suppression benefits.

Second, we found that achieving a conceptual 50% reduction point in wounding rate requires considerably more than a 50% control effort. Thus, not only is the relationship non-linear, it also skews toward the need for a disproportionately higher level of control effort to achieve desired reductions in lamprey populations. To suppress the lamprey population into the region between points 1 and 2 on Figure 3, control efforts needed to address more than 80% of the known sources of larval sea lamprey production in the basin. That same required level of control effort appeared consistently and repeatedly during the ECP, PCP, and the LTCP (Table 1). We do not suggest that the observed percentages constitute specific numeric management benchmarks, but we do think the long-term data reflect the existence of a threshold for required control effort, below which measurable reductions in the Lake Champlain lamprey population cannot be achieved.

Evaluation of sea lamprey control efforts is indirect where the larval and adult life history stages receive control while assessment of those efforts focuses on the juvenile (parasitic) stage. This indirect evaluation prevents immediate determinations of population suppression measures corresponding to applied effort. The cumulative effects (changes in wounding rates) of individual control efforts are also not observable until at least one year following implementation. These control program characteristics make comparisons between sea lamprey and other fish species controlled by removal and assessment of the same life history stages tenuous when looking for common relationships between control effort and population response. So while other long-term invasive fish control programs have modeled and quantified relationships between direct species removal and measured population responses [37–39], we hesitate in seeking to relate our findings to theirs because of the differences in target species life histories, niche, and control and assessment methodologies.

Relating our findings to other sea lamprey control efforts are complicated by scale and management focus. The Great Lakes program has historically used different measures and models [40–42] to prioritize their allocation of limited resources to achieve the greatest benefits across their larger scale. Lake Champlain differs in that limits to control have historically been the result of socio-political issues rather than limited resources. This difference and the 20× smaller scale enables the Lake Champlain program to control a higher proportion of its lamprey-producing tributaries. Currently, the Great Lakes regularly controls 166 of their 450 (37%) lamprey-producing tributaries with lampricides [13]. Those 166 do represent a large portion (more than 37%) of the total basin-wide larval population, yet it compares to 19 of 26 (73%) lampricide treated tributaries in Lake Champlain. Despite the apparent advantage Lake Champlain has in proportional control effort, lamprey wounding rates on lake trout in lakes Superior, Michigan, Huron, and Erie have remained under 20 since at least the year 2000 and under five in Lake Ontario since 1985 [43]. This compares to Lake Champlain lake trout wounding rates that have remained above the management target of 25 since recording began in 1982 (Figure 2). There are many presumed reasons for this [20,21], yet aside from the causes, the differences in response relative to control highlights how control effort and population responses can differ widely between two similar programs that focus on the same target species. This leads us to conclude that our specific findings may have limited applicability to Lake Champlain or similar, smaller watersheds.

5. Conclusions

The differing phases of Lake Champlain sea lamprey control over 30 years offered an occasion to evaluate trends and anomalies during periods of cessation, adjustment, and improvement. With 26 current lamprey-producing tributaries in the basin, the potential for each to exert influence on the population forces managers to remain vigilant in assessing larval population densities and distributions. It also requires validation that implemented control efforts meet management expectations. We learned that fractional efforts do not correspond to fractional reductions and that the minimum effort required to successfully control sea lamprey falls closer to the maximum end of the range. In recent years, we began referring to our approach as "comprehensive" to imply that we have come to realize the need to address all sources of lamprey production in the basin. Ignoring even a few or one source of lamprey production can negate gains that have taken years to achieve.

As sea lamprey control continues to serve as a tool to facilitate the restoration and recovery of native fish stocks in Lake Champlain, further refinement of current methodologies and the development of new approaches are both needed to ultimately meet the management targets for sea lamprey population suppression. Continued reliance on thorough larval assessment is critical to keeping pace with expanding colonization of new tributaries. We also look to shift the assessment of the parasitic population from exclusive reliance on wounding rates to a more inclusive and direct measure of lamprey abundance used by the Great Lakes [44,45]. Having established the level of assessment and control effort required to achieve and sustain population suppression during the previous three decades of the Lake Champlain sea lamprey control program, we expect further reductions during the ensuing fourth decade to require additional approaches, not just additional effort.

Author Contributions: Conceptualization, B.Y., BJA., and S.S.; methodology, B.Y., BJA. and S.S.; formal analysis, B.Y., BJA. and S.S.; investigation, B.Y., BJA. and S.S.; writing—original draft preparation, B.Y.; writing—review and editing, BJA. and S.S.; visualization, B.Y., BJA. and S.S.; supervision, B.Y.; project administration, B.Y. All authors have read and agreed to the published version of the manuscript.

Funding: This research received no external funding. The Lake Champlain Sea Lamprey Control Program is federally funded by the U.S. Department of the Interior and the U.S. State Department. The States of New York and Vermont contribute in kind by providing staff and equipment to participate in control activities.

Institutional Review Board Statement: Not Applicable.

Informed Consent Statement: Not Applicable.

Data Availability Statement: Data evaluated here are products of the United States Fish and Wildlife Service, New York State Department of Environmental Conservation, and the Vermont Fish and Wildlife Department and are available upon request from the authors.

Acknowledgments: We would like to collectively recognize and thank the over 100 employees from the New York State Department of Environmental Conservation, Vermont Fish and Wildlife Department, and U.S. Fish and Wildlife Service, from leaders to seasonal field technicians, who all contributed to the success of the Lake Champlain sea lamprey control program over the past 30 years. We thank United States Senator Patrick J. Leahy (Vermont) for his leadership and commitment in advocating for the program over its course. The authors thank the United States Fish and Wildlife Service for supporting this programmatic review effort and allowing us the time needed to compile, review, evaluate, and publish this work. Finally, we thank the families who endure the long and unusual schedules and travel of those who work to control sea lamprey and for their continued support.

Conflicts of Interest: The authors declare no conflict of interest. The Federal and State government entities who funded the work presented here had no role in the design of the study; in the collection, analyses, or interpretation of data; in the writing of the manuscript, or in the decision to publish the results. The findings and conclusions in the article are those of the authors and do not necessarily represent the views of the USFWS.

References

1. Lake Champlain Fish and Wildlife Management Cooperative. *Strategic Plan for Lake Champlain Fisheries*; U.S. Fish and Wildlife Service: Essex Junction, VT, USA, 2020.
2. MacKenzie, C. *Lake Champlain Lake Sturgeon Recovery Plan*; Vermont Fish and Wildlife Department: Montpelier, VT, USA, 2016.
3. Bryan, M.B.; Zalinski, D.; Filcek, K.B.; Libants, S.; Li, W.; Scribner, K.T. Patterns of invasion and colonization of the sea lamprey (*Petromyzon marinus*) in North America as revealed by microsatellite genotypes: Sea lamprey population structure. *Mol. Ecol.* **2005**, *14*, 3757–3773. [CrossRef] [PubMed]
4. Waldman, J.R.; Grunwald, C.; Wirgin, I. Evaluation of the native status of sea lampreys in Lake Champlain based on mitochondrial DNA sequencing analysis. *Trans. Am. Fish. Soc.* **2006**, *135*, 1076–1085. [CrossRef]
5. Waldman, J.; Daniels, R.; Hickerson, M.; Wirgin, I. Mitochondrial DNA analysis indicates sea lampreys are indigenous to lake ontario: Response to comment. *Trans. Am. Fish. Soc.* **2009**, *138*, 1190–1197. [CrossRef]
6. Eshenroder, R.L. Comment: Mitochondrial DNA analysis indicates sea lampreys are indigenous to Lake Ontario. *Trans. Am. Fish. Soc.* **2009**, *138*, 1178–1189. [CrossRef]
7. Eshenroder, R.L. The role of the Champlain Canal and Erie Canal as putative corridors for colonization of Lake Champlain and Lake Ontario by sea lampreys. *Trans. Am. Fish. Soc.* **2014**, *143*, 634–649. [CrossRef]
8. Marsden, J.E.; Chipman, B.D.; Nashett, L.J.; Anderson, J.K.; Bouffard, W.; Durfey, L.; Gersmehl, J.E.; Schoch, W.F.; Staats, N.R.; Zerrenner, A. Sea lamprey control in Lake Champlain. *J. Great Lakes Res.* **2003**, *29*, 655–676. [CrossRef]
9. Langdon, R.W.; Ferguson, M.T.; Cox, K.M. *Fishes of Vermont*; Vermont Fish and Wildlife Department: Montpelier, VT, USA, 2006.

10. Marsden, J.E.; Langdon, R.W. The history and future of Lake Champlain's fishes and fisheries. *J. Great Lakes Res.* **2012**, *38*, 19–34. [CrossRef]
11. Fisheries Technical Committee. *A Strategic Plan for the Development of Salmonid Fisheries in Lake Champlain*; Lake Champlain Fish and Wildlife Management Cooperative, U.S. Fish and Wildlife Service: Essex Junction, VT, USA, 1977.
12. Brant, C. *Great Lakes Sea Lamprey: The 70 Year War on a Biological Invader*; University of Michigan Press: Ann Arbor, MI, USA, 2019. [CrossRef]
13. Siefkes, M.J.; Steeves, T.B.; Sullivan, W.P.; Twohey, M.B.; Li, W. Sea lamprey control: Past, present, and future. In *Great Lakes Fisheries Policy and Management: A Binational Perspective*; Taylor, W.W., Lynch, A.J., Leonard, N.J., Eds.; Michigan State University Press: East Lansing, MI, USA, 2013; pp. 651–704.
14. Wilkie, M.P.; Hubert, T.D.; Boogaard, M.A.; Birceanu, O. Control of invasive sea lampreys using the piscicides TFM and niclosamide: Toxicology, successes & future prospects. *Aquat. Toxicol.* **2019**, *211*, 235–252. [CrossRef]
15. Lavis, D.S.; Hallett, A.; Koon, E.M.; McAuley, T.C. History of and advances in barriers as an alternative method to suppress sea lampreys in the Great Lakes. *J. Great Lakes Res.* **2003**, *29*, 362–372. [CrossRef]
16. McLaughlin, R.L.; Hallett, A.; Pratt, T.C.; O'Connor, L.M.; McDonald, D.G. Research to guide use of barriers, traps, and fishways to control sea lamprey. *J. Great Lakes Res.* **2007**, *33* (Suppl. S2), 7–19. [CrossRef]
17. Miehls, S.; Sullivan, P.; Twohey, M.; Barber, J.; McDonald, R. The future of barriers and trapping methods in the sea lamprey (*Petromyzon marinus*) control program in the Laurentian Great Lakes. *Rev. Fish Biol. Fish.* **2020**, *30*, 1–24. [CrossRef]
18. Ebener, M.P.; Bence, J.R.; Bergstedt, R.A.; Mullett, K.M. Classifying sea lamprey marks on Great Lakes lake trout: Observer agreement, evidence on healing times between classes, and recommendations for reporting of marking statistics. *J. Great Lakes Res.* **2003**, *29*, 283–296. [CrossRef]
19. Firkus, T.J.; Murphy, C.A.; Adams, J.V.; Treska, T.J.; Fischer, G. Assessing the assumptions of classification agreement, accuracy, and predictable healing time of sea lamprey wounds on lake trout. *J. Great Lakes Res.* **2020**. [CrossRef]
20. Adams, J.V.; Jones, M.L.; Bence, J.R. Using simulation to understand annual sea lamprey marking rates on lake trout. *J. Great Lakes Res.* **2020**. [CrossRef]
21. Adams, J.V.; Jones, M.L. Evidence of Host Switching: Sea lampreys disproportionately attack chinook salmon when lake trout abundance is low in Lake Ontario. *J. Great Lakes Res.* **2020**. [CrossRef]
22. King, E.L., Jr. Classification of sea lamprey (*Petromyzon marinus*) attack marks on Great Lakes lake trout (*Salvelinus namaycush*). *Can. J. Fish. Aquat. Sci.* **1980**, *37*, 1989–2006. [CrossRef]
23. Jones, M.L.; Irwin, B.J.; Hansen, G.J.A.; Dawson, H.A.; Treble, A.J.; Liu, W.; Dai, W.; Bence, J.R. An operating model for the integrated pest management of Great Lakes sea lampreys. *Open Fish Sci. J.* **2009**, *2*, 59–73. [CrossRef]
24. Jones, M.L.; Bergstedt, R.A.; Twohey, M.B.; Fodale, M.F.; Cuddy, D.W.; Slade, J.W. Compensatory mechanisms in Great Lakes sea lamprey populations: Implications for alternative control strategies. *J. Great Lakes Res.* **2003**, *29*, 113–129. [CrossRef]
25. Haeseker, S.L.; Jones, M.L.; Bence, J.R. Estimating uncertainty in the stock-recruitment relationship for St. Marys River sea lampreys. *J. Great Lakes Res.* **2003**, *29*, 728–741. [CrossRef]
26. Dawson, H.A.; Jones, M.L. Factors affecting recruitment dynamics of Great Lakes sea lamprey (*Petromyzon marinus*) populations. *J. Great Lakes Res.* **2009**, *35*, 353–360. [CrossRef]
27. Jones, M.L.; Adams, J.V. Eradication of sea lampreys from the Laurentian Great Lakes is possible. *J. Great Lakes Res.* **2020**. [CrossRef]
28. Fisheries Technical Committee. *Salmonid-Sea Lamprey Management Alternatives for Lake Champlain*; Lake Champlain Fish and Wildlife Management Cooperative, U.S. Fish and Wildlife Service: Essex Junction, VT, USA, 1985.
29. Fisheries Technical Committee. *A Comprehensive Evaluation of an Eight Year Program of Sea Lamprey Control in Lake Champlain*; Lake Champlain Fish and Wildlife Management Cooperative, U.S. Fish and Wildlife Service: Essex Junction, VT, USA, 1999.
30. Fisheries Technical Committee. *A Long-Term Program of Sea Lamprey Control in Lake Champlain: Final Supplemental Impact Statement*; U.S. Fish and Wildlife Service: Essex Junction, VT, USA, 2001.
31. Jones, M.L. Toward improved assessment of sea lamprey population dynamics in support of cost-effective sea lamprey management. *J. Great Lakes Res.* **2007**, *33* (Suppl. S2), 35–47. [CrossRef]
32. Hansen, M.J.; Adams, J.V.; Cuddy, D.W.; Richards, J.M.; Fodale, M.F.; Larson, G.L.; Ollila, D.J.; Slade, J.W.; Steeves, T.B.; Young, R.J.; et al. Optimizing larval assessment to support sea lamprey control in the Great Lakes. *J. Great Lakes Res.* **2003**, *29*, 766–782. [CrossRef]
33. Derosier, A.L.; Jones, M.L.; Scribner, K.T. Dispersal of sea lamprey larvae during early life: Relevance for recruitment dynamics. *Environ. Biol. Fishes* **2007**, *78*, 271–284. [CrossRef]
34. U.S. Fish and Wildlife Service. *Proposed Changes to the Long-Term Sea Lamprey Control Program on Lake Champlain*; U.S. Fish and Wildlife Service: Essex Junction, VT, USA, 2008.
35. U.S. Fish and Wildlife Service. *Proposed Additions to the Final Supplemental Environmental Impact Statement (August 2001): A Long-Term Program of Sea Lamprey Control in Lake Champlain*; U.S. Fish and Wildlife Service: Essex Junction, VT, USA, 2017.
36. U.S. Fish and Wildlife Service. *Addition of Hoisington Brook to A Long-Term Program of Sea Lamprey Control in Lake Champlain: Final Supplemental Environmental Impact Statement (August 2001)*; U.S. Fish and Wildlife Service: Essex Junction, VT, USA, 2019.
37. Dux, A.M.; Hansen, M.J.; Corsi, M.P.; Wahl, N.C.; Fredericks, J.P.; Corsi, C.E.; Schill, D.J.; Horner, N.J. Effectiveness of lake trout (*Salvelinus namaycush*) suppression in Lake Pend Oreille, Idaho: 2006–2016. *Hydrobiologia* **2019**, *840*, 319–333. [CrossRef]

38. Healy, B.D.; Schelly, R.C.; Yackulic, C.B.; Smith, E.C.O.; Budy, P. Remarkable response of native fishes to invasive trout suppression varies with trout density, temperature, and annual hydrology. *Can. J. Fish. Aquat. Sci.* **2020**, *77*, 1446–1462. [CrossRef]
39. Syslo, J.M.; Brenden, T.O.; Guy, C.S.; Koel, T.M.; Bigelow, P.E.; Doepke, P.D.; Arnold, J.L.; Ertel, B.D. Could ecological release buffer suppression efforts for non-native lake trout (*Salvelinus namaycush*) in Yellowstone Lake, Yellowstone National Park? *Can. J. Fish. Aquat. Sci.* **2020**, *77*, 1010–1025. [CrossRef]
40. Koonce, J.F.; Eshenroder, R.L.; Christie, G.C. An economic injury level approach to establishing the intensity of sea lamprey control in the Great Lakes. *N. Am. J. Fish. Manag.* **1993**, *13*, 1–14. [CrossRef]
41. Christie, G.C.; Adams, J.V.; Steeves, T.B.; Slade, J.W.; Cuddy, D.W.; Fodale, M.F.; Young, R.J.; Kuc, M.; Jones, M.L. Selecting Great Lakes streams for lampricide treatment based on larval sea lamprey surveys. *J. Great Lakes Res.* **2003**, *29*, 152–160. [CrossRef]
42. Irwin, B.J.; Liu, W.; Bence, J.R.; Jones, M.L. Defining economic injury levels for sea lamprey control in the Great Lakes Basin. *N. Am. J. Fish. Manag.* **2012**, *32*, 760–771. [CrossRef]
43. Marsden, J.E.; Siefkes, M.J. Control of invasive sea lamprey in the Great Lakes, Lake Champlain, and Finger Lakes of New York. In *Lampreys: Biology, Conservation and Control*; Springer: Dordrecht, The Netherlands, 2019; pp. 411–479. [CrossRef]
44. Harper, D.L.M.; Horrocks, J.; Barber, J.; Bravener, G.A.; Schwarz, C.J.; McLaughlin, R.L. An evaluation of statistical methods for estimating abundances of migrating adult sea lamprey. *J. Great Lakes Res.* **2018**, *44*, 1362–1372. [CrossRef]
45. Adams, J.V.; Bravener, G.A.; Lewandoski, S.A. Quantifying Great Lakes sea lamprey populations using an index of adults. *J. Great Lakes Res.* **2020**, in press.

Article

Yellowstone Lake Ecosystem Restoration: A Case Study for Invasive Fish Management

Todd M. Koel [1,*], Jeffery L. Arnold [1,†], Patricia E. Bigelow [1], Travis O. Brenden [2], Jeffery D. Davis [3], Colleen R. Detjens [1], Philip D. Doepke [1], Brian D. Ertel [1], Hayley C. Glassic [4], Robert E. Gresswell [5], Christopher S. Guy [6], Drew J. MacDonald [7], Michael E. Ruhl [7,‡], Todd J. Stuth [8], David P. Sweet [9], John M. Syslo [4,§], Nathan A. Thomas [1], Lusha M. Tronstad [10], Patrick J. White [1] and Alexander V. Zale [6]

1. U.S. National Park Service, Yellowstone Center for Resources, Native Fish Conservation Program, P.O. Box 168, Yellowstone National Park, WY 82190, USA; jeff_arnold@nps.gov (J.L.A.); pat_bigelow@nps.gov (P.E.B.); colleen_detjens@nps.gov (C.R.D.); philip_doepke@nps.gov (P.D.D.); brian_ertel@nps.gov (B.D.E.); nathan_thomas@nps.gov (N.A.T.); pj_white@nps.gov (P.J.W.)
2. Quantitative Fisheries Center, Michigan State University, 375 Wilson Road, East Lansing, MI 48824, USA; brenden@anr.msu.edu
3. Yellowstone Forever, P.O. Box 117, Yellowstone National Park, WY 82190, USA; jddavis@yellowstone.org
4. Montana Cooperative Fishery Research Unit, Department of Ecology, Montana State University, P.O. Box 173460, Bozeman, MT 59717, USA; hcg0509@gmail.com (H.C.G.); john.syslo@noaa.gov (J.M.S.)
5. U.S. Geological Survey, Northern Rocky Mountain Science Center, 2327 University Way, Bozeman, MT 59715, USA; bgresswell@usgs.gov
6. U.S. Geological Survey, Montana Cooperative Fishery Research Unit, Department of Ecology, Montana State University, MSU–P.O. Box 173460, Bozeman, MT 59717-3460, USA; cguy@montana.edu (C.S.G.); zale@montana.edu (A.V.Z.)
7. Montana Institute on Ecosystems, Montana State University, MSU–P.O. Box 173490, Bozeman, MT 59717-3490, USA; drew_macdonald@nps.gov (D.J.M.); michael.ruhl@state.nm.us (M.E.R.)
8. Hickey Brothers Research, LLC, 4083 Glidden Drive, Sturgeon Bay, WI 54235, USA; stuthfishing@charter.net
9. Wyoming Council of Trout Unlimited, P.O. Box 3008, Cody, WY 82414, USA; davidps992@gmail.com
10. Wyoming Natural Diversity Database, University of Wyoming, 1000 East University Avenue, Department 3381, Laramie, WY 82071, USA; tronstad@uwyo.edu

* Correspondence: todd_koel@nps.gov; Tel.: +1-307-344-2281

† Current affiliation: U.S. National Park Service, Science and Resource Management, Glen Canyon National Recreation Area & Rainbow Bridge National Monument, P.O. Box 1507, Page, AZ 86040, USA.

‡ Current affiliation: New Mexico Department of Game & Fish, Fisheries Management Division, 1 Wildlife Way, Santa Fe, NM 87507, USA.

§ Current affiliation: NOAA Fisheries, Pacific Islands Fisheries Science Center, 1845 Wasp Blvd., Bldg. 176, Honolulu, HI 96818, USA.

Received: 17 May 2020; Accepted: 9 June 2020; Published: 12 June 2020

Abstract: Invasive predatory lake trout *Salvelinus namaycush* were discovered in Yellowstone Lake in 1994 and caused a precipitous decrease in abundance of native Yellowstone cutthroat trout *Oncorhynchus clarkii bouvieri*. Suppression efforts (primarily gillnetting) initiated in 1995 did not curtail lake trout population growth or lakewide expansion. An adaptive management strategy was developed in 2010 that specified desired conditions indicative of ecosystem recovery. Population modeling was used to estimate effects of suppression efforts on the lake trout and establish effort benchmarks to achieve negative population growth ($\lambda < 1$). Partnerships enhanced funding support, and a scientific review panel provided guidance to increase suppression gillnetting effort to >46,800 100-m net nights; this effort level was achieved in 2012 and led to a reduction in lake trout biomass. Total lake trout biomass declined from 432,017 kg in 2012 to 196,675 kg in 2019, primarily because of a 79% reduction in adults. Total abundance declined from 925,208 in 2012 to 673,983 in 2019 but was highly variable because of recruitment of age-2 fish. Overall, 3.35 million lake trout were killed by suppression efforts from 1995 to 2019. Cutthroat trout abundance remained below target levels, but relative condition increased, large individuals (> 400 mm) became more abundant, and individual weights doubled, probably because of reduced density. Continued actions to suppress lake trout will facilitate further recovery of the cutthroat trout population and integrity of the Yellowstone Lake ecosystem.

Keywords: adaptive management; cutthroat trout; ecosystem restoration; nonnative fish suppression; national park; lake trout; native species recovery; *Oncorhynchus*; predatory fish invasion; *Salvelinus*; trophic cascade; wilderness preserve

---

1. Introduction

Apex predatory fishes introduced to freshwaters of the United States Intermountain West are invasive because they can spread within lakes or through interconnected river networks and pose a high risk to native species [1,2]. Native fish species richness in this region is naturally low and made up largely of non-predatory guilds [3]. Although introduced predatory fish often provide enhanced sport fishing opportunities [4], they prey upon vulnerable natives resulting in reductions of native species abundance or complete extirpation [5–7]. Predation losses in some areas have contributed to listing of native fishes as threatened or endangered under the Endangered Species Act (ESA) [8–10]. Introduction of a novel apex predator to a freshwater ecosystem may also result in cascading changes whereby inverse patterns in abundance, productivity, or biomass of populations or communities emerge across links in the aquatic food web [1,11]. Given that invasive predatory fishes have been introduced to all large lakes and rivers in the Western United States [12–14], mitigating negative effects from these introductions is a widespread problem faced by resource managers.

Because complete eradication or containment is generally not feasible in large aquatic systems [15,16], programs have been implemented to suppress invasive fish populations and relieve predation pressure on sympatric native species or desired, introduced sportfish populations [17–19]. Suppression programs for other waters are being contemplated [20–22]. However, proposed suppression programs are often challenged by constituents of the popular nonnative fisheries that have become established [23,24] or are complicated by presence of ESA-listed species that might be harmed by the suppression actions [25]. Lack of species-selective removal methods [15] and uncertainty in outcomes of the removal programs are common obstacles. In addition, because complete, system-wide eradication of an invasive fish is probably unattainable, a long-term commitment is required to maintain suppression actions (and funding to support them) to ensure the invasive population does not rebound [26].

The natural variation in abiotic conditions and complexity of biotic interactions within large aquatic ecosystems make outcomes of conservation actions to suppress invasive fish uncertain. These uncertainties may be accounted for, however, if an active adaptive management strategy is adopted [27,28]. In taking this approach, conservation actions are treated as deliberate, large-scale experimental manipulations and the results of these actions increase knowledge about the system and decrease uncertainty in management outcomes. Alternative approaches are incorporated into the monitoring design and evaluated as experimental treatments with expectations (hypotheses) in outcomes [29–31]. Future management decisions are adjusted based on new knowledge about the resource being managed [32]. Because introduced apex predatory fish directly result in loss of prey fish and indirectly force altered, cascading interactions throughout food webs [33], their removal is predicted to allow recovery of the prey species and a return of food-web interactions and other ecosystem services to their natural state [34]. An active adaptive management approach allows for incorporation of the response uncertainties within these complex aquatic ecosystems.

Further adding to the complexities of predicting ecological responses to invasive fish removal are anthropogenic alterations within watersheds. Agriculture, cattle grazing, mining, power generation, timber harvest, and urbanization are common disturbances in the United States Intermountain West. Multiple, interacting invaders may also occur and contribute to altered ecological interactions and the complexity of responses to management actions [35,36]. Understanding the effects of predatory fish introduction and assessing outcomes that are specifically driven by suppression actions are challenging because of these concurrent, confounding factors. The majority of these challenges are minimal within large federally-protected wildlands in the United States, including national parks and wilderness areas, where habitats are strictly preserved to support fish life history, diversity, population persistence, intact food webs, and natural ecological function. Ecological recovery of populations is more likely in areas with relatively little anthropogenic disturbance and few other invaders [36] than where confounding anthropogenic factors exist. Studies assessing the long-term benefits of invasive, predatory fish suppression in protected natural areas may therefore be more informative than those in more complex, anthropogenically confounded systems elsewhere.

2. Study Area and Focal Species

Yellowstone Lake is a large aquatic system on the Yellowstone Plateau (2357 m in elevation) with a highly protected watershed (>3200 km^2) located within Yellowstone National Park and the Bridger-Teton Wilderness of Wyoming (Figure 1, Video S1). As such, invasive fish are the only large-scale impact *sensu* [37] on the lake; its waters remain physically and chemically pristine. Yellowstone Lake is the largest alpine (above 2000 m) lake in North America and has a surface area of 34,000 ha, 239 km of shoreline, mean depth of 48 m, maximum depth of 137 m, and volume of 1.5×10^{10} m^3 [38,39]. Powerboat access is limited to only two locations, at Bridge Bay and Grant marinas, and most of the shoreline lies in protected (federally proposed) wilderness. Thermal structure of the lake is typically unstable with a weak and variable thermocline at a depth of 12–15 m during July to September (Figure A1). Surface water temperatures rarely exceed 18 °C [40,41]. Specific conductance is typically <100 μS/cm [42]. The lake freezes over by late December and can remain frozen until late May (Figure 2) or early June. In winter, ice about 1 m thick under deep (>1 m) snow covers much of the lake except where shallow water covers active hot springs. Roads are not cleared of snow, and access to the lake during winter is restricted to over-snow vehicles. These logistical constraints and safety risks preclude work during winter months.

The fish assemblage in Yellowstone Lake includes only two natives, Yellowstone cutthroat trout (see Table A1 for all scientific names) and the less abundant minnow, longnose dace. Ancestral Yellowstone cutthroat trout, hereafter cutthroat trout, are thought to have accessed the upper Yellowstone River and Yellowstone Lake from the upper Snake River via natural connections across the Continental Divide [43,44] following glacial recession about 14,000 years ago [45]. Cutthroat trout then evolved as the sole salmonid and dominant fish within the lake and its connected river network. During spring (May–July), cutthroat trout spawn in up to 68 tributaries around Yellowstone Lake, move downstream to spawn in the Yellowstone River below Fishing Bridge, or make long distance spawning migrations upstream into the remote headwaters of the upper Yellowstone River [46]. Because of their lacustrine-adfluvial life history strategy [40,47,48], they transport lake-derived nutrients into numerous tributary streams [49] where they are important prey for grizzly bears [50], black bears [51], and numerous avian predators [52,53]. Although some cutthroat trout fry may remain in the natal stream for 1–2 years, most move into Yellowstone Lake within several months of hatching. In the lake, juvenile cutthroat trout are pelagic and feed on zooplankton [54]. Adults occupy the epilimnion at depths < 20 m but are most frequently found in the littoral zone where they feed on benthic macroinvertebrates and zooplankton. Because cutthroat trout are commonly found in shallow waters of Yellowstone Lake, they are preferred prey of river otters [55], osprey, bald eagles, and several colonial waterbirds [41,56–58].

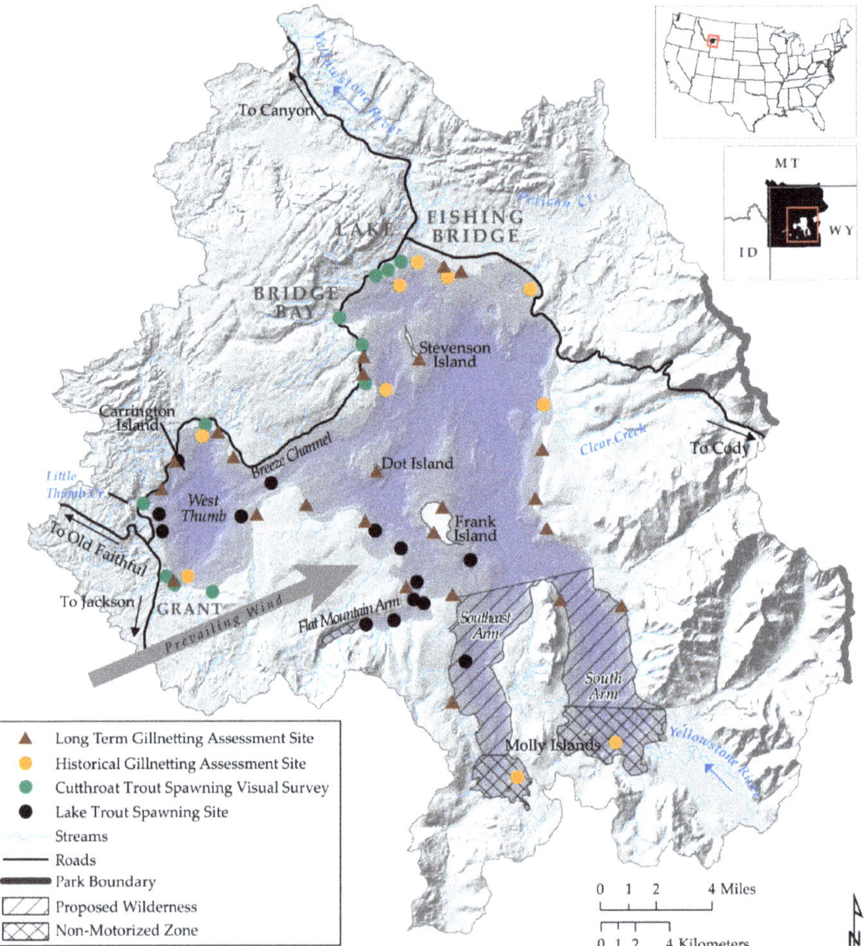

Figure 1. Yellowstone Lake within Yellowstone National Park in Northwestern Wyoming, USA, indicating locations of long-term gillnetting assessment sites for annual lake-wide monitoring of cutthroat trout and lake trout, historical gillnetting assessment sites that were sampled for cutthroat trout (prior to 2010), tributaries visually surveyed for spawning cutthroat trout each spring, and verified lake trout spawning sites [59]. Although 14 lake trout spawning sites are known in Yellowstone Lake, others probably exist [60]. Prevailing southwest (247°) winds and lake fetch during the autumn spawning period may preclude successful spawning along the eastern shore [61,62].

Along with their ecological importance, cutthroat trout are also historically significant. Early explorers of the Yellowstone Lake area noted their unique beauty and abundance. Soon after the establishment of Yellowstone National Park in 1872, cutthroat trout played an important role locally for subsistence, and nationally for recreation, as anglers were drawn from the Eastern United States for the angling experience [63,64]. Initial activities of the newly formed U.S. Commission on Fish and Fisheries focused on Yellowstone Lake. With the development of methods to propagate and move fish species, 310 million cutthroat trout eggs were shipped during 1901–1956 across the United States and elsewhere [65]. Other non-native fishes were introduced to Yellowstone Lake, presumably by anglers, including lake chub, longnose sucker, and redside shiner [65]. Although rarely studied, these fishes were new additions to the food web and likely altered the aquatic and terrestrial ecosystems by feeding on plankton and macroinvertebrates [54,66] and serving as prey for piscivorous birds [56,67] and mammals [55]. There was no evidence these fishes negatively affected the native cutthroat trout [40,68,69]. Yellowstone National Park prohibited stocking non-native fish into park waters as early as 1936 [70].

By the 1950s, following half a century of liberal angler harvest and egg collections by the U.S. Bureau of Sport Fisheries, abundance of the cutthroat trout population of Yellowstone Lake was declining and showing imminent signs of collapse [40]. Numbers of cutthroat trout migrating into tributary spawning streams were declining, and lakewide angler catch rates were low. A paradigm shift in National Park Service (NPS) management then occurred to one with an ecological basis [71] resulting in a redefinition of the role of cutthroat trout in Yellowstone Lake. Following restrictions on angler harvest and closure of the egg-collection operations, the population rebounded in the 1960s and 1970s and became so abundant that >70,000 cutthroat trout were counted spawning in a single spawning tributary (Clear Creek; Figure 1) during the spring of 1979 [72]. Biologists estimated 3.5 million (95% CI: 1.9–11.2 million) cutthroat trout (>350 mm total length) inhabited Yellowstone Lake at that time, and the consumers of these fish, such as bears, otters, ospreys, and bald eagles, were numerous near the lake. The ecosystem reflected its natural, pre-Euro-American condition [71].

Figure 2. Yellowstone Lake in Yellowstone National Park on 26 May 2019. Yellowstone Lake is the largest alpine (above 2000 m) lake in North America and has a surface area of 34,000 ha, 239 km of shoreline, mean depth of 48 m, maximum depth of 137 m, and volume of 1.5×10^{10} m^3 [38,39].

3. Predatory Fish Invasion and Initial Management Response

The perception of Yellowstone Lake as a secure refuge for cutthroat trout changed abruptly on 30 July 1994, when a nonnative lake trout was caught from the lake by an angler on a guided fishing trip [73]. Additional lake trout were caught soon afterwards causing grave concern, because their potential to negatively affect native trout had previously been well-documented in other large lakes in the Western United States (e.g., Lake Tahoe) [6,7]. An NPS press release dated 11 August 1994, described the discovery of lake trout in Yellowstone Lake, outlined ecological consequences that could result from establishment of this highly piscivorous, invasive fish species, and offered a US$10,000 reward for information leading to the arrest and conviction of the person(s) responsible for illegally stocking the fish. The NPS immediately implemented a must-kill regulation to prevent angler-caught lake trout from being returned to Yellowstone Lake alive. An illegal stocking of lake trout was assumed because natural movement into Yellowstone Lake from waters of the upper Snake River (in which they had previously been established) was not thought possible [74]. Regardless of the mode of introduction, lake trout were present and were already well on their way to establishing themselves as a new apex predator in Yellowstone Lake.

The native range of lake trout in North America includes Alaska, Canada, the Great Lakes, and parts of New England [75]. In their native range, lake trout fill an important ecological niche as an apex predator in food webs of lakes [76] and support valuable fisheries [77]. Lake trout are a deep-water dwelling, cold-adapted (<10 °C) predatory species that do not serve as an ecological substitute for cutthroat trout in Yellowstone Lake. Lake trout spawn within the lake and do not use tributary streams, making them inaccessible to native piscivorous avian and terrestrial wildlife. Additionally, lake trout can be extremely long-lived (30+ years, if unexploited) [78], grow longer than any other charr, can weigh more than 27 kg [44], and are capable of capturing prey at least half their body length (Figure 3) [79]. Lake trout in Yellowstone Lake mature at an earlier age (males age 4) than other populations in the Western U.S., probably because of their fast growth rates [7]. Fecundity is high, with a 5-kg female capable of producing 6000–8000 eggs in a single spawning event [79,80]. In addition, the early life history (pre-recruit) survival of lake trout in Yellowstone Lake is estimated to be 4–6 times greater than in their native range (discussed below) [81].

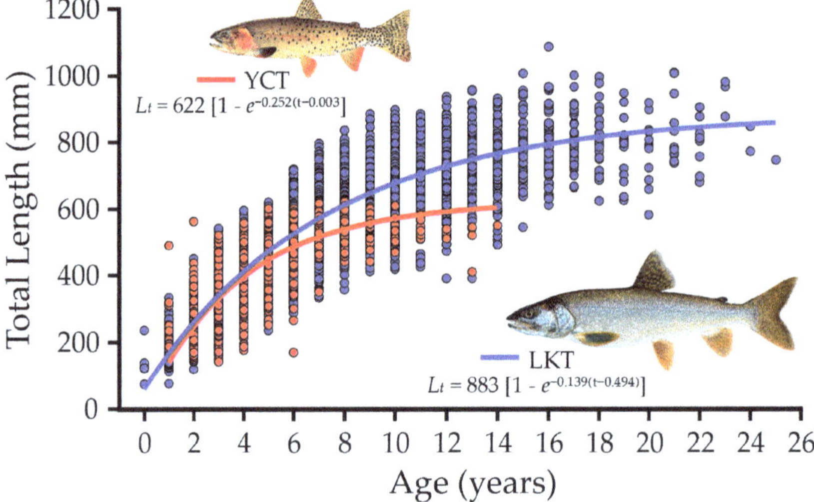

Figure 3. Total lengths-at-age (L_t) of native cutthroat trout (YCT; n = 1350) and invasive lake trout (LKT; n = 6387) collected by gillnetting on Yellowstone Lake during 1998 to 2019, with von Bertalanffy growth functions. Asymptotic mean lengths of cutthroat trout and lake trout were 622 mm (95% CI: 600–648) and 883 mm (95% CI: 869–898), respectively. Cutthroat trout with a maximum age of 14 years had a mean predicted length of 604 mm and lake trout with a maximum age of 25 years had a mean predicted length of 854 mm

The NPS convened a panel of experts from throughout the United States and Canada to assess the consequences of lake trout presence in Yellowstone Lake in 1995 [82]. At that time, in the absence of knowledge of the behavior and habits of lake trout in Yellowstone Lake, the primary recommendations of the science panel were to develop a program for limiting their expansion coupled with careful monitoring and application of adaptive management strategies [83]. The 1995 panel concluded that, despite a high level of uncertainty, the probability of eliminating lake trout was low and that the introduced predator would reduce the cutthroat trout population in Yellowstone Lake (Figure 4). At the same time, the group suggested that lake trout abundance could, with a high degree of probability, be limited by initiating an aggressive control program using gillnetting. Because complete eradication of lake trout was unlikely, a long-term commitment would be required to control lake trout abundance. It was agreed that the cutthroat trout population would decline even if lake trout could be suppressed, but a lake trout suppression program could reduce the expected loss of cutthroat trout by 50% or more. Most of the information needed to increase the effectiveness of initial control measures could be obtained from the control program itself, but some modification of the existing mid-September cutthroat trout gillnetting assessment program would be required to also evaluate changes in the lake trout population [83].

Figure 4. Twelve cutthroat trout from the stomach of a lake trout (approximately 3 kg) gillnetted immediately following ice-off from Yellowstone Lake in May 2017. During 8 months of the year (mid-October through mid-June) there is no thermal cause for separation of lake trout from cutthroat trout. Predation pressure by lake trout under the ice, which has never been studied, is likely high.

Lake trout suppression began in 1995 primarily by control (targeted) gillnetting at depths > 20 m to avoid cutthroat trout, and secondarily by experimental (exploratory) gillnetting designed to gain information on lake trout distribution seasonally throughout the lake. By 1999, two key spawning areas in the West Thumb of Yellowstone Lake had been identified by telemetry—Carrington Island and Solution Hump—where high concentrations of lake trout occurred during autumn (Figure 1). By 1999, nearly 15,000 lake trout were killed by gillnetting, with progressively more netted each year despite limited NPS resources (Figure A3). Retrospective population modeling estimated that 10,000 or more lake trout were probably in Yellowstone Lake in 1994 when they were first discovered [79]. Several years of bioenergetics research provided estimates of lake trout predation on the cutthroat trout population [79,84], and it became apparent that a dedicated program was required to curtail further lake trout population growth. However, enough funding to support such an effort had not been obtained.

Yellowstone National Park biologists then successfully competed for and were awarded an NPS Natural Resource Management grant in 1999 specifically for the development of a lake trout suppression program in 1999–2001. The supplemental funding allowed hiring a seasonal NPS crew dedicated to suppressing lake trout. With these new resources and redirected park funds, the park developed a comprehensive program for suppression of lake trout that consisted of intensive control gillnetting, gillnetting for annual monitoring, and focused gill netting of spawning lake trout. This program resulted in the removal of more than 340,000 lake trout by 2008. However, as the enhanced suppression program continued, gillnet catch rates continued to increase, suggesting that lake trout population abundance was continuing to increase and the population was expanding spatially across Yellowstone Lake. In addition, long-term monitoring of the cutthroat trout population suggested a concurrent decline to levels lower than ever previously recorded [85]. After more than a decade of sustained lake trout gillnetting and no evidence of cutthroat trout recovery, a comprehensive scientific appraisal of the on-going program was warranted.

4. Development of a Conservation Strategy that Embraces Uncertainty

The lake trout-induced collapse of cutthroat trout became increasingly apparent to the public during the 2000s as angler catch rates declined severely on Yellowstone Lake and the Yellowstone River both upstream and downstream of the lake (Figure 1) [46,51]. The loss was particularly noteworthy because cutthroat trout were the only sport fish available to anglers in this ecosystem. The decline also affected backcountry outfitters and anglers seeking the migratory, spawning fish in distant reaches of the upper Yellowstone River south of Yellowstone Lake within the Bridger-Teton Wilderness of Wyoming [46]. In addition to the precipitous decline of cutthroat trout, the novel piscivore altered plankton assemblages within the lake and reduced nutrient transport to tributary streams [49,86]. Effects extended across the aquatic-terrestrial ecosystem boundary as grizzly bears and black bears in the area necessarily sought alternative foods [41,87]. Nest density and success of ospreys greatly declined, and bald eagles shifted their diet to compensate for the loss of cutthroat trout [41,52]. An urgent need for action to reverse these declining trends was recognized by resource managers, scientists, and a wide range of constituents of Yellowstone Lake fishery and wildlife resources.

4.1. Scientific Review Panel

Yellowstone National Park requested assistance from an independent scientific review panel in August 2008 to critically evaluate the effectiveness of the lake trout suppression program in Yellowstone Lake [88]. The panel was tasked with evaluating the effectiveness of the lake trout suppression program, reviewing emerging technological opportunities for suppressing lake trout, and providing alternatives for the future direction of the program in the context of the primary mission of the NPS, which is to preserve unimpaired the natural and cultural resources and values of the National Park System for the enjoyment, education, and inspiration of this and future generations. To that end, the panel sought to ensure the long-term persistence of native cutthroat trout and the natural function of the Yellowstone Lake ecosystem.

The panel concluded that suppression gillnetting during 1995 to 2008 had not curtailed lake trout population growth, and that the cutthroat trout population had declined severely; however, the cutthroat trout population was not completely lost and the ecosystem could be restored with immediate, aggressive action. Because overharvest had caused collapse of lake trout populations throughout their native range, the panel thought that intensified suppression gillnetting could drive the lake trout population of Yellowstone Lake into decline. Although specific gillnetting effort benchmarks that would result in lake trout decline could not be determined, the panel recommended a doubling of the 28,000 units of annual effort (unit of effort = 100-m net nights) expended at that time. Because an immediate increase in gillnetting effort of that magnitude was beyond the capacity of NPS resources, the panel recommended incorporation of private sector (commercial, professional) gillnetters, an approach which was successful for lake trout suppression on Lake Pend Oreille in

Northern Idaho [89]. The panel also recommended reinitiating lake trout telemetry studies to determine movement patterns, locate spawning habitats, and inform the suppression gillnetting. Additionally, development of novel suppression alternatives to gillnetting and experimentation to assess their effectiveness was also supported by the panel. Specific recommendations of the 2008 scientific review panel, 15 years after initial lake trout discovery, were to:

1. Intensify existing lake trout suppression efforts for a minimum of 6 years.
2. Maintain and enhance cutthroat trout monitoring.
3. Develop a statistically robust lake trout monitoring program.
4. Develop a lake trout suppression plan with benchmarks for control to increase program effectiveness and ensure the conservation of the Yellowstone Lake ecosystem through the coming decades.

4.2. Planning and Environmental Compliance

These major actions to suppress lake trout and restore the Yellowstone Lake ecosystem required increased funding, heightened support by partners and stakeholders, detailed long-term planning, and National Environmental Policy Act (NEPA) compliance. The recommendations of the scientific review panel provided a well-defined need for fund-raising to support the increased efforts. The park supported incorporation of private sector gillnetters during a limited, pilot phase during 2009 to 2010 to determine their feasibility for operations on Yellowstone Lake. Hickey Brothers Research, LLC, a company from Door County, Wisconsin, with extensive commercial fishing experience on Lake Michigan, was awarded the NPS contract to initiate the expanded gillnetting. Combined with the continued NPS operations, the enhanced gillnetting effort resulted in eradication of nearly a quarter million lake trout during 2009 to 2010. Yellowstone National Park concurrently completed an environmental compliance process evaluating potential effects on park resources by an increased suppression program. Input from the public was sought on alternative management actions that would ensure the long-term recovery of cutthroat trout and restoration of natural ecosystem function. A Native Fish Conservation Plan/Environmental Assessment (EA) was made available for review and comment on 16 December 2010 [90]. Development of the plan included scientific review of current conservation efforts, projected changes in native fish status given known threats, a review of relevant emerging science and technology, and public and stakeholder input received during a public scoping process. The plan that emerged identified the following goals:

- Reduction in the long-term extinction risk for native fishes;
- Restoration and maintenance of the important ecological role of native fishes; and
- Creation of sustainable native fish angling and viewing opportunities for the public.

The plan proposed to conserve native fish from threats of lake trout and other nonnative species, disease, and climate-induced environmental change. It provided guidance for managing fisheries and aquatic resources over the following two decades. The plan described in detail the development of an adaptive management strategy (Figure 5) for implementing large-scale removal of lake trout on Yellowstone Lake by NPS netting crews and incorporation of private sector, contract netters, and called for the development and implementation of robust monitoring and continued scientific review through collaboration with partners. Assumptions made in the selection of reasonable alternatives for the Yellowstone Lake ecosystem included:

- A direct relationship exists between lake trout gillnetting effort and the number of lake trout captured.
- Cutthroat trout will recover naturally without large-scale supplementation (stocking) once lake trout are functionally removed.
- A significant lag (in time) in the recovery of cutthroat trout will occur following lake trout suppression (cutthroat trout recovery may not be immediately perceptible).
- Large-scale supplementation (stocking) of cutthroat trout from existing hatchery broods would be detrimental because it would reduce the genetic integrity of the population.

Table 1. Desired conditions, conservation actions, quantitative responses, and performance metrics for cutthroat trout (YCT) and lake trout (LKT) in the Yellowstone Lake ecosystem. Column labels A-D correspond to elements of the adaptive management strategy conceptual model (Figure 5).

Desired Conditions	Conservation Actions	Quantitative Responses	Performance Metrics
Primary • Restore YCT to pre-LKT abundances. • Free YCT of all stress by LKT.	• Gillnetting by NPS crews. • Private sector, contract gillnetting. • Incorporate new and emerging technologies.	• YCT long-term gillnetting assessment CPUE >40. • An average of >60 spawning YCT during visual surveys of 11 front country tributaries. • Restored angler YCT catch to at least 2.0 per hour. (1980s levels)	• Reduce LKT population growth rate (λ) to 0.75. • LKT long-term gillnetting assessment CPUE <0.01. • Reduce angler LKT catch per hour to <0.05.
	• Reconnect spawning tributary surface waters to the lake.		• Each year 75% of tributaries maintain a surface water connection with the lake.
Secondary • Restore YCT to abundance during early stages of LKT invasion. • Significantly reduce LKT stress on YCT.	• Gillnetting by NPS crews. • Private sector, contract gillnetting. • Incorporate new and emerging technologies.	• YCT long-term gillnetting assessment CPUE >26. • An average of >40 spawning YCT during visual surveys of 11 front country tributaries. • Restored angler YCT catch to at least 1.5 per hour. (1990s levels)	• Reduce LKT population growth rate (λ) to 0.85. • LKT long-term gillnetting assessment CPUE <0.1. • Reduce angler LKT catch per hour to <0.05.
	• Reconnect spawning tributary surface waters to the lake.		• Each year 75% of tributaries maintain a surface water connection with the lake.
Tertiary • Restore YCT to abundances during later stages of LKT invasion. • Moderately reduce LKT stress on YCT.	• Gillnetting by NPS crews. • Private sector, contract gillnetting. • Incorporate new and emerging • technologies.	• YCT long-term gillnetting assessment CPUE >12. • An average of >20 spawning YCT during visual surveys of 11 front country tributaries. • Restored angler YCT catch to at least 1.0 per hour. (2000s levels)	• Reduce LKT population growth rate (λ) to 0.95. • LKT in long-term gillnetting assessment is <0.5.
	• Reconnect spawning tributary surface waters to the lake.		• Each year 75% of tributaries maintain a surface water connection with the lake.
	• Reintroduce YCT to tributaries lacking use by spawners.		• Establish self-sustaining YCT spawning populations in tributaries.

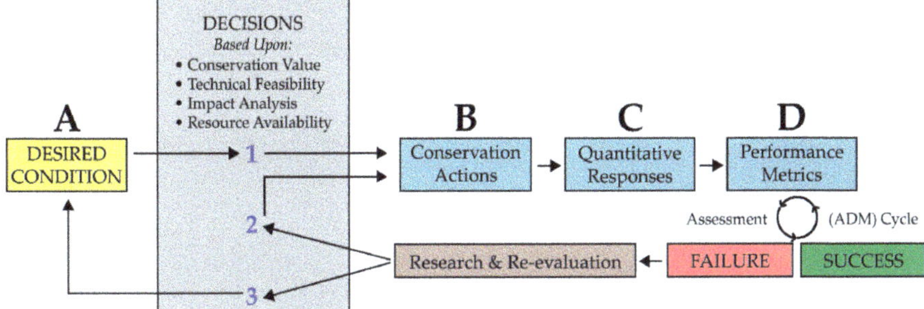

Figure 5. Adaptive management (ADM) strategy conceptual model for Yellowstone Lake ecosystem restoration. Desired condition (**A**), conservation actions (**B**), quantitative responses (**C**), and performance metrics (**D**) are defined in the text and listed in Table 1 [90].

Because > 100,000 lake trout were being killed annually by 2010, the plan included an alternative analysis of the marketing and sale (or the donation) of gillnetted lake trout for human consumption. The thought was that proceeds from the sale of lake trout could be used to supplement the increased funding needed for the program. However, a large portion of the catch was not suitable for human consumption because the soak times of suppression gillnets were often 7 nights. In addition, handling time and care, including holding fish on ice and transport, would greatly reduce time available for suppression gillnetting. Gillnetting effort would decrease by about an estimated 50% if lake trout were processed for human consumption, requiring a doubling of the number of boats and crews to maintain the same gillnetting suppression effort. Markets and food banks were far from Yellowstone Lake, resulting in high shipment costs. Moreover, the enabling legislation for Yellowstone National Park does not allow for the sale of its natural resources. Sale or donation of lake trout was rejected because of these significant issues. The lake trout carcasses were to instead be deposited in deep (>65 m) regions of the lake to retain their nutrients within the lake ecosystem.

4.3. Conceptual Ecosystem Model and Hypothesized Linkages

A conceptual model (Table A3) was developed to assist in identifying issues confronting the Yellowstone Lake ecosystem and to clarify which aspects of the ecosystem would likely respond as a result of management actions. The conceptual model illustrated the complex relations among agents of change, stressors on native fish, and ecosystem responses. Agents of change were sources of stressors on native fish when they operated outside the range of natural variability; they included natural processes and events as well as human activities. Ecosystem responses were defined as measurable and detectable changes or trends in the quality or integrity of ecosystem structure, function, or processes.

Agents of change appropriate to the Yellowstone Lake ecosystem were organized into five broad categories (Table A3):

- regional physical/chemical forces;
- biological introductions;
- angling;
- park infrastructure and operations; and
- local physical/chemical forces.

The degree to which each agent of change contributed to a problem was considered and a list of potential stressors on native fish was compiled. Each stressor was matched to general management issues within the 2006 NPS Management Policies [91]. The preliminary list of ecosystem responses was also grouped into five broad categories (Table A3):

- biogeochemical cycling;
- productivity/biomass change;
- fish functional role;
- fish life history strategy; and
- avian/terrestrial fish consumers.

For example, the lake trout invasion was an agent of change within the Yellowstone Lake ecosystem that resulted in stressors on native fish in the form of fewer cutthroat trout recruited to the spawning population, direct mortality, predation losses, and loss due to competition/displacement (Table A3). These stressors can also result in ecosystem responses such as changes in nutrient transport, primary and secondary production, fish functional roles and life history strategies, and impacts on avian and terrestrial fish consumers. The conceptual model was not intended to represent a comprehensive account of the entire ecosystem but rather was a framework implicating known or hypothesized agents of change that stress native fish and result in negative ecosystem responses. The goal of the model was to illustrate relationships between and among agents of change and key ecosystem processes and variables. It served to demonstrate the complexity of interactions within the Yellowstone Lake ecosystem, many of which are unknown. Multiple agents of change can lead to multiple stressors, resulting in multiple ecosystem responses.

The underlying hypotheses for the preferred alternative of the Native Fish Conservation Plan were that a gillnetting-driven reduction in lake trout would result in cutthroat trout recovery, and this recovery would, in turn, result in positive responses by piscivorous wildlife (Table A3). In addition, the cascading changes within the lake that followed the cutthroat trout decline (e.g., shifts in zooplankton, phytoplankton, and nutrient transport) would revert to pre-lake trout conditions.

4.4. Desired Conditions

Primary, secondary, and tertiary desired conditions for the Yellowstone Lake ecosystem were described in the Native Fish Conservation Plan (Table 1) [90]. Complete eradication of lake trout, the most significant agent of change, was the primary desired condition for Yellowstone Lake. However, the secondary condition was initially set as the management target because available lake trout suppression methods were incapable of achieving the primary desired condition. The tertiary condition would become the management target if implementation of conservation actions did not achieve the secondary desired condition. Failure to achieve at least the tertiary condition would be considered a failure to meet the objectives of the plan. Although cutthroat trout are expected to naturally recover following lake trout decline, conservation actions of all desired conditions included ensuring spawning tributary connectivity to Yellowstone Lake during drought years [51,90,92] and reintroduction (stocking) of cutthroat trout to tributaries lacking use by spawners if deemed necessary to maintain the tertiary desired condition.

4.4.1. Primary Desired Condition

The primary desired condition was characterized by cutthroat trout restored to pre-lake trout abundances, and free from all stress by lake trout. This condition would be achieved by a 100% eradication of lake trout or a suppression of lake trout to the point where the species had no measurable impact on the ecology of Yellowstone Lake. Quantitative responses to characterize this condition would include full recovery of cutthroat trout abundance to the averages observed during the five years prior to lake trout discovery (1987–1991; 40 per 100-m net night during long-term gillnetting assessments (relative abundance monitoring, see below); 60 observed during visual spawning surveys; angler catch rate of 2.0 per hour; Table 1). Performance metrics were a lake trout population growth rate (λ) ≤ 0.75; catch-per-unit-effort (CPUE) = 0.01 per 100-m net night during long-term gillnetting assessments; and angler catch rate < 0.05 per hour. Lake trout abundance would be extremely low and difficult to detect in this condition.

4.4.2. Secondary Desired Condition

The secondary desired condition would be characterized by restoration of cutthroat trout to abundances present during the early stages of lake trout invasion, indicating significantly reduced lake trout stress on cutthroat trout. This condition would be achieved by significantly reducing lake trout abundance in Yellowstone Lake. Quantitative responses to characterize this condition would include recovery of cutthroat trout abundance to the averages observed during the five years following lake trout discovery (1995–1999; 26 per 100-m net night during long-term gillnetting assessments; 40 observed during visual spawning surveys; angler catch rate of 1.5 per hour; Table 1). Performance metrics were a lake trout population growth rate (λ) $\leq$ 0.85; CPUE = 0.1 per 100-m net night during annual long-term gillnetting assessments; and angler catch rate < 0.1 per hour.

4.4.3. Tertiary Desired Condition

The tertiary desired condition would be characterized by cutthroat trout restored to abundances during the later stages of lake trout invasion, indicating moderately reduced lake trout stress on cutthroat trout. This condition would be achieved by slightly reducing lake trout abundance in Yellowstone Lake. Quantitative responses to characterize this condition would include maintaining cutthroat trout abundance at the average observed prior to lake-wide expansion by lake trout (2001–2005; 12 per 100-m net night during long-term gillnetting assessments; 20 observed during visual spawning surveys; angler catch rate of 1.0 per hour; Table 1). Performance metrics were a lake trout population growth rate (λ) $\leq$ 0.95; CPUE = 0.5 per 100-m net night during annual long-term gillnetting assessments; and angler catch rate < 0.5 per hour.

5. Stakeholder Involvement and Fundraising to Support Conservation Actions

Following completion of the Native Fish Conservation Plan/EA [90], a Finding of No Significant Impact (FONSI) was signed by the NPS Intermountain Region Director in June 2011. That year, Yellowstone National Park entered into a 5-year contract with Hickey Brothers Research, LLC, to increase lake trout gillnetting suppression effort lakewide. In addition, research to improve suppression efficiency began in earnest as adult lake trout were surgically implanted with acoustic tags to determine broad scale movement patterns and locate key spawning sites. A plan with clearly articulated objectives and benchmarks for Yellowstone Lake ecosystem restoration and feedback from annual scientific panel reviews provided strength for acquiring funding to support the program. Missing however, was a process to better incorporate anglers, conservation groups, and the general public in on-the-ground actions to conserve native fish as described in the plan.

5.1. Yellowstone Fly Fishing Volunteer Program

The *Yellowstone Fly Fishing Volunteer Program* was initiated in 2002 to acquire information about fish populations throughout the park without requiring Yellowstone biologists to travel to sample the populations themselves using electrofishing or other sophisticated gear [93]. The volunteers fly-fished to gather and archive information and biological samples that park biologists would otherwise not be able to collect. In addition to providing valuable data, samples, and assistance to the fisheries program, volunteer fly fishers have played an important role with the public by interacting positively with park biologists and the public and demonstrating their passion for native fish and the importance of protecting these species. Volunteer fly fishers have promoted an understanding of the Yellowstone Lake ecosystem restoration program and generated greater awareness of the current issues facing Yellowstone's native fish. These passionate and informed supporters have been an important contribution to the success of our program.

5.2. Yellowstone Lake Workgroup

A consortium of conservation groups met with NPS officials in 2011 with the intent of becoming partners in addressing the threats to the Yellowstone Lake fishery and ecosystem. From that meeting and subsequent discussions, a Memorandum of Understanding (MOU) was developed that formalized a cooperative relationship among participants to ensure the ecosystem was protected, maintained, and managed to achieve established goals. Signatories to the MOU were Trout Unlimited National, Wyoming Council, Montana Council, and Idaho Council; National Parks Conservation Association; Greater Yellowstone Coalition; Yellowstone Park Foundation (subsequently named Yellowstone Forever); and Yellowstone National Park. These stakeholders began meeting semi-annually and created a formal *Yellowstone Lake Workgroup* that acted as a sounding board to review lake trout suppression activities, population monitoring activities and trends, telemetry research results, new suppression technologies, and other fisheries-related science. They also initiated positive public outreach and education and authored publications directed at the general public and potential donors, including a publication with answers to frequently asked questions about the science supporting management of Yellowstone Lake [94]. The group has responded to public concerns about fish conservation actions when applicable. The *Yellowstone Lake Workgroup* has been actively involved in fundraising and has raised over US$1 million to directly support cutthroat trout restoration in Yellowstone Lake. The majority of the funds have been spent on (1) telemetry studies to determine lake trout seasonal movement patterns and location of spawning areas, (2) studies assessing the reproductive potential, cycles, and timing of lake trout spawning, and (3) studies to identify and optimize alternative suppression technologies aimed at lake trout embryos. In addition, members of the *Yellowstone Lake Workgroup* have provided volunteer labor to support the work on Yellowstone Lake.

5.3. Yellowstone Forever Fund-Raising Partnership

As the need for lake trout suppression effort increased, so did the need for increased funding to support it. Funding for the Yellowstone Lake ecosystem restoration was ≤ US$500,000 until 2009, when it was increased to support expanded suppression effort by contracted gillnetters. *Yellowstone Forever*, which was the official fund-raising partner of Yellowstone National Park, approved a grant in 2012 to initiate strong, annual support of the program. This support, funded by private donations, increased total program funding (donated and NPS) from US$1,000,000 in 2011 to more than US$2,000,000 by 2013 (Figure 6A). These funds allowed a rapid increase in suppression gillnetting effort by the contracted crews to meet (and surpass) annual gillnetting benchmarks, while concurrently enhancing long-term monitoring, population modeling, and applied research to improve program efficiency.

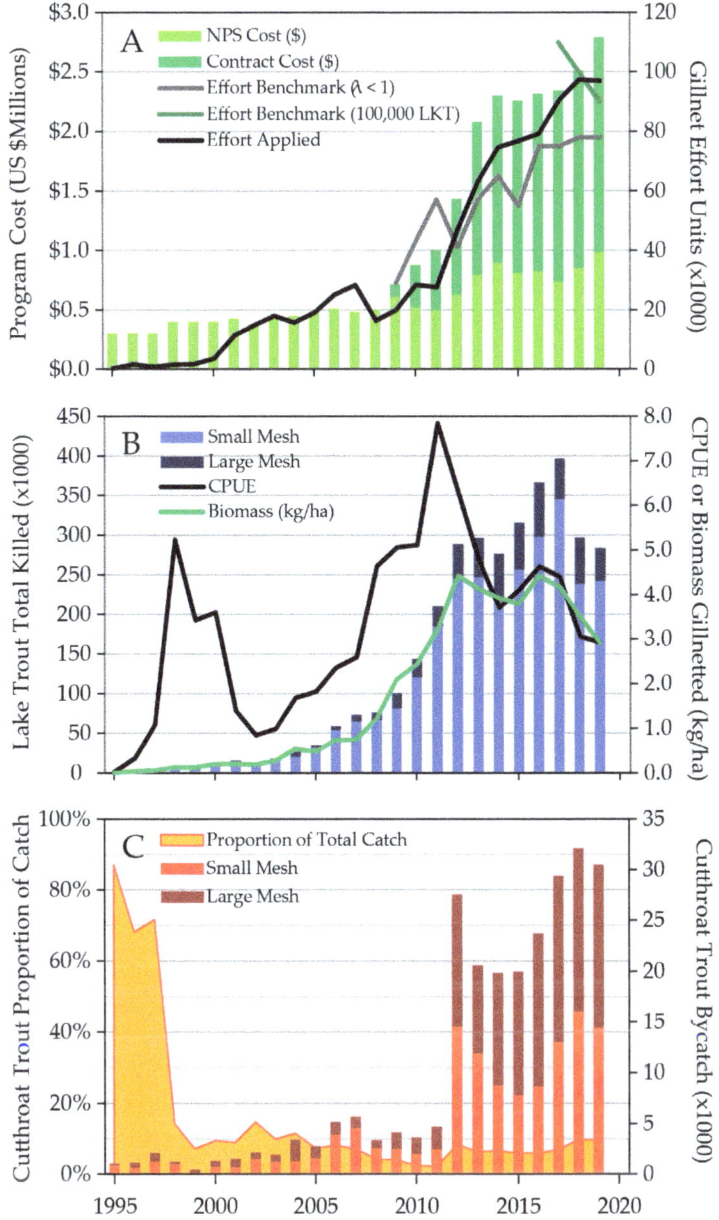

Figure 6. Costs, effort, and catch of the Yellowstone Lake ecosystem restoration program, 1995–2019, including (**A**) the total program cost (US $Millions) of National Park Service (NPS) and contracted operations, gillnetting effort unit (100-m net nights) benchmarks estimated by statistical-catch-at-age modeling to achieve λ < 1 (2009–2019) or to reduce lake trout (LKT) total abundance to 100,000 fish (2017–2019), and the actual total annual gillnetting effort applied, (**B**) total numbers of lake trout killed by small and large mesh gillnets, catch-per-unit-effort (CPUE) by all mesh sizes combined, and total biomass of lake trout gillnetted, and (**C**) cutthroat trout proportion (%) of total gillnet catch and number of bycatch in small and large mesh gillnets.

6. Historical Development of the Gillnetting Program

Gillnets have remained the primary lake trout suppression gear on Yellowstone Lake throughout the first 25 years of the program because they were the preferred gear in commercial fisheries and could overharvest lake trout populations in their native range [83,95]. At first (1995–1999), lake trout distribution or abundance in Yellowstone Lake was poorly understood, crews were inexperienced in population suppression gillnetting, and boats lacked specializations for lifting and processing long gillnets. Summer (June–September) gillnet sets were short (< 200 m length) and mostly lifted by hand from relatively small (6–8 m) research and monitoring vessels. Despite these constraints, NPS crews annually processed an average of 1164 100-m net nights and killed an average of 2965 lake trout (14,823 total lake trout killed during 1995–1999, mean CPUE = 2.1 per 100-m net night, mean biomass caught = 0.07 kg/ha; Figure 6B) with nearly all of the effort focused in the West Thumb region, where the original discovery of lake trout occurred and catch rates were highest (Figure 7).

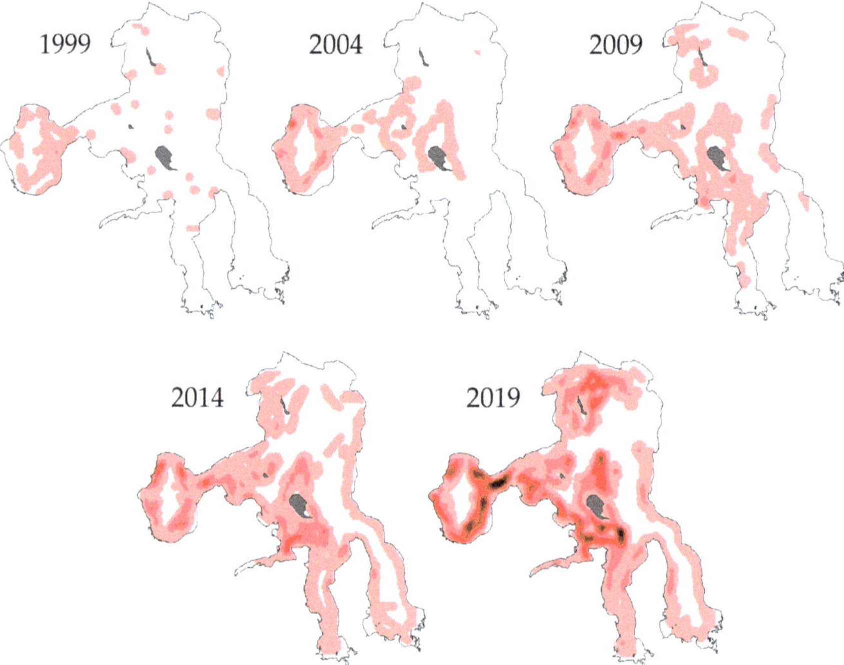

Figure 7. Spatial expansion and increase in intensity of suppression gillnetting effort on Yellowstone Lake during years of rapid lake trout population growth (1999, 2004, and 2009) and years where distribution of gillnets and overall effort was great enough to curtail population growth (i.e., $\lambda < 1$; 2014 and 2019). Darker shades of red indicate lake areas with greater intensity of gillnetting effort.

National Park Service crews increased gillnetting effort and total lake trout killed after acquisition of a large (10 m) enclosed gillnetting boat in 2001. Both the new boat and an existing boat (8 m) were outfitted with hydraulic net lifters, thereby greatly increasing their ability to retrieve long gangs of nets (550 m). Crews used sonar and detailed lake bathymetry [96] to locate concentrations of lake trout and extended gillnetting throughout the possible field season from late-May through mid-October in 2000–2004 to increase gillnetting effort more than 10-fold throughout the Breeze Channel and into the Main Basin of Yellowstone Lake (Figure 7). An average of 12,675 100-m net nights were processed during this period resulting in the kill of 17,072 lake trout annually (85,861 total lake trout killed during 2000–2004, mean CPUE = 1.7 lake trout per 100-m net night, mean biomass caught = 0.28 kg/ha; Figure 6B).

Gillnet effort and lake trout killed increased again during 2005–2009 as experienced NPS crews focused solely on suppression. Netting effort expanded across Yellowstone Lake (Figure 7) to follow the increasing lake trout population that, despite the 10+ years of control efforts, was experiencing exponential growth. A pilot study of contracted gillnet services initiated in 2009 brought a third specialized gillnetting vessel to the lake and increased gillnetting effort by 3 weeks in 2009. An average of 21,769 100-m net nights were processed and 68,761 lake trout were killed annually (343,807 total lake trout killed during 2005–2009, mean CPUE = 3.3 lake trout per 100-m net night; mean biomass caught = 1.05 kg/ha; Figure 6B).

The gillnetting effort was increased during 2010 to 2014 to a level that curtailed further lake trout population growth [81] as a result of (1) completion of the Native Fish Conservation Plan/EA, (2) estimation of gillnetting benchmarks by population modeling (see below), (3) growing support of stakeholders, (4) significant private donor funding support, and (5) guidance by the scientific review panel. Five large, specialized gillnetting vessels expanded effort and spatial distribution of gillnets across Yellowstone Lake (Figure 7). Contiguous gillnet lengths were increased to >3 km by the contract crews and were set in a serpentine pattern along bottom contours to maximize catch of lake trout, which swam parallel to net panels. The increased effort focused on large adult lake trout, especially during the autumn spawning period, while not reducing effort targeting smaller lake trout [97] (Figure 6B). The increased focus on adult lake trout along with telemetry of acoustic-tagged fish (see below) resulted in the discovery of several additional lake trout spawning sites that were subsequently targeted [59,98]. During this period of rapid effort expansion, an average of 48,073 100-m net nights were processed and 249,466 lake trout were killed annually (1,247,332 total lake trout killed during 2010–2014, mean CPUE = 5.6 per 100-m net night; mean biomass caught = 3.62 kg/ha; Figure 6B).

Large deepwater trap nets were set for extended periods (months) in fixed locations in 2010–2013 to complement gill netting and maximize capture of large adult lake trout (and minimize cutthroat trout bycatch) [42]. The trap nets also captured live lake trout for critical telemetry studies (see below). Although the trap nets caught nearly 33,000 additional adult lake trout during a total of 2810 net nights over four years, the fish were of the same size classes caught by large-mesh gillnets. Moreover, the trap nets, which were 9–15 m high, had complex leads > 300 m long, were held in place by heavy (50 kg) anchors, and required highly trained contracted crews to set and maintain. Use of the trap nets was discontinued after 4 years because the time and cost of their use were high relative to that required to achieve similar catches with gill nets.

The suppression program was expanded during 2015 to 2019 with the purchase of an additional NPS gillnetting vessel resulting in a total of six large specialized boats (Figure A4) and crews (NPS and contract) with substantial fishing experience (Video S2). Although all size classes of lake trout were targeted, effort continued to be focused on removal of large, adult lake trout. Net inventories were increased greatly to accommodate the increase in effort and included a broad range of mesh sizes to target the changing population and maximize catches throughout the suppression season. Twine (monofilament) diameters were reduced, which increased catch efficiencies. Gillnets were distributed across most of the lake that was <60 m deep, the depths that had proven to be most productive (Figure 7). Proportionally, effort continued to be highly focused on the West Thumb, Breeze Channel, and Main Basin regions near Frank Island where catches remained the highest (Figure 1). Experienced, professional gillnetting crews processed an average of 88,124 100-m net nights resulting in an annual average kill of 331,783 lake trout and a reduced CPUE (1,658,917 total lake trout killed during 2015 to 2019, mean CPUE = 3.8 lake trout per 100-m net night; mean biomass caught = 3.77 kg/ha; Figure 6B).

Bycatch of cutthroat trout has occurred throughout the gillnetting program. Both lake trout and cutthroat trout occupy shallow-water habitats during spring and autumn despite spatial segregation during summer stratification (late-July through mid-September) when lake trout seek cold, deep lake areas. Targeted gillnetting of lake trout while avoiding cutthroat trout bycatch has therefore been challenging during spring and autumn. Immediately following the discovery of lake trout during 1995–1997, 68%–87% of annual catches were cutthroat trout (Figure 6C). As lake trout abundances increased and crews developed better methods to target them, the bycatch of cutthroat trout declined to <15%. Annual bycatch increased to about 20,000–30,000 cutthroat trout beginning in 2012, concurrent with rapid expansion of gillnetting effort and cutthroat trout recovery. Bycatch makes up only about 6%–10% of the total catch but is an unfortunate consequence of the use of gill nets as the primary lake trout suppression tool. However, the number of cutthroat trout saved by the killing of 100,000s of lake trout annually far exceeds bycatch losses given that each lake trout consumes an estimated 41 cutthroat trout annually [79].

7. Lake Trout Population Modeling and Gillnetting Effort Benchmarks

The most robust approach for evaluating the success of the lake trout suppression program was a combination of long-term monitoring and population modeling [99–101]. We estimated lake trout abundances and mortality through time by integrating gillnetting effort, harvest data, and standardized monitoring data (long-term gillnetting assessments, described below) in a statistical catch-at-age (SCAA) assessment model [80,81,102]. The results of the SCAA model were used to forecast the amount of gillneting effort required to achieve a given level of mortality [103,104]. Our goal, established in 2010, was to reduce the abundance of lake trout to their mid-1990s levels (about 100,000 fish), when they probably had little effect on the native cutthroat trout population [90]. However, a major uncertainty in reaching that goal was the amount of gillnetting effort needed. Population modeling and analyses of lake trout suppression data collected over several years were used to address this question and to assess suppression program success. The three important metrics assessed were (1) total annual mortality, (2) population abundance, and (3) population growth rate (λ) of lake trout.

Annual gillnetting effort benchmarks were estimated iteratively using the SCAA model beginning in 2009, when 29,000 100-m net nights were predicted to be needed to reduce $\lambda < 1$ within 5 years (Figure 6A). This initial benchmark was estimated using age-0 and age-1 survival rates from the native range of lake trout [80]. Subsequent analyses of local data indicated that pre-recruit survival rates in Yellowstone Lake were much higher than in the native range and the model was adjusted accordingly. The benchmark increased to 57,000 units in 2011 and stabilized at 75,000 units from 2016 to 2019. Because annual lake trout harvests remained high, we estimated the effort required to reduce lake trout abundance to the goal of 100,000 fish identified in the Native Fish Conservation Plan. This target ranged from 90,000 to 110,000 100-m net nights between 2017 and 2019, respectively, which appeared to be converging with estimates of effort required to maintain $\lambda < 1$ (Figure 6A). Fortunately, the combined NPS and contracted suppression gillnetting crews were able to achieve > 90,000 effort units in those years.

8. Lake Trout Population Response to Suppression Gillnettting

Annual monitoring metrics indicated that suppression gillnetting successfully reduced lake trout abundances (ages 3+) and biomass. The catch rate of lake trout in annual long-term gillnetting assessments declined from 2011 to 2019 ($p = 0.032$), with a high of 4.9 per 100-m net night in 2014 and a low of 2.0 in 2018 (Table A4, Figure 8), but remained above established desired conditions (Table 1). Large adult lake trout (> 400 mm) consistently made up 13% to 29% of the catch. Proportions of the catch ≤ 280 mm (primarily age-2 individuals) were <50% in all years except 2019 when small lake trout made up 63% of the catch (Figure A5). Small, juvenile lake trout consistently represented a large proportion of the annual catch, suggesting that recruitment remained strong.

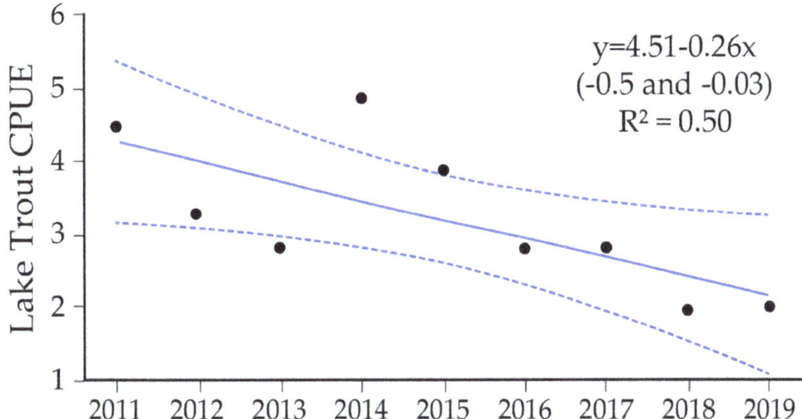

Figure 8. Catch-per-unit-effort (CPUE) of lake trout during annual long-term gillnetting assessments on Yellowstone Lake, 2011–2019. The blue line represents a simple linear regression model with 95% confidence intervals (dashed lines). Numbers within parentheses are the upper and lower 95% confidence limits of the slope parameter estimate, indicating a temporal decline in the CPUE response variable because the interval does not include zero.

Total lake trout abundance estimates (age 2 and older, at the beginning of the year, derived from the SCAA model, Table A5) declined ($p = 0.038$) from 925,208 in 2012 to 673,983 in 2019 (Figure 9A) after suppression gillnetting effort increased sufficiently to curtail lake trout population growth (i.e., $\lambda < 1$). However, high among-year variation in estimated total abundance ($R^2 = 0.54$) was apparently driven by highly variable recruitment of age-2 fish to the population. Abundances of age-2 lake trout during 2012 to 2019 ranged from a low of 318,640 in 2013 to a high of 480,961 in 2015 and no significant relationship existed between abundance and year ($p = 0.834$, Figure 10A). Abundances of age 3–5 ($p = 0.011$) and age 6+ ($p < 0.001$) lake trout declined from 416,814 and 57,722 in 2012 to 197,681 and 12,345 in 2019, respectively (Figure 10B,C). These declines resulted in a 54% decrease in total lake trout biomass ($p < 0.001$) from 432,017 kg in 2012 to 196,675 kg in 2019 (Figure 9B).

The lake trout population growth rate (λ) from 2017 to 2018 was 0.75 (95% CI: 0.65–0.85), which met our primary desired condition for this performance metric (Table 1) [81]. However, it grew to 1.18 (95% CI: 0.95–1.40) from 2018 to 2019 because of high age-2 recruitment. Even though total abundances of older lake trout declined between 2012 and 2017, abundance of age-2 lake trout increased by 69% between 2018 and 2019 (Figure 10A) because of the high year-class strength of the 2017 cohort, which was detected in 2019 after it recruited to the gear. This increase in recruitment, despite reductions in adult abundances, indicated a compensatory response by the lake trout population. Lake trout length at maturity did not change between 1995 and 2019; female lake trout matured at 515 mm (95% CL: 503–525) and male lake trout matured at 431 mm (95% CL: 423–444) (Figure A6). However, relative condition (K_n) of large lake trout (400+ mm), increased during this period, from 102.8 during 1995–1999 to 111.8 during 2005–2019 (Table A6, Figure A7). Nevertheless, estimated total egg production declined from 51.1 million in 2010 to only 15.8 million in 2017 (Figure 11). Accordingly, age-2 recruitment was maintained despite reductions in abundances of adults and egg production.

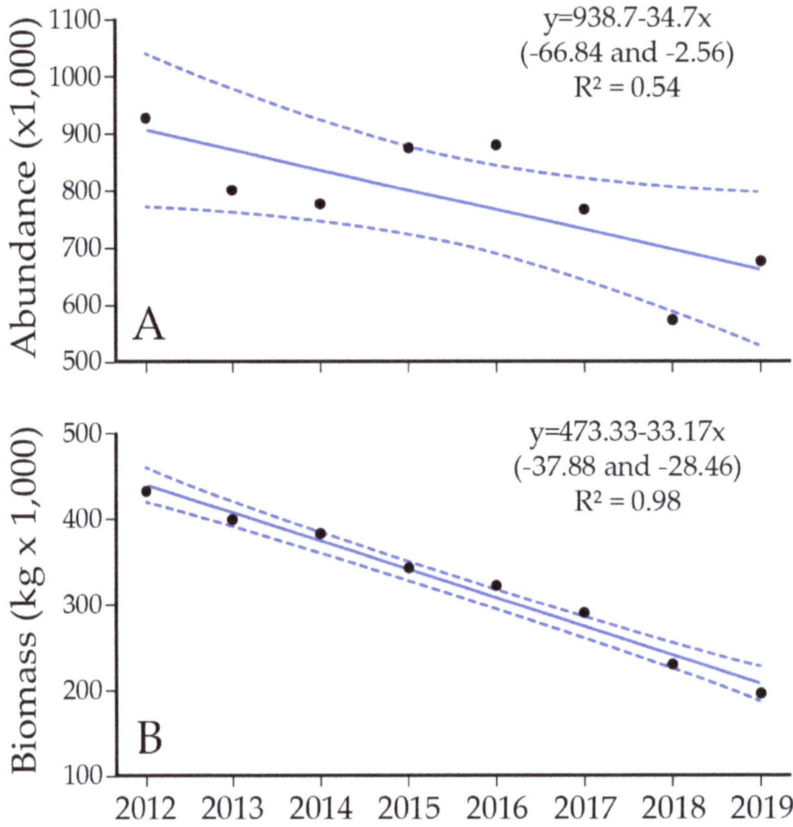

Figure 9. Total (**A**) abundance and (**B**) biomass of age-2 and older lake trout at the start of the year from 2012 through 2019 estimated using a statistical catch-at-age (SCAA) model [81]. Blue lines represent simple linear regression models with 95% confidence intervals (dashed lines). Numbers within parentheses are the upper and lower 95% confidence limits of the slope parameter estimate, indicating temporal declines in these response variables because intervals do not include zero.

Mandatory angler-harvest of lake trout, which was implemented immediately upon lake trout discovery, is substantial at about 5% of the total killed by all methods (angling and suppression gillnetting) combined in recent years. Moreover, angler harvest is additive to total mortality and comes at no cost to the program. Annual angler harvest increased from an estimated 500 lake trout in 1995 to 2900 in 1999 and averaged about 17,000 after 2002. Angler catch rate declined from 0.5 fish per hour in 2012 to 0.2 per hour in 2019 but remained above desired conditions for this performance metric (<0.05 lake trout per hour; Table 1). Historically, more than 50% of lake trout caught by anglers were large (>450 mm), but only 32%–34% were >450 mm during 2017 to 2019, reflecting the reduced abundance caused by suppression gillnetting. Anglers have consistently caught fewer lake trout than cutthroat trout in Yellowstone Lake.

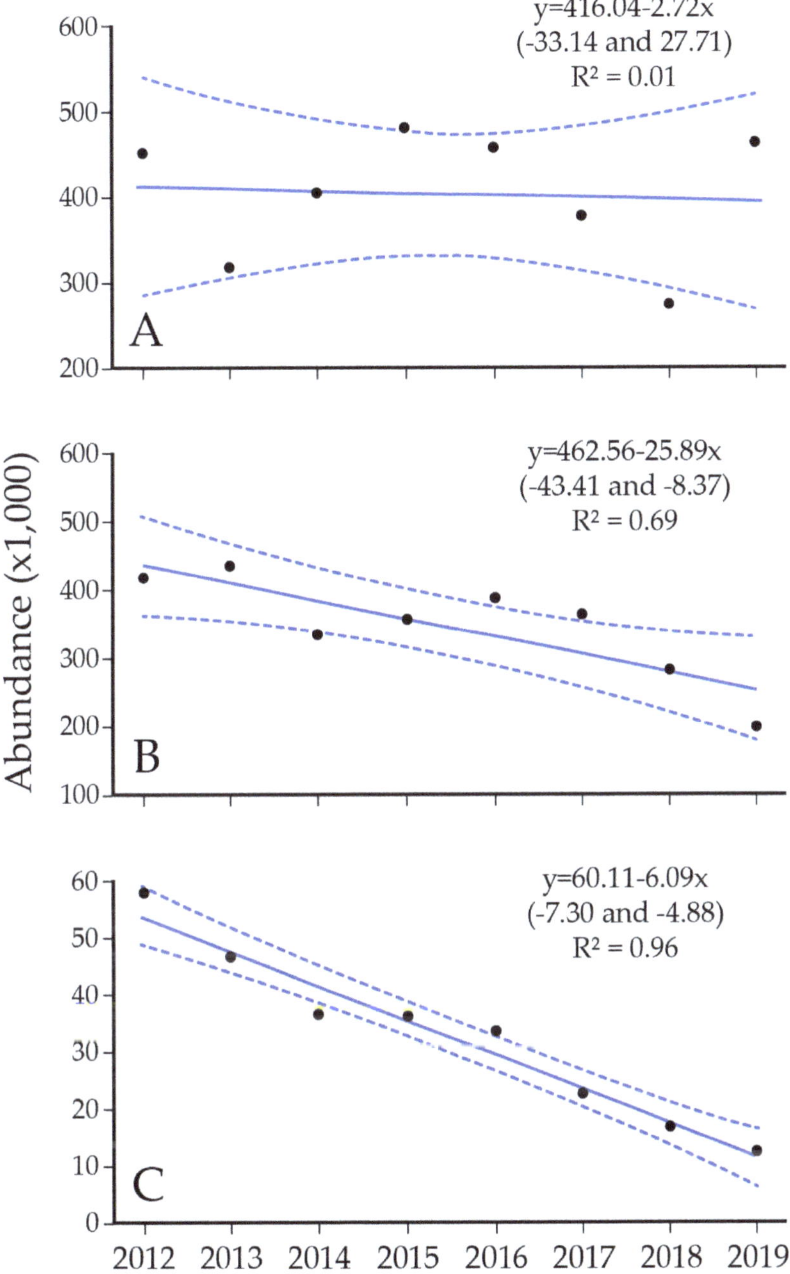

Figure 10. Abundances of (**A**) age-2, (**B**) age-3 to age-5, and (**C**) age-6+ lake trout at the start of the year from 2012 through 2019 estimated using a statistical catch-at-age (SCAA) model [81]. Blue lines represent simple linear regression models with 95% confidence intervals (dashed lines). Numbers within parentheses are the upper and lower 95% confidence limits of the slope parameter estimate. There was no temporal change in age-2 abundance because the interval includes zero. Age-3 to age-5 abundance and age-6+ abundance declined significantly.

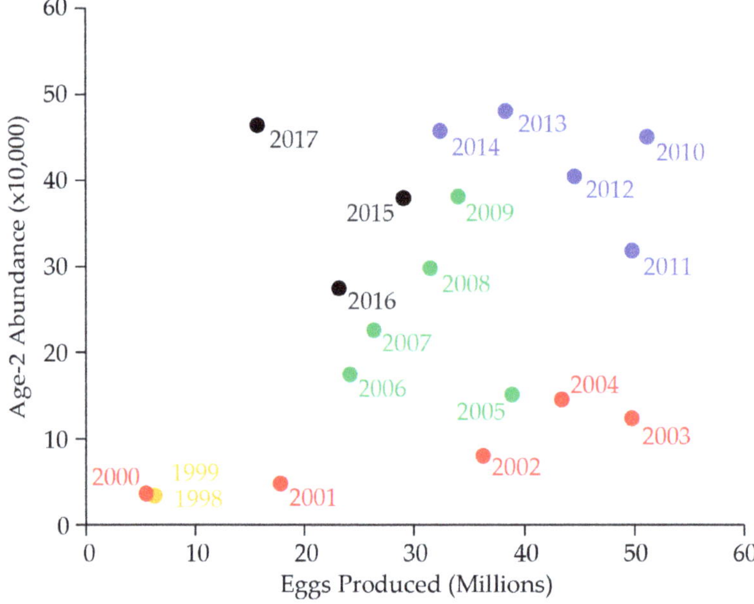

Figure 11. Age-2 lake trout abundances during 1998 to 2017 and the number of spawned eggs that produced these recruits two years previously estimated by a statistical catch-at-age (SCAA) assessment model [81]. Different colors represent 5-year periods of the lake trout suppression program described in text. Although fewer eggs were produced annually since lake trout population growth was curtailed in 2012, similar (or often greater) abundances of age-2 lake trout were recruited to the population, suggesting a compensatory response.

9. Cutthroat Trout Population Response to Lake Trout Suppression

The average cutthroat trout CPUE (during historical gillnetting assessments; Figure 1) was 37.8–48.7 per 100-m net night in the late-1980s and 41.9 in 1994, the year lake trout were first discovered (Figure 12A). Cutthroat trout CPUE then declined to 19.3 by 2004 (average of 8% reduction per year) following more than a decade of predation pressure by lake trout. The lowest lake-wide cutthroat trout gillnetting CPUE was 10.9 in 2010, and other monitoring metrics (see below) also reached a minimum.

Cutthroat trout abundance declined precipitously until suppression efforts reached sufficient levels to reduce lake trout abundances in 2012 [41]. The number of cutthroat trout caught during annual long-term gillnetting assessments varied subsequently, with mean CPUE ranging from a low of 12.5 per 100-m net night in 2011 to highs of 27.3 and 26.4 in 2014 and 2018, respectively (Table A4, Figure 13). These CPUEs met established secondary desired conditions for cutthroat trout (CPUE > 26; Table 1; Figure 12A). Size structure of the cutthroat trout population also varied during this period. The proportion of the long-term gillnetting assessment catch ≤ 280 mm (primarily age-2 individuals) was extremely low (16%) in 2011 when lake trout predation pressure was high (Figure A8). Although large (> 400 mm) individuals continued to dominate the cutthroat trout population during 2012 to 2019 when abundances of age 3+ lake trout declined, cohorts of smaller cutthroat trout subsequently became a more common component of the population, indicating greater avoidance of lake trout predation.

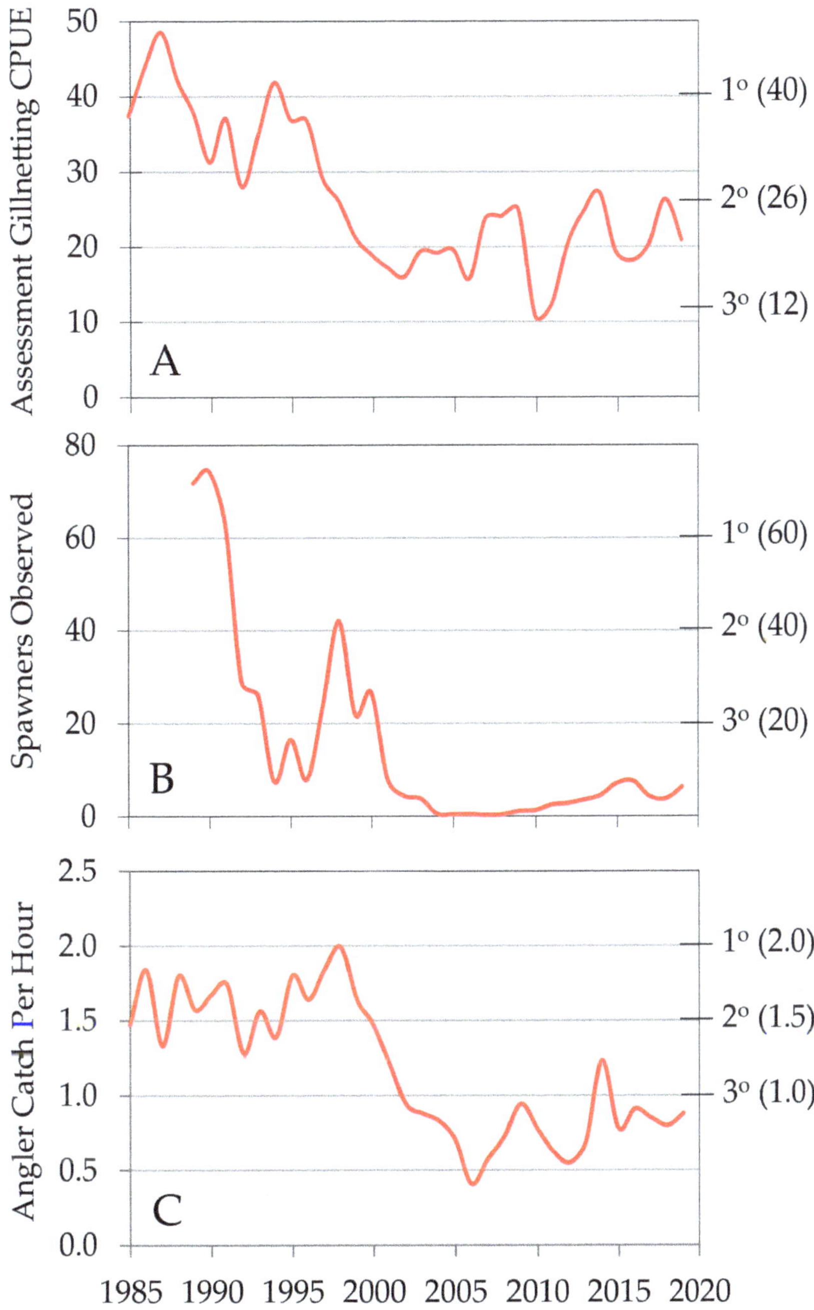

Figure 12. Cutthroat trout quantitative response variables monitored to assess the effects of conservation actions in the adaptive management strategy for Yellowstone Lake included the (**A**) catch-per-unit-effort (100-m net night) during within-lake gillnetting assessments, (**B**) observed during visual surveys of spawning streams, and (**C**) caught per hour by lake anglers, 1985–2019. Primary (1°), secondary (2°), and tertiary (3°) desired conditions are from the Native Fish Conservation Plan (Table 1) [90].

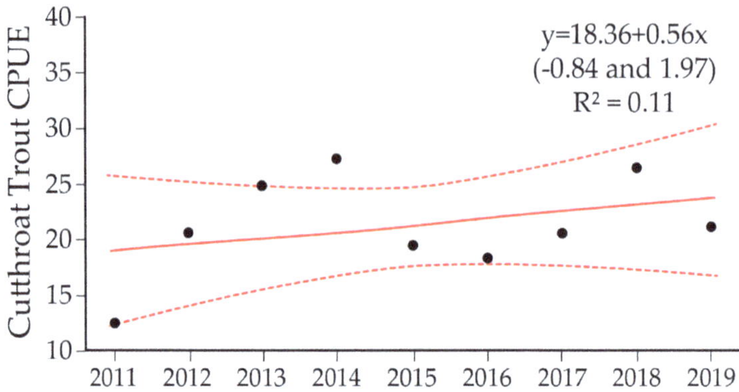

Figure 13. Catch-per-unit-effort (CPUE) of cutthroat trout during annual long-term gillnetting assessment monitoring on Yellowstone Lake, 2011–2019. The red line represents a simple linear regression model with 95% confidence intervals (dashed lines). Numbers within parentheses are the upper and lower 95% confidence limits of the slope parameter estimate, indicating no temporal change in the CPUE response variable because the interval includes zero.

Spawning cutthroat trout abundances decreased significantly following the lake trout invasion as judged by visual surveys of 11 spawning streams conducted annually since 1989 (Figure 1) [51,105]. An average of 74 cutthroat trout was observed during each stream survey in 1990, compared to ≤1 cutthroat trout during 2004 to 2010 (Figure 12B). Subsequently, spawning cutthroat trout increased slightly to a mean of 7.5 and 6.2 per survey in 2016 and 2019, respectively. Abundance increased in Little Thumb Creek, a tributary in the West Thumb near Grant (Figure 1), where more than 50 cutthroat trout were seen during a single visit in 2013, and more than 100 were seen during visits in 2014 and 2015. The number of fish observed in Little Thumb Creek after 2016 was about 80% of the total fish counted each spring at all of the visually-assessed spawning tributaries combined. Although the increased number of fish observed in Little Thumb Creek is encouraging, counts remained far below the primary or secondary desired conditions (means of 60 and 40 spawning cutthroat trout, respectively, observed per visit to all 11 of the visually-assessed spawning tributaries combined).

An estimated 68,000 cutthroat trout were caught by Yellowstone Lake anglers in 2019 at a catch rate of 0.9 fish per hour (Figure 12C). This catch rate was below the primary (2.0) and secondary (1.5) desired conditions (Table 1), which were the catch rates experienced during the 1980s and early 1990s. However, the average size of cutthroat trout caught in 2019 (440 mm) was much larger than caught prior to the lake trout invasion (380 mm). Although fewer fish were caught by anglers, the quality of the fish (from an anglers perspective) was much greater than in the past.

Lake trout predation was associated with a long-term shift in cutthroat trout lengths from dominance by small (100–280 mm) and midsized (290–390 mm) individuals to dominance by large individuals (400+ mm) in annual gillnetting assessments. The mean CPUE of small and midsized cutthroat trout declined from 18.6 per 100-m net night and 15.1, respectively, in the 1980s to just 6.9 and 3.9, respectively in the 2010s (Table A7, Figure 14A). Concurrently, the mean CPUE of large cutthroat trout nearly doubled, from 7.5 in the 1980s to 14.6 in the 2010s. Lake trout also caused increases in individual weights and condition of cutthroat trout. Although the average weight of small cutthroat trout slightly declined, the average weight of midsized and large cutthroat trout increased from 408.0 g and 682.8 g, respectively, in the 1980s to 463.4 g and 1418.6 g, respectively, in the 2010s (Table A7, Figure 14B). Relative weights (condition factors) of individual cutthroat trout also increased during this period. Mean relative weights of small, midsize, and large cutthroat trout were 58.8, 56.5, and 55.8, respectively, in the 1980s and increased to 68.4, 70.4, and 67.7, respectively, in the 2010s (Table A7, Figure 14C). Lower densities of cutthroat trout with higher individual weights and conditions should have higher fecundity [106], which should aid further recovery.

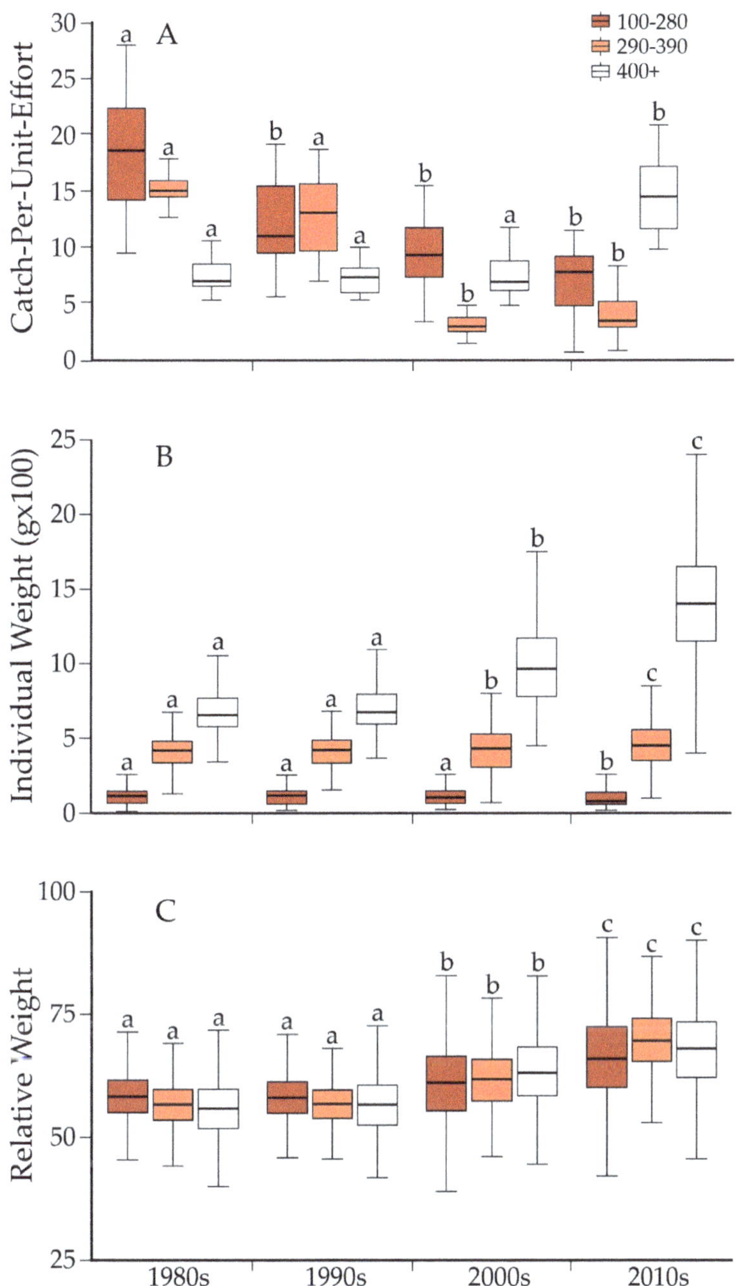

Figure 14. Mean (**A**) catch-per-unit-effort (100-m net night), (**B**) individual weight (g × 100), and (**C**) relative weight of each of three length groups (mm) of cutthroat trout from annual gillnetting assessments during each decade (1980–2019) on Yellowstone Lake. Means marked with the same letters (a–c) within each length group indicate no statistical difference among decades (α = 0.05).

10. Ecological Response to Lake Trout Suppression

Lake trout induced stress on cutthroat trout caused trophic shifts over four decades across multiple trophic levels both within and outside of Yellowstone Lake [41]. Hypothesized outcomes of lake trout reduction and cutthroat trout recovery were the return of trophic levels to natural, pre-lake trout conditions (Table A3). We therefore monitored components of the aquatic and terrestrial ecosystems to document the cascading changes that occurred after lake trout suppression.

The introduction of lake trout added a fourth trophic level [86] and resulted in cascading interactions within the aquatic food web of Yellowstone Lake, including shifts in cutthroat trout prey consumption and the biomass and individual lengths of zooplankton. When cutthroat trout were abundant in the 1980s, large-bodied cladocerans made up 80% of their diet [79]. After cutthroat trout declined in the 2000s, cladocerans made up only 11% of their diet, and cutthroat trout more frequently consumed amphipods, which made up 79% of their diet by 2011 [107]. After cutthroat trout abundance declined and predation on large zooplankton was reduced, the biomass of small-bodied copepods declined and the biomass of large-bodied zooplankton increased within the pelagic zone [41]. In addition, the average length of large-bodied zooplankton increased. Increased grazing by the dominant large-bodied cladocerans following the decline of cutthroat trout resulted in lower phytoplankton biomass and increased water clarity [41]. Although zooplankton and phytoplankton communities can be early indicators of ecological change (because of rapid turnover rates), we did not observe any shifts of plankton to a pre-lake trout condition resulting from the suppression program.

Lake trout invasion caused substantial indirect effects [108] that extended to native avian and terrestrial animals, such as ospreys, eagles, and bears, because spawning cutthroat trout were an important high-energy food for them. Ospreys are obligate piscivores that cannot switch to alternative food sources in the absence of fish. Osprey nest densities declined concurrently with declines in cutthroat trout prey, from an average of 38 nests during 1987 to 1991 to 11 during 2004 to 2008 [52] and only 2 during 2015 to 2019 [41]. Nesting success during 1987 to 1991 averaged 59% but declined to zero during 2008 to 2011 when no young ospreys were fledged from Yellowstone Lake nests. An average of 13% of osprey nests successfully fledged young during 2015 to 2019. We did not observe an increase in osprey nest density resulting from the suppression program, but expected a lagged response to increased cutthroat trout abundance.

Four to six bald eagle nests were typically present on Yellowstone Lake during the 1960s and 1970s. The nest count increased to an average of 11 during 2004 to 2008 [52] but then declined to 8 nests by 2015 to 2019 [41]. A steady long-term decline in bald eagle nest productivity occurred over two decades concurrently with the lake-wide decline in cutthroat trout. During 1985 to 1989, 56% of bald eagle nests on Yellowstone Lake successfully fledged young; however, nest success declined to zero in 2009 when prey fish abundance was low. Bald eagles are opportunistic feeders and increased consumption of alternative prey, including scavenging carnivore-provided carcasses and winterkill. Bald eagles were also observed preying on common loons and trumpeter swan cygnets, which have declined recently in Yellowstone National Park.

Grizzly and black bears are opportunistic feeders with flexible diets that consumed other foods available in the Yellowstone Lake area when cutthroat trout abundance was low [51,87]. No bear activity was found on surveyed spawning streams in 2008, 2009, or 2011 after cutthroat trout declined [41]. Compared to estimates obtained from 1997 to 2000, the number of grizzly bears visiting spawning streams a decade later (2007–2009) decreased by 63%, and the number of black bears decreased by 64 to 84% [109]. Bear activity on spawning tributaries subsequently increased in response to the slight recovery of spawning cutthroat trout [41].

11. Applied Research to Inform Decision Making

Management actions to restore the Yellowstone Lake ecosystem were supported by science through a strong program of applied research. Primary university partners were Michigan State University, Montana State University, University of Montana, University of Vermont, University of Wyoming, and Utah State University. Agency research partners included Montana Department of Fish, Wildlife and Parks, U.S. Geological Survey, U.S. Fish and Wildlife Service, U.S. Department of Agriculture, and Wyoming Game and Fish Department.

Research identified Lewis Lake or other waters in the upper Snake River system as the source of lake trout introduced to Yellowstone Lake [110,111] and identified the potential impacts the new predatory trophic level could have on the cutthroat trout and the ecology of the lake and tributary spawning streams [79,86,112]. As the cutthroat trout population declined, additional causal factors were investigated including *Myxobolus cerebralis* Hofer, 1903 (the causative agent of whirling disease), which was discovered in cutthroat trout from Yellowstone Lake in 1998 [113,114] and caused localized losses of cutthroat trout in Pelican Creek and the Yellowstone River downstream (Figure 1) [115–118]. Environmental factors influencing variation in cutthroat trout year class strength were also investigated. Persistent drought conditions were considered a strong contributing factor leading to cutthroat trout decline from the 1980s to the 2000s [51,92]. Studies also documented large-scale migrations, spawning locations (natal origins), and lake-wide movements of cutthroat trout [46,119].

After more than a decade of suppression gillnetting did not stop lake trout population growth, SCAA modeling of the lake trout population was conducted in the late 2000s to estimate lake trout demographics and establish gillnetting effort benchmarks that would result in $\lambda < 1$ [80,102]. An annual long-term monitoring protocol (long-term gillnetting assessment) was developed with sufficient power to detect changes in lake trout abundances over time [120]. Mark-recapture of tagged lake trout estimated population abundances and independently validated modeled lake trout abundance estimates [42]. The population model was updated annually and was critical for evaluating the efficacy of lake trout suppression relative to Native Fish Conservation Plan performance metrics [81,90].

Identification of lake trout spawning areas and movement patterns was critical for increasing the efficiency of the suppression program. Spawning habitat models, based on wave energy theory and geomorphology, suggested conditions for lake trout spawning habitat are patchy and exist in <4% of the lake [62]. Acoustically-tagged "Judas" fish were used to document seasonal movement patterns, habitat use (including spawning), and guide gillnetting efforts. For example, telemetry of lake trout conducted over several years (2011–2016) using fixed array receivers revealed extensive movements throughout the lake, important migration corridors, and spawning habitats [98,121]. Active mobile telemetry (boat-mounted receivers) further refined lake trout spawning site locations [60] and provided gillnetting crews with real-time locations and depths of lake trout aggregations, which increased gillnetting catches [122]. Subsequently, putative lake trout spawning sites were searched for presence of gametes, and all verified sites were investigated using remotely operated vehicles or scuba diving to delineate substrate boundaries, estimate substrate sizes and depths, and document substrate types [59]. The early life history of lake trout in Yellowstone Lake was assessed to better understand potential vulnerabilities to alternative (complementary) suppression methods that target embryos or fry [123].

Ongoing research focused on the development of alternative suppression methods that target embryos on spawning sites and complement gillnetting (see below). Because the diets of lake trout and cutthroat trout shifted to dominance by amphipods following the cutthroat trout decline [107], research was conducted to better understand lake ecology related to large-scale carcass deposition (about 300,000 carcasses per year in lake areas deeper than 65 m) and treatment of spawning sites with carcasses and organic pellets during autumn (see below). Ongoing efforts are underway to increase efficiency of monitoring cutthroat trout spawning abundances in tributaries using eDNA technology [124], detect congregations of lake trout using airborne lidar [125], and to further document effects of the lake trout and suppression actions on the lake and associated terrestrial ecosystems, including indirect effects on bears, birds, and river otters [41].

12. Need for Complementary Methods that Target Multiple Life Stages

The integrated pest management (IPM) approach to controlling invasive species uses a variety of suppression methods to target multiple life stages of invaders to maintain abundance levels below those causing harm [126–128]. The IPM approach has been most widely used in terrestrial systems to control agricultural pests [129,130]; however, a growing body of scientific evidence suggests that control of multiple life stages is required to suppress invasive fish populations over long time scales [131–133]. The suppression of sea lamprey in the Laurentian Great Lakes is an example of IPM in which chemical treatments, pheromone attractants, migration barriers, and other methods have been successfully used in combination [134–136]. The Yellowstone Lake gillnetting program curtailed lake trout population growth, but required extremely high levels of effort [41,81]. Because lake trout population growth rates are most sensitive to changes in age-0 survival [18,137,138], we developed and experimentally assessed methods to reduce it, with the intent of implementing an IPM approach on Yellowstone Lake.

The search for suppression methods that could complement gillnetting began at Montana State University in 2004 by a mechanical and chemical engineering senior design team [139] and continued in 2008 with a comprehensive literature review [140]. Later, the effectiveness of high-pressure water [42], electricity [141], suction-dredging [142], tarping (suffocation), and use of lake trout carcasses [143,144] and organic (plant-based) pellets [59,145,146] were evaluated for increasing lake trout embryo mortality *in situ* on Yellowstone Lake spawning sites. Additionally, chemical compounds (rotenone, sodium chloride, calcium carbonate, gelatin), sedimentation, fish carcasses, and organic (plant- and fish-based) pellets were evaluated in the laboratory [147]. To date, the most promising method for increasing lake trout embryo mortality was to intentionally degrade interstitial water quality. Treatment of spawning sites with lake trout carcasses or organic (soy and wheat gluten) pellets (Figure A9) induced organic decomposition and decreased dissolved oxygen concentrations at the substrate surface and 20 cm below the surface [59,144,145]. Biological oxygen demand of the decomposing organic materials caused dissolved oxygen to decline to 0 mg/L immediately after treatments and caused high embryo mortality within 200 h.

Fourteen lake trout spawning sites have been located over the past 25 years by gillnetting, sonar, telemetry, and shoreline visual surveys [59]. These sites vary in size, depth, substrate type, and thermal characteristics (Figure 15), and are generally located near western shorelines in areas of relatively low fetch during the autumn spawning period (when prevailing winds are from the southwest, Figure 1). Because only a few weeks are available to safely work on Yellowstone Lake following the peak of lake trout spawning each autumn, we expanded the embryo suppression research to include a comprehensive treatment of a spawning site with organic pellets by helicopter (with long line and seeder/spreader) to better understand the logistical constraints that may be faced when attempting large-scale, multi-site applications in the future. During an October 2019 experimental treatment, all of the rocky substrate at the Carrington Island spawning site (0.5 ha; Figures 1 and A9) was treated with 18,000 kg of organic pellets in less than one day (Video S3). Fry traps placed at Carrington Island in spring 2020 captured no lake trout fry, indicating all lake trout embryos were likely killed by the treatment and recruitment from the site was completely eliminated. Relative to the expansive lake areas intensively gillnetted over a 22-week season (>60 km of gill nets set daily), lake trout embryo suppression targets relatively small sites during a period of 2–3 weeks in autumn where the majority of a future year class is concentrated. Broad-scale application of pellets in autumn may reduce lake trout recruitment and enhance population suppression as part of an IPM approach targeting multiple lake trout life stages because the area of the 14 verified spawning sites is only 11.4 ha (0.03% of lake surface area) [59].

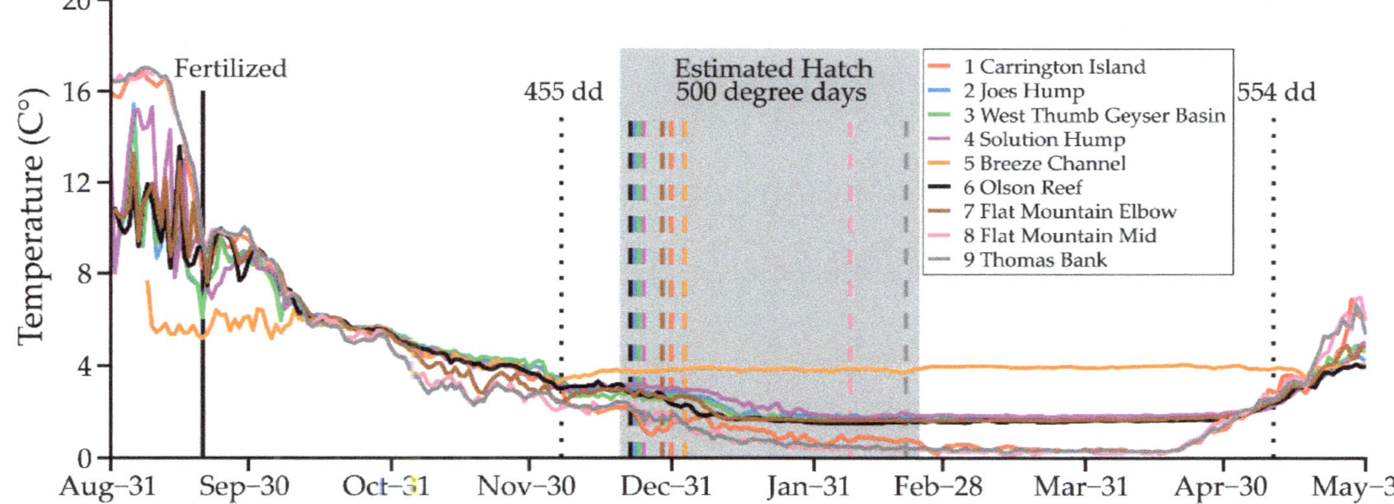

Figure 15. Water temperatures at the substrate of lake trout spawning sites in Yellowstone Lake during embryo fertilization, incubation, and hatching, September 2017–May 2018. Lake trout embryos require 455–554 degree days (dd) to hatch at 8 °C in the laboratory [147]. Dates of hatching at each site after 500 dd are within shaded gray. Although cooling temperatures slow development [148], embryos fertilized by 20 September may hatch prior to Yellowstone Lake becoming ice covered in late-December, many months earlier than within their native range in the Great Lakes [149].

13. Discovery of Nonnative Cisco in Yellowstone Lake

A cisco (likely *Coregonus artedi* Lesueur, 1818), not native to Yellowstone Lake, was caught during lake trout gillnetting operations north of Stevenson Island during August 2019 (Figure 1). The cisco was a live immature female, age-3, caught in 50 m depth. Otolith microchemistry analysis indicated it was probably hatched in Yellowstone Lake, meaning that parents and siblings are probably present. Someone illegally introduced them because no possible natural pathway exists for this species to reach Yellowstone Lake. The nearest possible source populations are in Tiber and Ft. Peck [150] reservoirs in the Missouri River drainage of Northern Montana, at travel distances of 560 km (6+ hrs) and 720 km (7+ hrs), respectively. We will implement monitoring for cisco population expansion using multimesh gillnetting of the lake pelagic zone, sampling for larvae, sampling for cisco eDNA, and by examining the stomachs of gillnetted lake trout. Cisco coevolved with lake trout and are their preferred prey in their native range within the Laurentian Great Lakes. If cisco become abundant in Yellowstone Lake, they could compete directly with cutthroat trout for zooplankton and other food resources, while providing additional prey for lake trout. Genetic analyses are being completed to confirm the fish caught in 2019 as *C. artedi* and possibly determine the source population for the illegal stocking.

14. Discussion

Yellowstone National Park is the site of successful restoration programs for iconic wildlife populations of American bison, gray wolves, grizzly bears, and trumpeter swans. Restoring cutthroat trout and the natural ecology of Yellowstone Lake, however, has been the most challenging restoration effort faced in the park's long history. The park was able to implement a suppression program that killed 3.35 million lake trout because scientific evidence strongly supported it, a must-harvest angler regulation was applied, numerous press releases and reports highlighted the urgent need, and a strong constituency for the lake trout fishery (that could have opposed suppression) never formed [151]. Moreover, the mission of the NPS is to preserve unimpaired the natural and cultural resources and values of the national park system for the enjoyment, education, and inspiration of this and future generations [152]. Cutthroat trout supported a fishery at Yellowstone Lake with historical importance and underpinned what had been the most intact naturally-functioning ecosystem in the lower 48 states. Therefore, allowing invasive predatory lake trout to persist and further degrade these nationally significant resources was not acceptable.

14.1. Suppressing an Invasive Population Below Carrying Capacity

Yellowstone Lake was perfectly suited for invasion and proliferation of predatory lake trout. The lake offered large expanses of deep, unoccupied habitat, devoid of co-evolved enemies [153], and rich in available resources [154] such as cutthroat trout forage that facilitated their establishment and expansion. Lack of thermal segregation (Figure A1) exacerbated interactions between lake trout and cutthroat trout throughout much of the year (mid-October through mid-June). Most of the lake (64%; 21,810 ha) is <60 m deep [155], depths at which lake trout thrive. The lake is mesotrophic and more productive than most lakes within the native range of lake trout. Lake trout in Yellowstone Lake therefore grew rapidly [7], matured early, and had high fecundity [80] (r-selected traits) [156,157]. Suppression of this invasive lake trout population, which never reached carrying capacity (K) [80,158], was therefore problematic. Managers use a total biomass yield threshold of 0.5 kg ha^{-1} yr^{-1} to avoid population collapse of lake trout fisheries in oligotrophic lakes in their native range [77,159]; yield densities of 95% of 145 populations in North America were <3.42 kg ha^{-1} yr^{-1} [95]. We did not drive the population into decline until biomass yield (suppression) reached 4.4 kg ha^{-1} yr^{-1} in 2012 (Figure 6B). Therefore, gillnetting effort applied during the first 17 years (1995–2011; 68%) of the 25-year suppression program was insufficient to force the lake trout population into decline (i.e., $\lambda < 1$; Figure 6B).

Shifts in habitat use by the expanding lake trout population probably increased their survival and enhanced the complexity of our gillnetting suppression efforts. Although gillnetting effort greatly increased during 1995 to 2011, lake trout abundance increased in advance and the population expanded lake-wide (Figure 7). The lake trout population probably used the most productive habitats early in the invasion [160] because the species is adaptable and able to colonize new environments that satisfy their basic habitat requirements [161]. Discrete 'islands' of rocky substrates provided spawning opportunities to the expanding population as it radiated outward from the West Thumb. Island biogeography theory predicts that habitat use is driven by a balance between colonization and extirpation [162]. Predation is a potent force driving species sorting along environmental gradients in freshwater habitats [163]. High predation risk in one habitat may cause a shift to another habitat where risk is lower. Our 'predatory' gillnetting efforts on spawning sites discovered early in the program (e.g., Carrington Island) to target spawning adults may have forced straying and pioneering of new spawning sites in more remote locations of the lake that were unknown or not targeted. Site fidelity and selection pressure would cause greater use of the most successful spawning habitats by each successive generation. Successful spawning at a site may have been caused by better habitat quality (e.g., favorable thermal characteristics; Figure 15) or simply our lack of awareness of it, which precluded suppression gillnetting there.

Spatially disproportionate suppression effort may have afforded lake trout refuge in remote regions of Yellowstone Lake. Suppression gillnetting crews actively sought lake trout and placed gillnets to maximize catches throughout each season. These efforts expanded unhindered across Yellowstone Lake as the lake trout population expanded (Figure 7) but were limited in the Flat Mountain, South, and Southeast arms, which are within proposed wilderness (Figure 1); 30% (6650 ha) of the 21,810 ha of Yellowstone Lake most suitable for gillnetting (<60 m deep) is within the proposed wilderness. Moreover, large lake trout move into the arms to prey upon juvenile cutthroat trout that emigrate from the upper Yellowstone River and other tributaries during late summer. Boat speeds within the arms were restricted to < 8 km hr^{-1}, and only nonmotorized boats (kayaks and canoes) were allowed in the far southern ends of the arms (including the delta of the upper Yellowstone River). Gillnetting in the arms was also deterred by long travel times and low catches, such that in 2013 only 14% of the total lake-wide gillnetting effort was applied there. The proposed wilderness was functioning as an aquatic (freshwater) protected area [164–166] limiting harvest of lake trout. We therefore completed a wilderness Minimum Requirements Analysis (MRA) in 2014 to suspend boating rules in the arms and allow aggressive targeting and gillnetting of lake trout lake-wide.

14.2. *Why are Lake Trout Resilient to Suppression Gillnetting on Yellowstone Lake?*

High early life history survival may buffer the effects of suppression gillnetting on the lake trout population. Sustainable exploitation of a fishery requires a reproductive surplus that can be removed [167–170], harvest that exceeds this surplus can cause population collapse [171–173]. Survival of lake trout in their native range is regulated *sensu* [174] during early life stages [175,176] but survival of pre-recruits in Yellowstone Lake may be 4–6 times higher [81] thereby requiring a 67% increase in gillnetting mortality at later stages to reduce population abundance [80,81]. The unoccupied habitat of Yellowstone Lake may provide lake trout a juvenile-survival advantage similar to that afforded to spawning common carp, arguably the most harmful invasive fish in the world, in predator-free, winterkill-prone shallow lakes [177,178]. Interstitial embryo predators such as sculpin *Cottus* spp. and crayfish *Orconectes* spp. [179,180] and fry predators such as rock bass and yellow perch [181] that are common in the native range of lake trout do not exist in Yellowstone Lake, which is naturally species-depauperate because of its isolation and elevation. Larval lake trout can therefore stay on spawning sites later into the summer, feed more, and achieve greater maximum lengths before dispersing [123]. The lack of predation on pre-recruit lake trout in Yellowstone Lake may afford the population an 'ecological release' [81,182], thereby buffering it against our suppression efforts.

High mortality of suppressed age classes may enhance survival of pre-recruit lake trout and add to population resilience through an an overcompensatory response to gillnetting mortality. Subjecting a life stage of a population to mortality can increase the abundance of other life stages and the total population [174]. Density-dependent processes can thereby confound removal efforts [183], resulting in positive population-level effects [132]. For example, intensive removal of age-0, juvenile, and adult smallmouth bass for 7 years from a north-temperate lake in New York, USA, reduced population biomass but increased population abundance, primarily by increasing juvenile abundance [184]. Size-selective mortality (i.e., uneven mortality across life stages) *sensu* [185] can elicit a similar response. For example, a large decrease in abundance of adult Eurasian perch resulted in a corresponding increase in juveniles [186]. Reduced competition among adult survivors increased somatic and reproductive growth, and juvenile survival was higher after release from cannibalism, collectively resulting in overcompensation. Similarly, the age composition of the Yellowstone Lake population shifted to predominantly younger fish as we increased gillnetting effort and targeted adult lake trout. Age-2 fish composed 26%–43% of total abundance during 1998 to 2004, but increased to 48%–55% during 2014 to 2018 [81] and 69% in 2019. Large-mesh gillnetting for 8 years (2012–2019) successfully reduced adult (age 6+) lake trout abundance by 79% and reduced total population biomass. However, abundance of age-2 fish did not change appreciably (Figure 10A). In Yellowstone Lake, per capita recruitment of lake trout at low levels of spawner abundance (Figure 11), pre-recruit survival [81], and maturation of age-4 and older fish (Figure A6) are all high. All of these characteristics can result in overcompensation whereby population abundance increases in response to harvest [183,185]. Moreover, variation in adult biomass (gradient from high to low, 2012–2019) may have gradually reduced competition, increased fecundity, or allowed a shift from alternate-year to annual spawning. Cannibalism of embryos and other early life stages may also be reduced, further enhancing pre-recruit survival. Although the actual mechanisms are unknown, such compensatory responses may impede attempts to curtail lake trout population growth in Yellowstone Lake.

14.3. Transition to Suppression of Multiple Lake Trout Life Stages

The realization that high survival of pre-recruit lake trout may offset increased mortality of older age classes has heightened interest in an IPM approach targeting multiple life stages with complementary suppression methods on Yellowstone Lake. Specifically, intense treatment of lake trout spawning sites with carcasses or organic pellets may mimic habitat degradation in their native range to increase mortality of embryos or fry, or both, and thereby decrease recruitment, especially as fewer adults spawn at fewer sites. Location and characterization of primary spawning sites and assessment of the quantity and quality of embryo-deposition habitat and hatching success will be critical. Methods that focus on early life-history stages to disrupt and reduce lake trout reproduction and recruitment should prove effective if spawning continues to be concentrated in shallow (<20 m) lake areas.

Considerable uncertainty and built-in time lags deter significant reductions in suppression gillnetting effort. Currently, an estimated 95,000–100,000 gillnetting effort units will be required annually for 5 years to achieve a 90% probability of reducing lake trout abundance to 100,000 fish [81]. However, we expect that a combination of gillnetting and embryo suppression will probably be used to maintain the lake trout population below target levels, and population-level effects of embryo suppression cannot be distinguished from the effects of on-going gillnetting that targets lake trout adults. The SCAA model estimates that 55,000 effort units will be required to maintain suppression after the target abundance of 100,000 fish is met, but the estimate includes uncertainty, and any resurgence in lake trout (caused by premature or excessive reduction of suppression gillnetting) would not be detected until they recruit to our monitoring and suppression gillnets at age 2. A management correction would not occur until the following year, giving the lake trout population a full 3 years of recovery. Therefore, a reduction of suppression gillnetting should only be made with extreme caution and vigilant monitoring.

15. Conclusions

Lake trout are being harvested from Yellowstone Lake at a greater rate than ever before anywhere on Earth [187]. This ecosystem restoration program illustrates that predatory fish invasions can be managed and controlled over large areas, even if total eradication may not be feasible. The process, however, requires a long-term commitment, is laced with uncertainty, and requires a great deal of collaboration. Program development, learning, demonstrating need, and building capacity to implement suppression actions at a large scale all require considerable time. Our adaptive management approach allows the program to move forward and implement conservation actions despite uncertainty in outcomes. Continuous learning from assessments and feedback obtained during annual science panel reviews are used to adjust lake trout suppression or other actions to progress towards desired conditions. This approach is used due to the varied environments and stressors (e.g., whirling disease and drought) impacting cutthroat trout in Yellowstone Lake, and the fact that some uncertainty exists in the possible responses by cutthroat trout and lake trout to future management actions. For example, although science-based findings indicate that lake trout population growth in Yellowstone Lake has been curtailed, uncertainty remains in the estimates of the number of years that high levels of suppression will need to be maintained to reduce the population to target levels (100,000 fish). Similarly, the rate of cutthroat trout recovery after the population is released from overriding lake trout impacts is also uncertain, as are the responses of avian and terrestrial wildlife. In the future, due to their use of shallow lake areas and dependence upon tributary streams, cutthroat trout may be more greatly harmed by climate-induced change than lake trout, which solely inhabit the comparatively stable, deep lake environment. The presence of cisco as a new, additional invader that functions as prey for lake trout (co-occurring exotic prey and exotic predator) *sensu* [35] further complicates matters. Lake trout suppression will become even more critical as these new threats emerge. Modeling of the cutthroat trout population is currently being conducted to better understand demographics and potentially refine objectives to include measures of stock biomass because lake trout predation resulted in a shift in cutthroat trout size structure to dominance by large fish (Video S4). Performance metrics will continue to be refined and monitored to track system responses to lake trout suppression, and the results will continue to be used to make adaptations and adjust management actions each year.

16. Materials and Methods

16.1. Lake Trout Suppression Netting

Up to six boats (Figure A4) were used to capture lake trout with sinking gillnets during late-May to mid-October 1995–2019 [97,188]. Suppression netting consisted of small-mesh (25 to 38 mm) and large-mesh (44 to 76 mm) bar measure gill nets targeting lake trout at depths typically greater than 20 m to reduce cutthroat trout bycatch. Nets were set shallower than 20 m at known spawning locations during peak spawning activity in autumn. Gill net soak time was typically three to four nights. Annual effort (effort unit = 100-m net per night) was 249 units in 1995 and increased to highs of 97,397 units and 96,971 units in 2018 and 2019, respectively (Figure 6A). Trap nets were also used during 2010 to 2013 to target large lake trout (i.e., >450 mm) [42,188]. Four to 10 trap nets were deployed at fixed locations throughout Yellowstone Lake each year. The netted lake trout were cut to puncture air bladders and then returned to deep (>65 m) regions of Yellowstone Lake.

16.2. Cutthroat Trout and Lake Trout Gillnet Assessment

Within Yellowstone Lake, cutthroat trout population metrics and individual characteristics (e.g., relative abundance, size structure, body condition) were assessed by standardized gillnetting programs. In mid-September during 1980 to 2010, gillnet surveys were conducted at 11 fixed sites throughout the lake (historical gillnetting assessment; Figure 1) [72,106]. At each site, five sinking experimental gill nets were set overnight perpendicular to shore in shallow water. Nets were set 100 m apart with the near-shore end about 1.5 m deep. Nets were 1.5 m in height and 38 m length, consisting of 7.6 m panels of 19- to 51-mm bar measure.

In 2011, a new protocol (long-term gillnetting assessment) was developed and implemented through 2019 to encompass monitoring of both cutthroat trout and lake trout. During the long-term gillnetting assessment program, 24 sites throughout the lake were sampled each year with a split-panel design to maximize spatial coverage and power for detecting temporal change [120]. Thirty-six sites were originally selected randomly with 12 designated to be revisited each year and the remaining 24 split into two panels of 12 sites each that were revisited every other year on an alternating basis. The sampling occurred after establishment of the lake thermocline during early August with a total of six experimental gill nets per site (Figure 1). At each site, a small-mesh and large-mesh sinking gill net were set overnight at each of three depth strata [epilimnion (3 to 10 m), metalimnion (10 to 30 m), and hypolimnion (>40 m)]. Small-mesh gill nets were 2 m in height and 76 m length, consisting of 13.7-m panels of 19- to 51-mm bar measure. Large-mesh gillnets were 3.3 m in height and 68.6 m length, consisting of 13.7-m panels of 57- to 89-mm bar measure. Gill nets were set perpendicular to shore and nets within a stratum were set parallel 100 m apart. Only the gillnets set in the shallow stratum were used to assess cutthroat trout. Gillnets at all three depth strata were used to assess lake trout.

Both the mid-September historical gillnetting assessment and August long-term gillnetting assessment were conducted for a period of 4 years to ensure that cutthroat trout mean CPUE and size structure were similar and a continuous dataset among the years of both monitoring programs could be compared, 1980–2019 (Figure 12). Discontinuing the historical gillnetting assessment of cutthroat trout in 2010 allowed additional time for NPS crews to focus on lake trout suppression activities during the critical autumn spawning period.

Relative weight for individual cutthroat trout was calculated using the equation $Wr = (W/Ws)*100$, where W = measured weight and Ws = standard weight predicted from a cutthroat trout (lentic) standard weight equation $\log10(Weight) = a + b * \log10(Length)$, where $a = -5.192$ and $b = 3.086$ [189]. One-way analysis of variance (ANOVA) was used to compare mean CPUE, individual weight, and relative weight for each length class among the four decades ($\alpha = 0.05$). If a statistical difference among decade means was detected, a post hoc Tukey's honestly significant difference multiple comparison procedure was used to test for differences between decade means. Data were manipulated using the "dplyr" package [190] in Program R and analyzed using Program R [191].

Relative condition (Kn) for individual lake trout was calculated using the equation $Kn = (W/W')*100$, where W = measured weight and W′ = predicted weight of a fish of the same length from a lake trout average weight-length equation $\log10(Weight) = a + b * \log10(Length)$, where $a = -5.589$ and $b = 3.210$ [95]. One-way analysis of variance (ANOVA) was used to compare mean Kn for each length class among the five time periods ($\alpha = 0.05$). If a statistical difference among means was detected, a post hoc Tukey's honestly significant difference multiple comparison procedure was used to test for differences between means of each time period.

Lake trout length at 50% maturity (L50) was estimated from lake trout (female n = 1766; male n = 2812) captured in gillnets in Yellowstone Lake during 1996 to 2019. Maturity stages (i.e., immature or mature) were assigned macroscopically, proportion of mature lake trout by 10 mm length bins was calculated, and sex-specific logistic regression models were fit to the proportional data. Length at 50% maturity was estimated from the model's inflection point and confidence intervals (95%) of L50 were estimated with bootstrapping using the bootCase() function from the "car" package [192] in program R [191]. Data were manipulated using the "dplyr" package [190] in Program R and analyzed using Program R.

16.3. Cutthroat Trout Tributary Spawner Assessment

Visual surveys for spawning cutthroat trout and bear activity were conducted annually during 1989 to 2019 on 9 to 11 tributaries located along the western side of Yellowstone Lake between Lake and Grant (Figure 1) [51,105]. Spawning reaches were delineated on each tributary, and the standardized reaches were walked in an upstream direction once each week from May to July. Observed cutthroat trout were counted, and the activity by black bears and grizzly bears was estimated by noting the presence of scat, parts of consumed trout, fresh tracks, and/or bear sightings. The average number of spawning cutthroat trout observed per visit was obtained by dividing the total observed (in all 9 to 11 tributaries combined, through the entire spawning period) by the number of surveys conducted.

16.4. Cutthroat Trout and Lake Trout Angler Catch

Because of its remote location, largely roadless (proposed wilderness) shoreline, and the short period of time (approximately 4 months annually) that local supportive facilities are open (campgrounds, gas stations, marinas), the angler effort on Yellowstone Lake is extremely low as compared to other large lakes in the Western U.S. We estimated angler effort and success via a report card distributed to all anglers when purchasing a special use permit for fishing in the park [72]. Annually, approximately 4000 anglers (10% of all park anglers) have voluntarily completed and returned cards for analysis. More than 9000 anglers fished Yellowstone Lake in 2019.

16.5. Lake Trout Population Modeling and Gillnetting Effort Benchmarks

The Yellowstone Lake lake trout population is assessed annually using an integrated SCAA assessment model that incorporates time-series of data from both suppression netting and long-term gillnetting assessment programs [80,102]. The SCAA model is age structured and encompasses ages ranging from age 2 to age 17, with the last age class an aggregate group including all fish age 17 and older. Age 2 is the age of recruitment in the SCAA model because younger age lake trout are not frequently captured in suppression or assessment gillnets. The SCAA model uses time-series of observations from the suppression gillnet program, the suppression trapnet program that ran from 2010 to 2013, and the long-term gillnetting assessment. The SCAA model generates predictions of abundances at age of the lake trout population based on model-based estimates of recruitment levels, abundances-at-age in the first assessment year, and underlying mortality levels for the different fishery components in operation and assumed natural mortality levels. Conditional on the predicted abundances at age, the SCAA model predicts suppression and assessment netting harvest and age-composition of harvest, which are compared to observed values. Predictions from the SCAA model can then be combined with other population descriptors (e.g., length at age, length-weight relationships, maturation at age, fecundity) to estimate the stock-recruitment relationship for the lake trout population, total and age-specific fishery yield, total and age-specific population biomass. In combination, the analyses and modeling results provided a robust time series prediction of the lake trout population, which in turn, can be used to gauge the success of the suppression program in decreasing lake trout abundance in Yellowstone Lake and strategize future efforts.

Annually, the SCAA model is used to estimate the amount of gillnetting effort required to cause an abundance decline in the lake trout population (i.e., $\lambda < 1$) and, in recent years (2017–2019), to achieve an abundance goal of 100,000 lake trout. These estimates have been critical for the restoration program as they have dictated the numbers of crews, boats, nets, and other gear (and therefore funding) needed to achieve suppression targets. The amount of housing required for the crews, which is extremely limited in the Yellowstone Lake area, was also driven by the suppression targets established from the assessment modeling. Annually, information gained from monitoring and suppression gillnetting was used to lengthen the time series of the data components that feed into the SCAA model. In turn, the annual gillnetting effort benchmarks evolve as the model updates the most recent estimates of abundances and mortalities (Figure 6A).

16.6. Monitoring for Ecological Response

A goal of the Yellowstone Lake ecosystem restoration is to restore the natural ecological role of native cutthroat trout. The lake trout-induced stress on cutthroat trout caused trophic shifts over the past four decades across multiple trophic levels both within and outside of Yellowstone Lake [41].

Hypothesized outcomes of lake trout reduction and cutthroat trout recovery is that these altered trophic levels will revert to their natural, pre-lake trout conditions (Table A3). To document the cascading changes that may occur due to lake trout suppression, we monitor several components of the aquatic and terrestrial ecosystems.

Aquatic ecological monitoring occurred during ice-free seasons and included measurements of zooplankton density, biomass, and size from samples collected at four lake regions (Main Basin, West Thumb, South Arm, and Southeast Arm; Figure 1) [41]. Phytoplankton biomass was estimated using chlorophyll a, and light transmission was measured using a Secchi disk in West Thumb. The thermal structures of Yellowstone Lake (e.g., isotherm depths) were measured in the West Thumb using a multiparameter sonde (Hydrolab Surveyor). Temperature was also measured routinely at the lake's surface. Lake surface levels, ice-on and -off dates, and outlet discharge (Yellowstone River at Fishing Bridge) were also obtained annually.

Avian and terrestrial consumers of cutthroat trout were annually monitored to document potential recovery. The number of breeding pairs and nesting success was determined for bald eagle and osprey populations each breeding season by surveying all forested areas up to 1 km from the Yellowstone Lake shoreline, connected tributaries, and forested islands using a fixed-wing Super Cub airplane [41]. Bear use of spawning cutthroat trout was documented during visual surveys for spawning cutthroat trout (as described above) on 9 to 11 tributaries located along the western side of Yellowstone Lake. Activity by black bears and grizzly bears was estimated by noting the presence of scat, parts of consumed trout, fresh tracks, and/or bear sightings along spawning stream corridors.

16.7. *Permits and Ethical Aspects*

This study was performed under the auspices of Montana State University Institutional Animal Care and Use Protocol 2018-68. Any use of trade, product, or firm names is for descriptive purposes only and does not imply endorsement by the U.S. Government.

Supplementary Materials: The following are available online, Video S1 (Native cutthroat trout and the Yellowstone Lake Ecosystem, http://doi.org/10.5281/zenodo.3820758); Video S2 (Gillnetting invasive lake trout http://doi.org/10.5281/zenodo.3829258); Video S3 (Organic pellet application to Carrington Island spawning site http://doi.org/10.5281/zenodo.3829479); and Video S4 (Angling for restored native cutthroat trout http://doi.org/10.5281/zenodo.3829613).

Author Contributions: Conceptualization, T.M.K., M.E.R.; methodology, J.L.A., P.E.B., C.R.D., P.D.D., B.D.E., R.E.G., D.J.M., T.J.S., L.M.T., A.V.Z.; formal analysis, T.M.K., P.E.B., T.O.B., C.S.G., J.M.S., N.A.T.; investigation, J.L.A., P.E.B., C.R.D., P.D.D., B.D.E., R.E.G., D.J.M., T.J.S., L.M.T., A.V.Z.; supervision, P.J.W.; project administration, T.M.K., P.J.W.; funding acquisition, T.M.K., J.D.D., D.P.S.; writing—original draft preparation, T.M.K.; writing—review and editing by all authors. All authors have read and agreed to the published version of the manuscript.

Funding: This research was funded by Yellowstone Forever, grant number G-022; George B. Storer Foundation; Greater Yellowstone Coalition; Idaho Council of Trout Unlimited; Montana Council of Trout Unlimited; Montana Fish, Wildlife and Parks; Montana State University; National Parks Conservation Association; National Park Foundation; University of Wyoming; U.S. Fish and Wildlife Service; U.S. Geological Survey; Whirling Disease Initiative, Montana Water Center; Wyoming Council of Trout Unlimited; Wyoming Game and Fish Department; Wyoming Wildlife and Natural Resource Trust; and the U.S. National Park Service, Yellowstone National Park.

Acknowledgments: Special thanks to the nearly five hundred seasonal NPS biological science technicians, Student Conservation Association (SCA) interns, contract gillnetting captains and crewmembers, and long-term volunteers that contributed to restoration of Yellowstone Lake over the past 25 years, This project greatly benefitted from management by Wayne Brewster, Jennifer Carpenter, David Hallac, Lynn Kaeding, Daniel Mahony, James Selgeby, S. Thomas Olliff, and John Varley, with strong support by park superintendents Michael Finley, Suzanne Lewis, Daniel Wenk, and Cameron Sholly. Partners advancing restoration science include Lindsey Albertson, Julie Alexander, Michelle Briggs, Kerry Gunther, Billie Kerans, Dominique Lujan, Thomas McMahon, Silvia Murcia, Alex Poole, James Ruzycki, Douglas Smith, Kole Stewart, and Jacob Williams. Along with NPS and contract staff the SCA program (www.theSCA.org) has provided immeasurable support for recovering cutthroat trout. The Rocky Mountains Cooperative Ecosystems Studies Unit (www.cfc.umt.edu/cesu) facilitated numerous agreements supporting critical research. We thank all Scientific Review Panel members 1995–2019—recent members include Michael Hansen, Michael Jones, Christopher Luecke, Ellen Marsden, Patrick Martinez, Jason Stockwell, Jack Williams, and Daniel Yule. Christopher Downs provided comments that greatly improved this manuscript. We greatly thank Allison Klein for producing all manuscript figures.

Conflicts of Interest: The authors declare no conflict of interest. The funders had no role in the design of the study; in the collection, analyses, or interpretation of data; in the writing of the manuscript, or in the decision to publish the results

Dedication: This case study is dedicated to the memory of Jacqueline J. Koel, loving mother of Todd Koel, who sadly lost her decade-long battle with Parkinson's disease during preparation of this manuscript.

Appendix A

Table A1. Common and scientific names of birds, fishes, and mammals referred to in text for locations within Yellowstone National Park and elsewhere.

Location	Taxon	Common Name	Scientific Name
Yellowstone			
	Birds	bald eagle	*Haliaeetus leucocephalus* (Linnaeus, 1766)
		common loon	*Gavia immer* (Brunnich, 1764)
		osprey	*Pandion haliaetus* (Linnaeus, 1758)
		trumpeter swan	*Cygnus buccinator* Richardson, 1832
	Fishes	cisco [1]	*Coregonus artedi* Lesueur, 1818
		lake chub	*Couesius plumbeus* (Agassiz, 1850)
		lake trout	*Salvelinus namaycush* (Walbaum, 1792)
		longnose sucker	*Catostomus catostomus* (Forster, 1773)
		redside shiner	*Richardsonius balteatus* (Richardson, 1836)
		Yellowstone cutthroat trout [2]	*Oncorhynchus clarkii bouvieri* (Jordan and Gilbert, 1883)
	Mammals	American bison	*Bison bison* (Linnaeus, 1758)
		American black bear	*Ursus* americanus Pallas, 1780
		gray wolf	*Canis lupus* Linnaeus, 1758
		grizzly bear	*Ursus arctos* (Linnaeus, 1758)
		North American river otter	*Lontra canadensis* (Schreber, 1777)
Elsewhere			
	Fishes	common carp	*Cyprinus carpio* Linnaeus, 1758
		Eurasian perch	*Perca fluviatilis* Linnaeus, 1758
		rock bass	*Ambloplites rupestris* (Rafinesque, 1817)
		sea lamprey	*Petromyzon marinus* Linnaeus, 1758
		smallmouth bass	*Micropterus dolomieu* Lacepède, 1802
		yellow perch	*Perca flavescens* (Mitchill, 1814)

[1] Likely *C. artedi*, confirmation by genetic analysis in process. [2] Subspecies designation.

Table A2. Public support played an important role in the Yellowstone Lake ecosystem restoration program. Outreach occurred via multiple sources using a wide range of media types. Information exchange with the public and program involvement by students, volunteers, interns, agency partners, and conservation organization staff were critical for highlighting the consequences of lake trout and helping to secure funding. This timeline identifies key events or changes that occurred, the public outreach that resulted, and links to available literature or other information online.

Year	Key Changes or Events	Public Outreach	Available URL
1994	Lake trout first discovered by an angler	NPS press releases	https://www.latimes.com/archives/la-xpm-1994-11-22-mn-424-story.html
	US$10,000 reward offered for information leading to arrest	Lake trout wanted poster	http://doi.org/10.5281/zenodo.3873385
	Lake trout must-kill angling regulation implemented	YNP angling regulations	https://www.nps.gov/yell/planyourvisit/upload/20FishReg_web.pdf
1995	Scientific panel review guides management response	Report to the NPS Director	https://www.nps.gov/parkhistory/online_books/yell/trout_invasion.pdf
	Public awareness of lake trout effect on birds and mammals	Numerous articles	https://www.nps.gov/articles/ys-25-1-shorts.htm
1996	Lake trout telemetry locates spawning areas	Graduate research programs	https://www.researchgate.net/publication/34007273
	Student Conservation Association, beginning of long-term support	Internships on suppression crews	https://www.thesca.org/
	Yellowstone Science, century of fisheries research and management	Special issue for the public	https://www.nps.gov/yell/learn/upload/YS_4_4_sm.pdf
1997	Applied research documents lake trout effects on cutthroat trout	Peer-reviewed publication	https://doi.org/10.1890/1051-0761(2003)013{[}0023:EOILTO{]}2.0.CO;2
1998	Whirling disease discovered in cutthroat trout	Peer-reviewed publication	https://doi.org/10.1577/H05-031.1
	Yellowstone fishes: ecology, history, and angling in the park	Book released	https://www.nps.gov/yell/learn/bookstore.htm
1999	Angler creel survey conducted to estimate lake trout catch	Interviews with lake anglers	https://www.nps.gov/yell/learn/fishreports.htm
2000	Park visitor perceptions of lake trout evaluated	Yellowstone Science article	https://www.nps.gov/yell/learn/upload/YS_9_2_sm.pdf
2001	Gillnetting enhanced by specialized boat and dedicated crews	NPS technical report on-line	https://www.researchgate.net/publication/237614952
	Cutthroat trout catch-and-release only angling regulation	YNP angling regulations	https://www.nps.gov/yell/planyourvisit/upload/20FishReg_web.pdf
2002	YNP native fish conservation reports for each year	Reports produced for the public	https://www.nps.gov/yell/learn/fishreports.htm
	YNP fly fishing volunteer program initiated	Volunteers interact with NPS biologists	https://www.nps.gov/articles/ys-25-1-shorts.htm
2003	Extent of whirling disease effect on cutthroat trout documented	Graduate research programs	https://doi.org/10.3354/dao071191
2004	Montana State University, College of Engineering, Senior design	Investigate suppression alternatives	http://doi.org/10.5281/zenodo.3873394
2005	Drought effects on cutthroat trout documented	Peer-reviewed publication	https://doi.org/10.1577/1548-8446(2005)30{[}10:NLTRIY{]}2.0.CO;2

Table A2. *Cont.*

Year	Key Changes or Events	Public Outreach	Available URL
2006	Barbless hook only angling regulation implemented	YNP angling regulations	https://www.nps.gov/yell/planyourvisit/upload/20FishReg_web.pdf
2006	Yellowstone Science, cutthroat trout conservation	Special issue for the public	https://www.nps.gov/yell/learn/upload/YS_14_2_sm.pdf
2007	Regional drought highlights potential effects of climate change	YNP angling closures to park waters	https://www.ncdc.noaa.gov/sotc/drought/200708
2008	Scientific panel review following cutthroat trout decline	Report to YNP Superintendent	https://www.nps.gov/yell/planyourvisit/upload/gresswell_final_updated_1_2010.pdf
	Initiate research regarding alternate suppression techniques	Comprehensive literature review	http://files.cfc.umt.edu/cesu/NPS/MSU/2008/08Zale_YELL_trout%20embryos_lit%20review.pdf
	Weird but true! Researcher proposes jello to kill lake trout	Montana State University news release	https://nypost.com/2008/12/29/weird-but-true-2347/
2009	Contract gillnetting pilot phase begins	YNP native fish program report	https://www.nps.gov/yell/learn/fishreports.htm
	Public scoping meetings for plan/ environmental assessment	Meetings and on-line comment period	https://parkplanning.nps.gov/projectHome.cfm?projectID=30504
2010	Native fish conservation plan/EA completed	Adaptive management plan	https://parkplanning.nps.gov/projectHome.cfm?projectID=30504
	Lake trout statistical-catch-at-age model created	Graduate research programs	https://doi.org/10.1139/cjfas-2019-0306
	Aquatic trophic cascade documented	Peer-reviewed publication	https://doi.org/10.1577/T09-151.1
	Public meetings at YNP gateway communities	Annually for information exchange	https://www.nps.gov/yell/learn/news/19013.htm
2011	Scientific panel reviews implemented annually	Reports to YNP Superintendent	https://www.nps.gov/yell/planyourvisit/upload/Lake_Trout_Suppression_Workshop_LowRes_Accessible.pdf
	Contract gillnetting fully implemented annually	YNP native fish program report	https://www.nps.gov/yell/learn/fishreports.htm
	Lake trout telemetry studies reinitiated	Public interpretive website	https://www.usgs.gov/centers/norock/science/yellowstone-lake-acoustic-biotelemetry-project-home-page
	Yellowstone Lake Workgroup formed	Memorandum of Understanding	http://doi.org/10.5281/zenodo.3873432
2012	Yellowstone Forever US$1,000,000 grant began annually	Save the Yellowstone cutthroat trout	https://www.yellowstone.org/what-we-do/native-fish/
	Trout Unlimited bloggers tour for conservation writers	Articles published on-line	http://www.sippingemergers.com/2012/08/setting-stage.html?m=1
2013	Native fish conservation program website	Public interpretive website	https://www.nps.gov/yell/learn/management/native-fish-conservation-program.htm
2013	Fund-raising by donating a lake trout telemetry tag	Trout Unlimited public website	https://eastyellowstonetu.org/images/savetheyellowstonecutthroat_2.html

Table A2. *Cont.*

Year	Key Changes or Events	Public Outreach	Available URL
	Wyoming Wildlife and Natural Resource Trust grant US$771,000	Wyoming Trout Unlimited	https://wwnrt.wyo.gov/
	Grizzly bear predation links loss of native trout to migratory elk	Peer-reviewed publication	https://doi.org/10.1098/rspb.2013.0870
	University science partner press releases	Articles published on-line	http://www.montana.edu/news/mountainsandminds/article.html?id=11921
2014	Public outreach for project support enhanced	Lake trout suppression FAQ website	https://www.usgs.gov/centers/norock/science/faq-invasive-lake-trout-yellowstone-lake
2015	Trout Unlimited document supporting science	Lake trout suppression FAQ report	http://wyomingtu.org/wp-content/uploads/2014/03/Science-Supporting-Management-of-Yellowstone-Lake-Fisheries.pdf
2016	NPS centennial celebration	Heightened public awareness	https://yellowstone.co/pdfs/centennial2016.pdf
	Outdoor Writers Association of America Conference	Tour and articles about the program	https://owaa.org/2014/02/owaa-announces-location-2016-annual-conference/
	Lake trout carcasses used to kill embryos on spawning sites	Graduate research programs	https://doi.org/10.1002/nafm.10259
2017	Yellowstone Science, native fish conservation issue	Special issue for the public	https://www.nps.gov/articles/series.htm?id=87AF9117-08B4-46BB-97B9C28B37A85BB2
	Sound of science in Yellowstone audio series	Podcast available online	https://www.nps.gov/yell/learn/photosmultimedia/onefishtwofish.htm
2018	Journey through Yellowstone's aquatic ecosystems	Interpretive video available online	https://www.nps.gov/yell/learn/nature/fishaquaticspecies.htm
	Realization that lake trout may have invaded naturally	Several articles by the media	https://www.jhnewsandguide.com/news/environmental/could-lake-trout-swim-to-y-stone/article_0625cb12-8782-59bd-a942-93653e5328e5.html
	Effects of actions to suppress lake trout on lake ecology	Graduate research programs	http://www.mtcfru.org/
	Organic pellets used to kill embryos on spawning sites	Peer-reviewed publication	https://doi.org/10.1002/tafs.10208
2019	Fly Fishing Film Tour features cutthroat trout return	Spring public film series	https://www.tu.org/blog/fly-fishing-film-tour-features-the-return/
	Partnership support of the restoration program continues	Trout Unlimited on-line articles	https://www.tu.org/blog/lake-trout-on-the-decline-in-yellowstone-lake/
2019	Magazine articles highlighting suppression efforts	National Geographic and others	https://www.nationalgeographic.com/animals/2019/06/how-to-eradicate-yellowstone-lake-trout/
	Public radio programming	Montana Public Radio	https://www.mtpr.org/post/non-native-lake-trout-numbers-declining-yellowstone-officials-say

Table A2. *Cont.*

Year	Key Changes or Events	Public Outreach	Available URL
	Regional and national print media	Cody Enterprise and others	https://www.codyenterprise.com/news/local/article_bd6ba2a0-93b9-11ea-b91b-2f3aadaca730.html
	Lake trout-induced indirect ecological effects documented	Heightened international awareness	https://advances.sciencemag.org/content/5/3/eaav1139
2020	Full length documentary film on the lake restoration effort	Video segments with this article	https://zenodo.org/record/3829613#.XtVlU2hKiUk
	Program updates made available to the public	NPS press releases continue	https://www.nps.gov/yell/learn/news/19053.htm
	Lake trout suppression on-line information updated	Public interpretive website	https://www.nps.gov/yell/learn/nature/lake-trout.htm
	Yellowstone Forever fund-raising partnership continues	A race against time donor website	https://www.yellowstone.org/trout/
	Scientific panel review results highlighted	Numerous articles on-line	https://www.tu.org/blog/science-panel-excited-about-numbers-on-yellowstone-lake/
	Two Ocean Pass: Alternative hypothesis for lake trout invasion	Peer-reviewed publication	https://doi.org/10.3390/w12061629
	Environmental assessment to expand spawning site treatments	Public scoping on-line anticipated	
	Global Covid-19 pandemic	Restoration program continues	

Table A3. Conceptual ecosystem model for the Yellowstone Lake ecosystem restoration. Cells with "X" are hypothesized linkages between agents of change, stressors on native fish, and ecosystem responses. Italic text with asterisks (*) and darkly shaded cells indicate elements addressed by the restoration program.

AGENTS OF CHANGE	Increased water temperatures	Altered hydrologic events (timing of max. & min. flows	Altered hydrologic events (flow volume)	Sedimentation	**Lake level declines/tributary disconnect*	Physiological stress & reduced fitness	**Fewer produced or recruited to spawning population*	**Direct native fish mortality*	**Predation by non-native fish*	**Competition/ displacement*	**Loss of genetic integrity (hybridization)*
Regional Physical/Chemical Forces											
Increasing temperature (air)	X	X	X		X	X	X				
**Changing precipitation patterns (snowpack, runoff)*	X	X	X		X	X	X				
Wildland Fire Frequency Increased	X	X	X	X			X				
Biological Introductions											
**Historical fish stocking by management*							X	X	X	X	X
**Stocking of fish illegally (lake trout)*							X	X	X	X	
Aquatic nuisance species (New Zealand mudsnails)							X			X	
Disease dissemination (whirling disease)						X	X				
Angling											
Intentional illegal harvest							X	X			
Mis-identification resulting in harvest							X	X			
Catch & release mortality							X	X			
Park Infrastructure/Operations											
Fire suppression						X					
Backcountry trails & campsites				X		X					
Land management (stock use, herbicide treatments)				X		X					
Road improvements				X		X					
Water treatment facilities		X	X			X					
Local Physical/Chemical Forces											
Natural geothermal inputs	X					X					

Table A3. *Cont.*

ECOSYSTEM RESPONSE	Increased water temperatures	Altered hydrologic events (timing of max. & min. flows	Altered hydrologic events (flow volume)	Sedimentation	*Lake level declines/tributary disconnect*	Physiological stress & reduced fitness	*Fewer produced or recruited to spawning population*	*Direct native fish mortality*	*Predation by non-native fish*	*Competition/ displacement*	*Loss of genetic integrity (hybridization)*
Biogeochemical Cycling											
Nutrient flux/transport altered					X	X		X	X	X	
Productivity/Biomass Change											
Primary production (algae) availability reduced					X	X		X	X	X	
Secondary production (inverts/zooplankton) availability altered/reduced	X	X		X	X	X		X	X	X	
Secondary production (fish) availability altered/reduced	X	X		X	X	X		X	X	X	
Fish Functional Role as Secondary Consumers											
Shift from invertivore (invert consumer) to piscivore (fish consumer/predator)								X	X	X	
Fish Life History Strategy											
Shift in spawning timing	X	X									X
Disrupt migration and/or shift in spawning location			X	X	X		X				X
Habitat volume (niche) available reduced	X		X	X	X					X	
Avian/Terrestrial Tertiary Consumers											
Displacement of grizzly bears from spawning streams					X			X	X	X	
Decline in native trout use by ospreys and eagles					X			X	X	X	
Increased physiological stress on river otters					X			X	X	X	

Table A4. Mean catch-per-unit-effort (CPUE, 100-m net night) of cutthroat trout and lake trout during annual long-term gillnetting assessments on Yellowstone Lake, 2011–2019, with lower (Lwr) and upper (Upr) 95% confidence limits (CL).

	Cutthroat Trout			Lake Trout		
Year	**YCT CPUE**	**Lwr 95% CL**	**Upr 95% CL**	**LKT CPUE**	**Lwr 95% CL**	**Upr 95% CL**
2011	12.46	8.10	16.82	4.44	2.91	5.97
2012	20.53	16.38	24.68	3.28	2.20	4.36
2013	24.82	19.03	30.61	2.80	2.00	3.60
2014	27.30	22.19	32.41	4.86	3.39	6.33
2015	19.42	14.66	24.18	3.89	3.00	4.78
2016	18.28	14.18	22.38	2.79	1.94	3.64
2017	20.38	15.99	24.77	2.80	2.08	3.52
2018	26.44	20.50	32.38	1.96	1.44	2.48
2019	20.96	16.82	25.10	2.00	1.11	2.89

Table A5. Mean annual total abundance and biomass, and abundances of age-2, age-3 to age-5, and age-6+ lake trout at the start of the year from 2012 through 2019 with lower (Lwr) and upper (Upr) 95% confidence limits (CL) estimated using a statistical catch-at-age (SCAA) model [81].

Year	Lake Trout	Mean	Lwr 95% CL	Upr 95% CL
2012	Abundance Total	925,208	771,271	1,125,280
2013		798,996	652,429	965,909
2014		774,569	633,884	943,216
2015		872,537	700,626	1,071,490
2016		877,724	706,840	1,089,630
2017		764,868	594,113	936,550
2018		572,550	449,699	741,077
2019		673,983	493,012	976,925
2012	Abundance Age-2	450,672	364,999	560,782
2013		318,640	249,647	397,507
2014		404,864	315,551	496,958
2015		480,961	372,384	597,922
2016		457,865	353,324	571,199
2017		379,099	289,469	476,558
2018		274,512	207,561	373,254
2019		463,958	313,814	688,536
2012	Abundance Age-3 to 5	416,814	336,620	510,974
2013		433,844	353,892	538,683
2014		333,291	268,885	417,800
2015		355,516	280,185	448,171
2016		386,503	301,593	490,933
2017		363,205	271,289	454,582
2018		281,511	213,664	363,813
2019		197,681	146,741	282,689
2012	Abundance Age-6+	57,722	41,025	72,130
2013		46,512	31,696	60,086
2014		36,414	24,458	48,976
2015		36,060	25,506	50,249
2016		33,356	23,762	48,202
2017		22,563	14,961	32,736
2018		16,527	11,088	25,040
2019		12,345	7991	20,163
2012	Biomass Total	432,017	343,977	521,873
2013		398,020	314,683	488,371
2014		382,138	298,369	469,323
2015		342,273	269,421	428,070
2016		321,580	257,333	414,412
2017		290,201	223,262	369,658
2018		229,509	179,299	304,616
2019		196,675	147,708	282,073

Table A6. Mean relative condition (Kn) with lower (Lwr) and upper (Upr) 95% confidence limits (CL) for three length classes (mm) of lake trout captured in Yellowstone Lake during periods of lake trout population growth (1995–1999, 2000–2004, and 2005–2009) and periods of population decline (2010–2014 and 2015–2019).

Length Class (mm)	Time Period	Mean Kn	Lwr 95% CL	Upr 95% CL
200–280	1995–1999	120.31	117.00	123.61
	2000–2004	120.03	110.22	129.83
	2005–2009	121.36	119.22	123.50
	2010–2014	115.74	114.32	117.16
	2015–2019	116.26	114.96	117.56
290–390	1995–1999	112.06	111.52	112.60
	2000–2004	112.86	111.83	113.89
	2005–2009	116.59	114.10	119.08
	2010–2014	111.88	111.02	112.75
	2015–2019	113.93	112.90	114.95
400+	1995–1999	102.78	102.23	103.33
	2000–2004	102.04	100.23	103.84
	2005–2009	106.21	105.11	107.32
	2010–2014	107.38	106.45	108.31
	2015–2019	111.78	110.20	113.35

Table A7. Mean catch-per-unit-effort (100-m net night), individual weight (g), and relative weight with lower (Lwr) and upper (Upr) 95% confidence limits (CL) of each of three length groups (mm) of cutthroat trout from annual gillnetting assessments during each decade (1980–2019) on Yellowstone Lake.

Cutthroat Trout	Length Class (mm)	Decade	Mean	Lwr 95% CL	Upr 95% CL
Catch-Per-Unit-Effort	100–280	1980s	18.63	14.27	22.99
		1990s	11.95	8.25	15.64
		2000s	9.43	6.77	12.09
		2010s	6.88	4.34	9.42
	290–390	1980s	15.09	13.94	16.24
		1990s	12.76	9.81	15.71
		2000s	3.09	2.25	3.93
		2010s	3.89	2.22	5.57
	400+	1980s	7.52	6.30	8.75
		1990s	7.38	6.13	8.63
		2000s	7.45	5.97	8.93
		2010s	14.61	12.05	17.17
Mean Individual Weight (g)	100–280	1980s	114.28	112.76	115.80
		1990s	113.61	111.46	115.75
		2000s	111.20	108.60	113.80
		2010s	103.90	100.63	107.18
	290–390	1980s	408.05	405.07	411.03
		1990s	411.01	107.02	414.99
		2000s	425.95	414.21	437.70
		2010s	463.37	453.82	472.92
	400+	1980s	682.80	674.98	690.63
		1990s	710.92	701.58	720.26
		2000s	1004.89	987.98	1021.81
		2010s	1418.60	1405.55	1431.64
Relative Weight	130–280	1980s	58.84	58.46	59.22
		1990s	58.71	58.37	59.05
		2000s	62.37	61.47	63.28
		2010s	68.38	67.48	69.28
	290–390	1980s	56.53	56.35	56.71
		1990s	56.71	56.51	56.91
		2000s	61.56	60.77	62.36
		2010s	70.35	69.70	71.00
	400+	1980s	55.78	55.43	56.13
		1990s	56.36	55.97	56.74
		2000s	63.50	63.00	64.00
		2010s	67.74	67.44	68.04

Appendix B

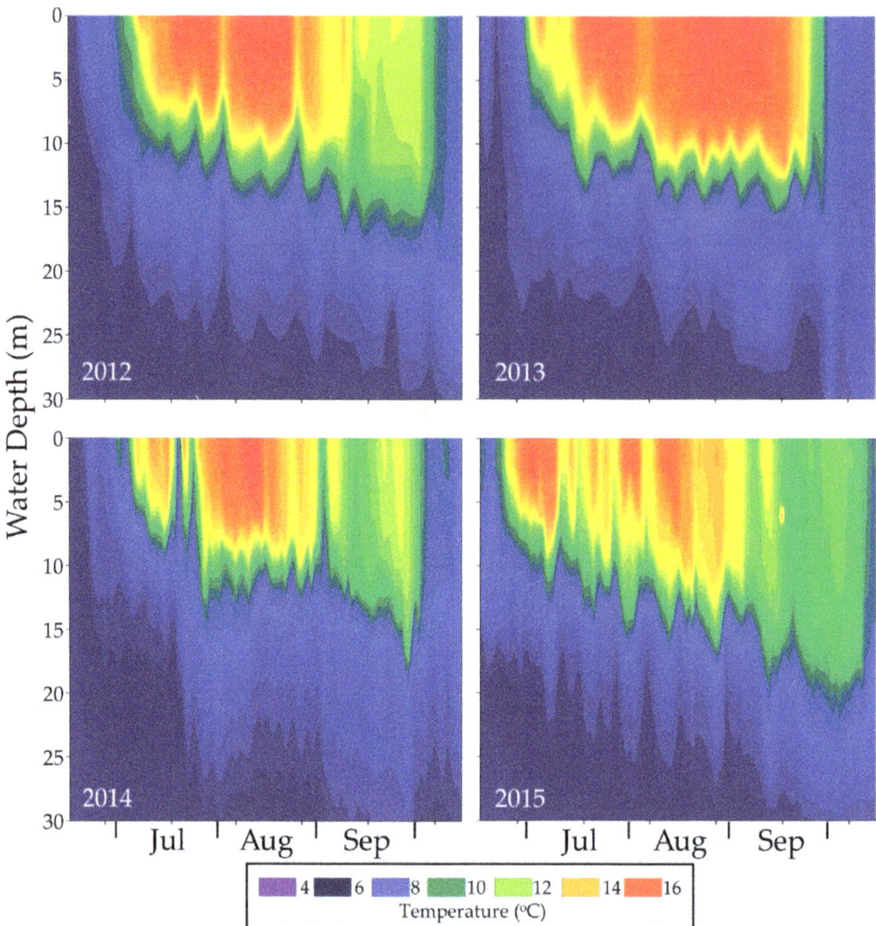

Figure A1. Depths of isotherms (°C) measured using a multiparameter sonde (Hydrolab Surveyor) in the West Thumb of Yellowstone Lake during a portion of the ice-free period, 2012–2015. The thermal structure of Yellowstone Lake is typically unstable with a weak and variable thermocline in July, August, and September. During nine months of each year, there is no thermal cause for segregation of invasive lake trout and native cutthroat trout. Only the upper 30 m are shown for better resolution of surface water temperatures.

Figure A2. Native cutthroat trout accessed the upper Yellowstone River and Yellowstone Lake from the upper Snake River via natural connections across the Continental Divide following glacial recession about 14,000 years ago. Cutthroat trout then evolved as the sole salmonid and dominant fish within the lake and its connected river network.

Figure A3. Invasive lake trout are a large-bodied, long-lived, and cold-adapted predatory species that became inadvertently introduced to Yellowstone Lake and were first discovered in 1994. They then became established as a new apex predatory trophic level within the lake. Because they are deep-water dwelling and do not use tributary streams, they are inaccessible to piscivorous avian and terrestrial wildlife and do not serve as an ecological substitute for native cutthroat trout.

Figure A4. Lake trout gillnetting boats on Yellowstone Lake included (**A–C**) National Park Service *Freedom*, *Hammerhead*, and *Cutthroat*, and (**D–F**) Hickey Brothers Research, LLC. *Kokanee*, *Patriot*, and *Northwester*.

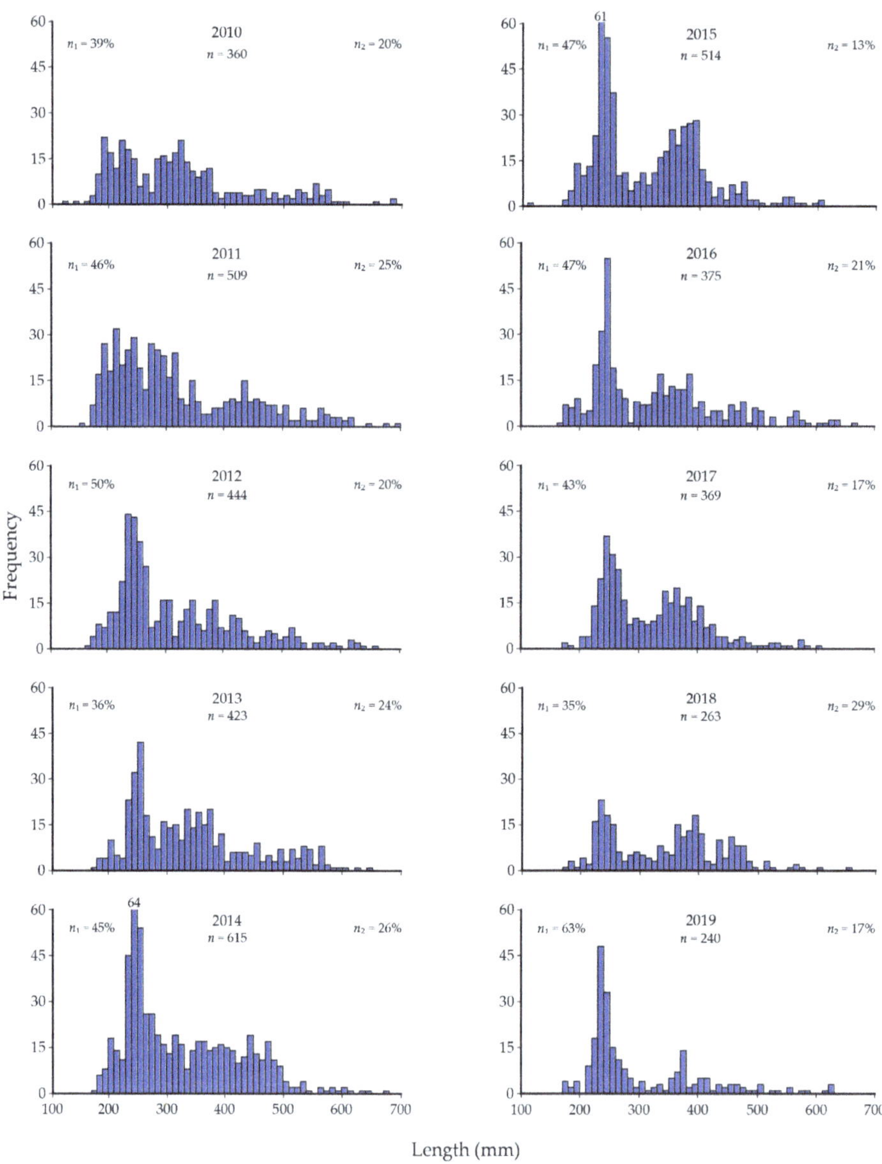

Figure A5. Length-frequency distributions of lake trout sampled during annual long-term gillnetting assessments on Yellowstone Lake with total number sampled (n), percentage $\leq$ 280 mm (n_1), and percentage $\geq$ 400 mm (n_2) each year, 2010–2019.

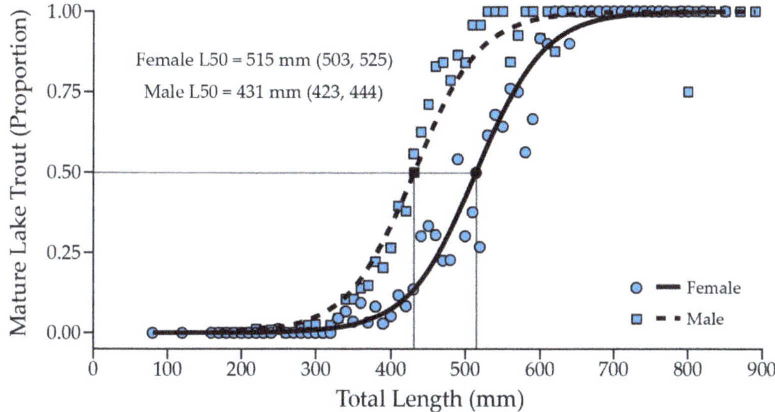

Figure A6. Proportions of mature female (n = 1766) and male (n = 2812) lake trout captured in Yellowstone Lake from 1996–2019 in 10-mm length bins with estimated length at which 50% of fish were mature (L50; black symbols at inflection points) and 95% confidence limits (in parentheses).

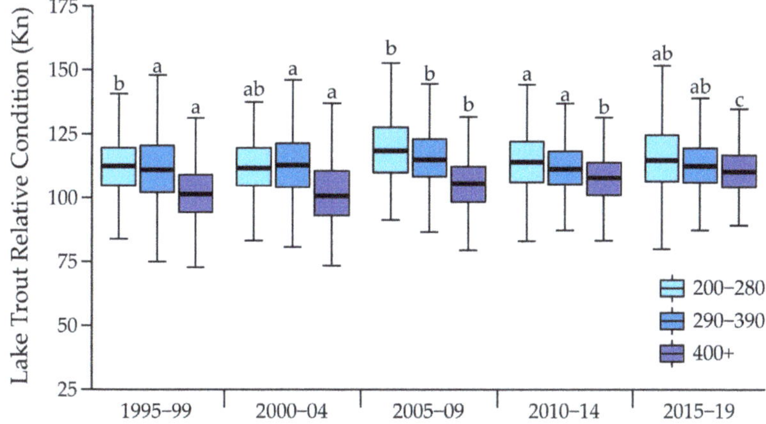

Figure A7. Relative condition (Kn) for three length classes (mm) of lake trout captured in Yellowstone Lake during periods of lake trout population growth (1995–1999, 2000–2004, and 2005–2009) and periods of population decline (2010–2014 and 2015–2019). Same letters (a–c) indicate no statistical difference in mean Kn between the five time periods for each length class. Over the past 25 years, the relative condition of large lake trout (400+ mm) has increased. Relative condition of smaller size classes did not appreciatively change.

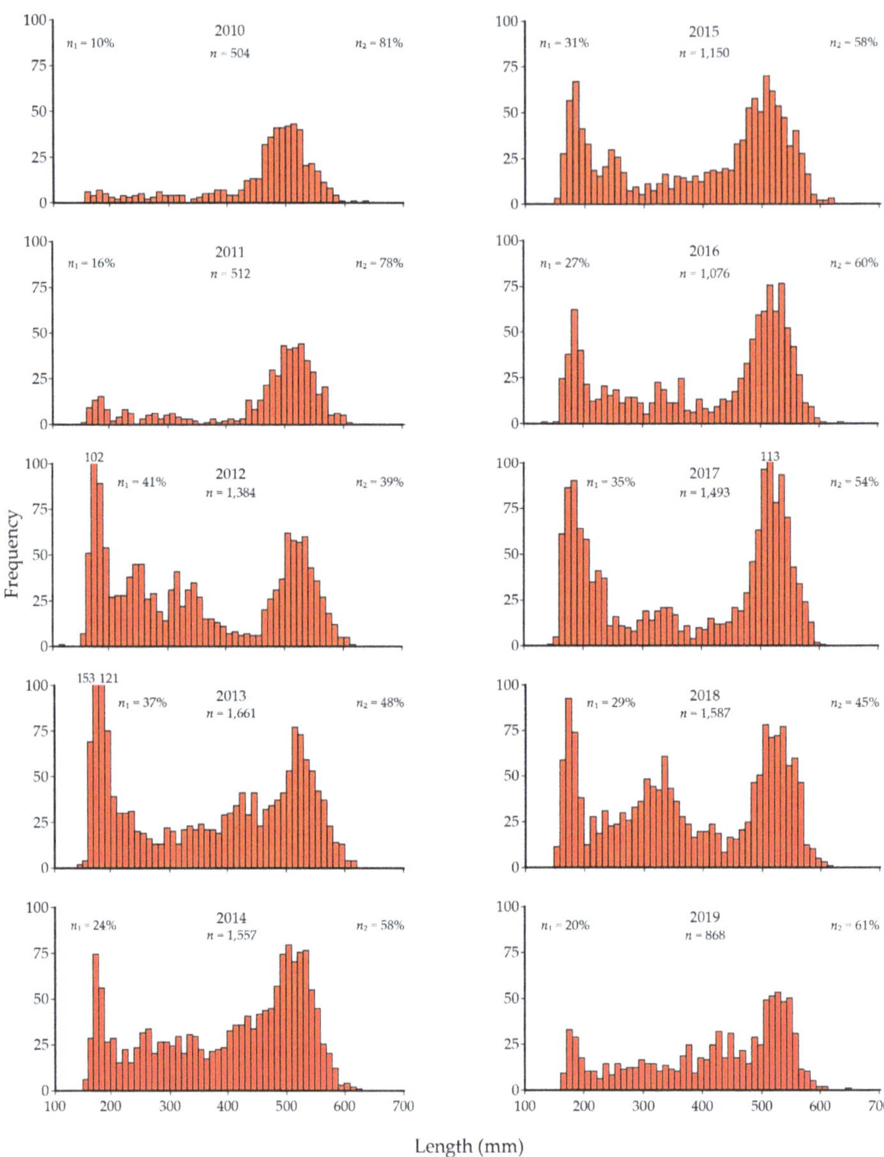

Figure A8. Length-frequency distributions of cutthroat trout sampled during annual long-term gillnetting assessments on Yellowstone Lake with total number sampled (n), percentage ≤ 280 mm (n_1), and percentage ≥ 400 mm (n_2) each year, 2010–2019.

Figure A9. (**A**) The angular-rock substrate surrounding Carrington Island in the West Thumb is prime lake trout spawning habitat in Yellowstone Lake. (**B**) Scuba divers inspect the spatial coverage of lake trout carcasses, and (**C**) organic (soy and wheat gluten) pellets spread by helicopter to induce decomposition, reduce dissolved oxygen concentrations, and increase mortality of lake trout embryos.

References

1. Gozlan, R.E.; Britton, J.R.; Cowx, I.; Copp, G.H. Current knowledge on non-native freshwater fish introductions. *J. Fish Biol.* **2010**, *76*, 751–786. [CrossRef]
2. Rahel, F.J.; Smith, M.A. Pathways of unauthorized fish introductions and types of management responses. *Hydrobiologia* **2018**, *817*, 41–56. [CrossRef]
3. Clarkson, R.W.; Marsh, P.C.; Stefferud, S.E.; Stefferud, J.A. Conflicts between native fish and nonnative sport fish management in the southwestern United States. *Fisheries* **2005**, *30*, 20–27. [CrossRef]
4. McMahon, T.E.; Bennett, D.H. Walleye and northern pike: Boost or bane to northwest fisheries? *Fisheries* **1996**, *21*, 6–13. [CrossRef]
5. Moyle, P.B.; Light, T. Biological invasions of fresh water: Empirical rules and assembly theory. *Biol. Conserv.* **1996**, *78*, 149–161. [CrossRef]
6. Vander Zanden, J.M.; Chandra, S.; Allen, B.C.; Reuter, J.E.; Goldman, C.R. Historical food web structure and restoration of native aquatic communities in the Lake Tahoe (California-Nevada) basin. *Ecosystems* **2003**, *6*, 274–288. [CrossRef]
7. Martinez, P.J.; Bigelow, P.E.; Deleray, M.A.; Fredenberg, W.A.; Hansen, B.S.; Horner, N.J.; Lehr, S.K.; Schneidervin, R.W.; Tolentino, S.A.; Viola, A.E. Western lake trout woes. *Fisheries* **2009**, *34*, 424–442. [CrossRef]
8. Olden, J.D.; Poff, N.L. Long–term trends of native and non–native fish faunas in the American Southwest. *Anim. Biodivers. Conserv.* **2005**, *28*, 75–89. Available online: https://www.raco.cat/index.php/ABC/article/view/56742 (accessed on 10 June 2020).
9. Sanderson, B.L.; Barnas, K.A.; Rub, A.M.W. Nonindigenous species of the Pacific Northwest: An overlooked risk to endangered salmon? *Bioscience* **2009**, *59*, 245–256. [CrossRef]
10. Coggins, L.G.; Yard, M.D.; Pine, W.E. Nonnative fish control in the Colorado River in Grand Canyon, Arizona: An effective program or serendipitous timing? *Trans. Am. Fish. Soc.* **2011**, *140*, 456–470. [CrossRef]
11. Carpenter, S.R.; Kitchell, J.F.; Hodgson, J.R. Cascading trophic interactions and lake productivity. *Bioscience* **1985**, *35*, 634–639. [CrossRef]
12. Nico, L.G.; Fuller, P.L. Spatial and temporal patterns of nonindigenous fish introductions in the United States. *Fisheries* **1999**, *24*, 16–27. [CrossRef]
13. Rahel, F.J. Homogenization of fish faunas across the United States. *Science* **2000**, *288*, 854–856. [CrossRef]
14. Schade, C.B.; Bonar, S.A. Distribution and abundance of nonnative fishes in streams of the western United States. *N. Am. J. Fish. Manag.* **2005**, *25*, 1386–1394. [CrossRef]
15. Britton, J.R.; Gozlan, R.E.; Copp, G.H. Managing non-native fish in the environment. *Fish Fish.* **2011**, *12*, 256–274. [CrossRef]
16. Rytwinski, T.; Taylor, J.J.; Donaldson, L.A.; Britton, J.R.; Browne, D.R.; Gresswell, R.E.; Lintermans, M.; Prior, K.A.; Pellatt, M.G.; Vis, C.; et al. The effectiveness of non-native fish removal techniques in freshwater ecosystems: A systematic review. *Environ. Rev.* **2018**, *27*, 71–94. [CrossRef]
17. Mueller, G.A. Predatory fish removal and native fish recovery in the Colorado River mainstem. *Fisheries* **2005**, *30*, 10–19. [CrossRef]
18. Fredenberg, C.R.; Muhlfeld, C.C.; Guy, C.S.; D'Angelo, V.S.; Downs, C.C.; Syslo, J.M. Suppression of invasive lake trout in an isolated backcountry lake in Glacier National Park. *Fish. Manag. Ecol.* **2017**, *24*, 33–48. [CrossRef]
19. Dux, A.M.; Hansen, M.J.; Corsi, M.P.; Wahl, N.C.; Fredericks, J.P.; Corsi, C.E.; Schill, D.J.; Horner, N.J. Effectiveness of lake trout (*Salvelinus namaycush*) suppression in Lake Pend Oreille, Idaho: 2006–2016. *Hydrobiologia* **2019**, *840*, 319–333. [CrossRef]
20. Dux, A.M.; Guy, C.S.; Fredenberg, W.A. Spatiotemporal distribution and population characteristics of a nonnative lake trout population, with implications for suppression. *N. Am. J. Fish. Manag.* **2011**, *31*, 187–196. [CrossRef]
21. Klein, Z.B.; Quist, M.C.; Rhea, D.T.; Senecal, A.C. Population characteristics and the suppression of nonnative burbot. *N. Am. J. Fish. Manag.* **2016**, *36*, 1006–1017. [CrossRef]
22. Kaus, D.J. Feasibility of Walleye Population Suppression in Buffalo Bill Reservoir, Wyoming. Master's Thesis, Montana State University, Bozeman, MT, USA, 2019.

23. Quist, M.C.; Hubert, W.A. Bioinvasive species and the preservation of cutthroat trout in the western United States: Ecological, social, and economic issues. *Environ. Sci. Policy* **2004**, *7*, 303–313. [CrossRef]
24. Shollenberger, H.; Dressler, E.; Mallinson, D.J. Invasive snakehead and introduced sport fish illustrate an environmental health paradox of invasive species and angler demand. *Case Stud. Environ.* **2019**, *3*, 1–10. [CrossRef]
25. Syslo, J.M.; Guy, C.S.; Cox, B.S. Comparison of harvest scenarios for the cost-effective suppression of lake trout in Swan Lake, Montana. *N. Am. J. Fish. Manag.* **2013**, *33*, 1079–1090. [CrossRef]
26. Hansen, M.J.; Guy, C.S.; Budy, P.; McMahon, T.E. Trout as native and nonnative species: A management paradox. In *Trouts and Char of the World*; Kershner, J.L., Williams, J.E., Gresswell, R.E., Lobón-Cerviá, J., Eds.; American Fisheries Society: Bethesda, MD, USA, 2019; pp. 645–684.
27. Walters, C.J.; Holling, C.S. Large-scale management experiments and learning by doing. *Ecology* **1990**, *71*, 2060–2068. [CrossRef]
28. McCarthy, M.A.; Possingham, H.P. Active adaptive management for conservation. *Conserv. Biol.* **2007**, *21*, 956–963. [CrossRef] [PubMed]
29. Holling, C.S. *Adaptive Environmental Assessment and Management*; Wiley: London, UK, 1978.
30. Walters, C.J. Is adaptive management helping to solve fisheries problems? *AMBIO J. Hum. Environ.* **2007**, *36*, 304–307. [CrossRef]
31. Walters, C.J. *Adaptive Management of Renewable Resources*; Macmillan: New York, NY, USA, 1986.
32. Runge, M.C. An introduction to adaptive management for threatened and endangered species. *J. Fish Wildl. Manag.* **2011**, *2*, 220–233. [CrossRef]
33. Estes, J.A.; Terborgh, J.; Brashares, J.S.; Power, M.E.; Berger, J.; Bond, W.J.; Carpenter, S.R.; Essington, T.E.; Holt, R.D.; Jackson, J.B.C.; et al. Trophic downgrading of planet Earth. *Science* **2011**, *333*, 301. [CrossRef]
34. Thom, R.; St Clair, T.; Burns, R.; Anderson, M. Adaptive management of large aquatic ecosystem recovery programs in the United States. *J. Environ. Manag.* **2016**, *183*, 424–430. [CrossRef]
35. Zavaleta, E.S.; Hobbs, R.J.; Mooney, H.A. Viewing invasive species removal in a whole-ecosystem context. *Trends Ecol. Evol.* **2001**, *16*, 454–459. [CrossRef]
36. Prior, K.M.; Adams, D.C.; Klepzig, K.D.; Hulcr, J. When does invasive species removal lead to ecological recovery? Implications for management success. *Biol. Invasions* **2018**, *20*, 267–283. [CrossRef]
37. Parker, I.M.; Simberloff, D.; Lonsdale, W.M.; Goodell, K.; Wonham, M.; Kareiva, P.M.; Williamson, M.H.; Von Holle, B.; Moyle, P.B.; Byers, J.E.; et al. Impact: Toward a framework for understanding the ecological effects of invaders. *Biol. Invasions* **1999**, *1*, 3–19. [CrossRef]
38. Kaplinski, M.A. Geomorphology and Geology of Yellowstone Lake, Yellowstone National Park, Wyoming. Master's Thesis, Northern Arizona University, Flagstaff, AZ, USA, 1991.
39. Remsen, C.C.; Maki, J.S.; Klump, J.V.; Aguilar, C.; Anderson, P.D.; Buchholz, L.; Cuhel, R.L.; Lovalvo, D.; Paddock, R.W.; Waples, J.; et al. Sublacustrine geothermal activity in Yellowstone Lake: Studies past and present. In *Yellowstone Lake: Hotbed of Chaos or Reservoir of Resilience, Proceedings of the 6th Biennial Scientific Conference on the Greater Yellowstone Ecosystem*; Anderson, R.J., Harmon, D., Eds.; Yellowstone Center for Resources and The George Wright Society: Yellowstone National Park, WY, USA, 2002.
40. Gresswell, R.E.; Varley, J.D. Effects of a century of human influence on the cutthroat trout of Yellowstone Lake. *Am. Fish. Soc. Symp.* **1988**, *4*, 45–52.
41. Koel, T.M.; Tronstad, L.M.; Arnold, J.L.; Gunther, K.A.; Smith, D.W.; Syslo, J.M.; White, P.J. Predatory fish invasion induces within and across ecosystem effects in Yellowstone National Park. *Sci. Adv.* **2019**, *5*, eaav1139. [CrossRef]
42. Koel, T.M.; Arnold, J.L.; Bigelow, P.E.; Detjens, C.R.; Doepke, P.D.; Ertel, B.D.; Ruhl, M.E. *Native Fish Conservation Program, Yellowstone Fisheries and Aquatic Sciences 2012–2014*; YCR-2015-01; National Park Service, Yellowstone Center for Resources: Yellowstone National Park, WY, USA, 2015. Available online: https://www.nps.gov/yell/learn/nature/upload/2012-2014_yellowstone_fisheries.pdf (accessed on 10 June 2020).
43. Jordan, D.S. A reconnaissance of streams and lakes of Yellowstone National Park, Wyoming in the interest of the U.S. Fish Commission. *Bull. U.S. Fish Comm.* **1891**, *9*, 41–63.
44. Behnke, R.J. *Trout and Salmon of North America*; Free Press: New York, NY, USA, 2002.
45. Licciardi, J.M.; Pierce, K.L. History and dynamics of the Greater Yellowstone Glacial System during the last two glaciations. *Quat. Sci. Rev.* **2018**, *200*, 1–33. [CrossRef]

46. Ertel, B.D.; McMahon, T.E.; Koel, T.M.; Gresswell, R.E.; Burckhardt, J.C. Life history migrations of adult Yellowstone cutthroat trout in the upper Yellowstone River. *N. Am. J. Fish. Manag.* **2017**, *37*, 743–755. [CrossRef]
47. Gresswell, R.E.; Liss, W.J.; Larson, G.L. Life-history organization of Yellowstone cutthroat trout (*Oncorhynchus clarkii bouvieri*) in Yellowstone Lake. *Can. J. Fish. Aquat. Sci.* **1994**, *51*, 298–309. [CrossRef]
48. Kaeding, L.R.; Boltz, G.D. Spatial and temporal relations between fluvial and allacustrine Yellowstone cutthroat trout, *Oncorhynchus clarkii bouvieri*, spawning in the Yellowstone River, outlet stream of Yellowstone Lake. *Environ. Biol. Fishes* **2001**, *61*, 395–406. [CrossRef]
49. Tronstad, L.M.; Hall, R.O., Jr.; Koel, T.M. Introduced lake trout alter nitrogen cycling beyond Yellowstone Lake. *Ecosphere* **2015**, *6*, 224. [CrossRef]
50. Felicetti, L.A.; Schwartz, C.C.; Rye, R.O.; Gunther, K.A.; Crock, J.G.; Haroldson, M.A.; Waits, L.; Robbins, C.T. Use of naturally occurring mercury to determine the importance of cutthroat trout to Yellowstone grizzly bears. *Can. J. Zool.* **2004**, *82*, 493–501. [CrossRef]
51. Koel, T.M.; Bigelow, P.E.; Doepke, P.D.; Ertel, B.D.; Mahony, D.L. Nonnative lake trout result in Yellowstone cutthroat trout decline and impacts to bears and anglers. *Fisheries* **2005**, *30*, 10–19. [CrossRef]
52. Baril, L.M.; Smith, D.W.; Drummer, T.; Koel, T.M. Implications of cutthroat trout declines for breeding ospreys and bald eagles at Yellowstone Lake. *J. Raptor Res.* **2013**, *47*, 234–245. [CrossRef]
53. Bergum, D.J.; Gunther, K.A.; Baril, L.M. Birds and mammals that consume Yellowstone cutthroat trout in Yellowstone Lake and its tributaries. *Yellowstone Sci.* **2017**, *25*, 86–89. Available online: https://www.researchgate.net/publication/318281009 (accessed on 10 June 2020).
54. Benson, N.G. *Limnology of Yellowstone Lake in Relation to the Cutthroat Trout*; Research Report 56; U.S. Department of the Interior, Fish and Wildlife Service, Bureau of Sport Fisheries and Wildlife: Washington, DC, USA, 1961. Available online: http://doi.org/10.5281/zenodo.3890470 (accessed on 10 June 2020).
55. Crait, J.R.; Ben-David, M. River otters in Yellowstone Lake depend on a declining cutthroat trout population. *J. Mammal.* **2006**, *87*, 485–494. [CrossRef]
56. Swenson, J.E. Prey and foraging behavior of ospreys on Yellowstone Lake, Wyoming. *J. Wildl. Manag.* **1978**, *42*, 87–90. [CrossRef]
57. Diem, K.L.; Pugesek, B.H. American white pelicans at the Molly Islands, in Yellowstone National Park: Twenty-two years of boom-and-bust breeding, 1966–1987. *Colonial Waterbirds* **1994**, *17*, 130–145. [CrossRef]
58. Walker, L.E.; Smith, D.W.; Albrechtsen, M.B.; Cassidy, B.J.; Shields, E.M.; Duffy, K. *Yellowstone Bird Project: Annual Report 2018*; YCR-2019-01; National Park Service, Yellowstone Center for Resources: Yellowstone National Park, WY, USA, 2019. Available online: https://www.nps.gov/yell/learn/nature/upload/2018-Bird-Report_web.pdf (accessed on 10 June 2020).
59. Koel, T.M.; Thomas, N.A.; Guy, C.S.; Doepke, P.D.; MacDonald, D.J.; Poole, A.S.; Sealey, W.M.; Zale, A.V. Organic pellet decomposition induces mortality of lake trout embryos in Yellowstone Lake. *Trans. Am. Fish. Soc.* **2020**, *149*, 57–70. [CrossRef]
60. Williams, J.R. Quantifying the Spatial Structure of Invasive Lake Trout in Yellowstone Lake to Improve Suppression Efficacy. Master's Thesis, Montana State University, Bozeman, MT, USA, 2019.
61. Flavelle, L.S.; Ridgway, M.S.; Middel, T.A.; McKinley, R.S. Integration of acoustic telemetry and GIS to identify potential spawning areas for lake trout (*Salvelinus namaycush*). *Hydrobiologia* **2002**, *483*, 137–146. [CrossRef]
62. Bigelow, P.E. Predicting Areas of Lake Trout Spawning Habitat within Yellowstone Lake, Wyoming. Ph.D. Thesis, University of Wyoming, Laramie, WY, USA, 2009.
63. Gresswell, R.E.; Liss, W.J. Values associated with management of Yellowstone cutthroat trout in Yellowstone National Park. *Conserv. Biol.* **1995**, *9*, 159–165. [CrossRef]
64. Byorth, J. Trout shangri—La: Remaking the fishing in Yellowstone National Park, Montana. *Mag. West. Hist.* **2002**, *52*, 38–47.
65. Varley, J.D.; Schullery, P.D. *Yellowstone Fishes: Ecology, History, and Angling in the Park*; Stackpole Books: Mechanicsburg, PA, USA, 1998.
66. Biesinger, K.E. Studies on the Relationship of the Redside Shiner (*Richardsonius balteatus*) and the Longnose Sucker (*Catostomus catostomus*) to the Cutthroat Trout (*Salmo clarki*) Population in Yellowstone Lake. Master's Thesis, Utah State University, Logan, UT, USA, 1961.

67. Davenport, M. *Piscivorous Avifauna on Yellowstone Lake, Yellowstone National Park*; U.S. Department of the Interior, National Park Service: Yellowstone National Park, WY, USA, 1974. [CrossRef]
68. Brown, C.J.D.; Graham, R.J. Observations on the longnose sucker in Yellowstone Lake. *Trans. Am. Fish. Soc.* **1954**, *83*, 38–46. [CrossRef]
69. Furey, K.M.; Glassic, H.C.; Guy, C.S.; Koel, T.M.; Arnold, J.L.; Doepke, P.D.; Bigelow, P.E. Diets of longnose sucker in Yellowstone Lake, Yellowstone National Park, U.S.A. *J. Freshw. Ecol.* **2020**. (under review).
70. Madsen, D.H. Protection of native fishes in the national parks. *Trans. Am. Fish. Soc.* **1937**, *66*, 395–397. [CrossRef]
71. Leopold, A.S.; Cain, S.A.; Cottam, C.M.; Gabrielson, I.N.; Kimba, T.L. *Wildlife Management in the National Parks*; Report to the Secretary of the Interior; Advisory Board on Wildlife Management: Washington, DC, USA, 1963; Available online: http://npshistory.com/publications/leopold_report.pdf (accessed on 10 June 2020).
72. Jones, R.D.; Gresswell, R.E.; Jennings, D.E.; Rubrecht, S.M.; Varley, J.D. *Fishery and Aquatic Management Program in Yellowstone National Park*; Tech. Rep. 1979; U.S. Fish and Wildlife Service: Yellowstone National Park, WY, USA, 1980. [CrossRef]
73. Kaeding, L.R.; Boltz, G.D.; Carty, D.G. Lake trout discovered in Yellowstone Lake threaten native cutthroat trout. *Fisheries* **1996**, *21*, 16–20. [CrossRef]
74. Koel, T.M.; Detjens, C.R.; Zale, A.V. Two Ocean Pass: An alternative hypothesis for invasion of Yellowstone Lake by lake trout, and implications for future invasions. *Water* **2020**, *12*, 1629. [CrossRef]
75. Scott, W.B.; Crossman, E.J. *Freshwater Fishes of Canada*; Bulletin 184; Fisheries Research Board of Canada: Ottawa, ON, Canada, 1973.
76. Ryder, R.A.; Kerr, S.R.; Taylor, W.W.; Larkin, P.A. Community consequences of fish stock diversity. *Can. J. Fish. Aquat. Sci.* **1981**, *38*, 1856–1866. [CrossRef]
77. Healey, M.C. The dynamics of exploited lake trout populations and implications for management. *J. Wildl. Manag.* **1978**, *42*, 307–328. [CrossRef]
78. Muir, A.M.; Blackie, C.T.; Marsden, J.E.; Krueger, C.C. Lake charr *Salvelinus namaycush* spawning behaviour: New field observations and a review of current knowledge. *Rev. Fish Biol. Fish.* **2012**, *22*, 575–593. [CrossRef]
79. Ruzycki, J.R.; Beauchamp, D.A.; Yule, D.L. Effects of introduced lake trout on native cutthroat trout in Yellowstone Lake. *Ecol. Appl.* **2003**, *13*, 23–37. [CrossRef]
80. Syslo, J.M.; Guy, C.S.; Bigelow, P.E.; Doepke, P.D.; Ertel, B.D.; Koel, T.M. Response of non-native lake trout (*Salvelinus namaycush*) to 15 years of harvest in Yellowstone Lake, Yellowstone National Park. *Can. J. Fish. Aquat. Sci.* **2011**, *68*, 2132–2145. [CrossRef]
81. Syslo, J.M.; Brenden, T.O.; Guy, C.S.; Koel, T.M.; Bigelow, P.E.; Doepke, P.D.; Arnold, J.L.; Ertel, B.D. Could ecological release buffer suppression efforts for non-native lake trout (*Salvelinus namaycush*) in Yellowstone Lake, Yellowstone National Park? *Can. J. Fish. Aquat. Sci.* **2020**, *77*, 1010–1025. [CrossRef]
82. Varley, J.D.; Schullery, P. *The Yellowstone Lake Crisis: Confronting a Lake Trout Invasion*; A Report to the Director of the National Park Service; Yellowstone Center for Resources, National Park Service: Yellowstone National Park, WY, USA, 1995. Available online: https://www.nps.gov/parkhistory/online_books/yell/trout_invasion.pdf (accessed on 10 June 2020).
83. McIntyre, J.D. Review and assessment of possibilities for protecting the cutthroat trout of Yellowstone Lake from introduced lake trout. In *The Yellowstone Lake Crisis: Confronting a Lake Trout Invasion. A Report to the Director of the National Park Service*; Varley, J.D., Schullery, P., Eds.; Yellowstone Center for Resources, National Park Service: Yellowstone National Park, WY, USA, 1995; pp. 28–33.
84. Ruzycki, J.R. Impact of Lake Trout Introductions on Cutthroat Trout of Selected Western Lakes of the Continental United States. Ph.D. Thesis, Utah State University, Logan, UT, USA, 2004.
85. Koel, T.M.; Arnold, J.L.; Bigelow, P.E.; Doepke, P.D.; Ertel, B.D.; Ruhl, M.E. *Yellowstone Fisheries and Aquatic Sciences: Annual Report, 2008*; YCR-2010-03; National Park Service, Yellowstone Center for Resources: Yellowstone National Park, WY, USA, 2010. Available online: https://www.nps.gov/yell/planyourvisit/upload/2008_fisheries_ar_final.pdf (accessed on 10 June 2020).
86. Tronstad, L.M.; Hall, R.O.; Koel, T.M.; Gerow, K.G. Introduced lake trout produced a four-level trophic cascade in Yellowstone Lake. *Trans. Am. Fish. Soc.* **2010**, *139*, 1536–1550. [CrossRef]
87. Middleton, A.D.; Morrison, T.A.; Fortin, J.K.; Robbins, C.T.; Proffitt, K.M.; White, P.J.; McWhirter, D.E.; Koel, T.M.; Brimeyer, D.G.; Fairbanks, W.S.; et al. Grizzly bear predation links the loss of native trout to the demography of migratory elk in Yellowstone. *Proc. R. Soc. B Biol. Sci.* **2013**, *280*, 20130870. [CrossRef]

88. Gresswell, R.E. *Scientific Review Panel Evaluation of the National Park Service Lake Trout Suppression Program in Yellowstone Lake, 25–29 August*; Final Report, YCR–2009–05; USGS Northern Rocky Mountain Science Center: Bozeman, MT, USA, 2009. Available online: https://www.nps.gov/yell/planyourvisit/upload/gresswell_final_updated_1_2010.pdf (accessed on 10 June 2020).
89. Hansen, M.J.; Horner, N.J.; Liter, M.; Peterson, M.P.; Maiolie, M.A. Dynamics of an increasing lake trout population in Lake Pend Oreille, Idaho. *N. Am. J. Fish. Manag.* **2008**, *28*, 1160–1171. [CrossRef]
90. Koel, T.M.; Arnold, J.L.; Bigelow, P.E.; Ruhl, M.E. *Native Fish Conservation Plan: Environmental Assessment*; U.S. Department of the Interior, National Park Service: Yellowstone National Park, WY, USA, 2010. Available online: https://parkplanning.nps.gov/projectHome.cfm?projectID=30504 (accessed on 10 June 2020).
91. U.S. Department of the Interior. *NPS Management Policies 2006*; National Park Service, U.S. Government Printing Office: Washington, DC, USA, 2006; ISBN 0-16-076874-8.
92. Kaeding, L.R. New climate regime started and further shaped the historic Yellowstone Lake cutthroat trout population decline commonly attributed entirely to nonnative lake trout predation. *Aquat. Ecol.* **2020**, *54*, 641–652. [CrossRef]
93. Detjens, C.R.; Voigt, W.; Voigt, J.; Koel, T.M. Fly fishing volunteers support native fish conservation in Yellowstone. *Yellowstone Sci.* **2017**, *25*, 82–84.
94. Trout Unlimited. *Science Supporting Management of Yellowstone Lake Fisheries: Responses to Frequently Asked Questions*; Wyoming Water Project, Trout Unlimited: Lander, WY, USA, 2014; Available online: http://wyomingtu.org/wp-content/uploads/2014/03/Science-Supporting-Management-of-Yellowstone-Lake-Fisheries.pdf (accessed on 10 June 2020).
95. Hansen, M.J.; Guy, C.S.; Bronte, C.R.; Nate, N.A. Life history and population dynamics. In *Lake Charr Salvelinus Namaycush: Biology, Ecology, Distribution, and Management*; Muir, A.M., Krueger, C.C., Hansen, M.J., Riley, S.C., Noakes, D.L.G., Eds.; Fish & Fisheries Series; Springer: New York, NY, USA. (in press)
96. Morgan, L.A.; Shanks, W.C.; Lovalvo, D.A.; Johnson, S.Y.; Stephenson, W.J.; Pierce, K.L.; Harlan, S.S.; Finn, C.A.; Lee, G.; Webring, M.; et al. Exploration and discovery in Yellowstone Lake: Results from high-resolution sonar imaging, seismic reflection profiling, and submersible studies. *J. Volcanol. Geotherm. Res.* **2003**, *122*, 221–242. [CrossRef]
97. Bigelow, P.E.; Doepke, P.D.; Ertel, B.E.; Guy, C.S.; Syslo, J.M.; Koel, T.M. Suppressing non-native lake trout in Yellowstone Lake. *Yellowstone Sci.* **2017**, *25*, 53–59. Available online: https://www.nps.gov/articles/suppressing-non-native-lake-trout-to-restore-native-cutthroat-trout-in-yellowstone-lake.htm (accessed on 10 June 2020).
98. Gutowsky, L.F.G.; Romine, J.G.; Heredia, N.A.; Bigelow, P.E.; Parsley, M.J.; Sandstrom, P.T.; Suski, C.D.; Danylchuk, A.J.; Cooke, S.J.; Gresswell, R.E. Revealing migration and reproductive habitat of invasive fish under an active population suppression program. *Conserv. Sci. Pract.* **2020**, *2*, e119. [CrossRef]
99. Allen, M.S.; Hightower, J.E. Fish population dynamics: Mortality, growth, and recruitment. In *Inland Fisheries Management in North America*, 3rd ed.; Hubert, W.A., Quist, M.C., Eds.; American Fisheries Society: Bethesda, MD, USA, 2010; pp. 43–79.
100. Quist, M.C.; Pegg, M.A.; DeVries, D.R. Age and growth. In *Fisheries Techniques*, 3rd ed.; Alexander, A.V., Parrish, D.L., Sutton, T.M., Eds.; American Fisheries Society: Bethesda, MD, USA, 2012; pp. 677–731.
101. Guy, C.S.; Brown, M.L. *Analysis and Interpretation of Freshwater Fisheries Data*; American Fisheries Society: Bethesda, MD, USA, 2007.
102. Syslo, J.M. Demography of Lake Trout in Relation to Population Suppression in Yellowstone Lake, Yellowstone National Park. Master's Thesis, Montana State University, Bozeman, MT, USA, 2010.
103. Morris, W.F.; Doak, D.F. *Quantitative Conservation Biology: Theory and Practice of Population Viability Analysis*; Sinauer Associates: Sunderland, MA, USA, 2002.
104. Cambray, J.A. Impact on indigenous species biodiversity caused by the globalisation of alien recreational freshwater fisheries. *Hydrobiologia* **2003**, *500*, 217–230. [CrossRef]
105. Reinhart, D.P. Grizzly Bear Habitat Use on Cutthroat Trout Spawning Streams in Tributaries of Yellowstone Lake. Master's Thesis, Montana State University, Bozeman, MT, USA, 1990.
106. Kaeding, L.R.; Koel, T.M. Age, growth, maturity, and fecundity of Yellowstone Lake cutthroat trout. *Northwest Sci.* **2011**, *85*, 431–444. [CrossRef]

107. Syslo, J.M.; Guy, C.S.; Koel, T.M. Feeding ecology of native and nonnative salmonids during the expansion of a nonnative apex predator in Yellowstone Lake, Yellowstone National Park. *Trans. Am. Fish. Soc.* **2016**, *145*, 476–492. [CrossRef]
108. Ripple, W.J.; Estes, J.A.; Schmitz, O.J.; Constant, V.; Kaylor, M.J.; Lenz, A.; Motley, J.L.; Self, K.E.; Taylor, D.S.; Wolf, C. What is a trophic cascade? *Trends Ecol. Evol.* **2016**, *31*, 842–849. [CrossRef]
109. Teisberg, J.E.; Haroldson, M.A.; Schwartz, C.C.; Gunther, K.A.; Fortin, J.K.; Robbins, C.T. Contrasting past and current numbers of bears visiting Yellowstone cutthroat trout streams. *J. Wildl. Manag.* **2014**, *78*, 369–378. [CrossRef]
110. Stott, W. *Molecular Genetic Characterization and Comparison of Lake Trout from Yellowstone and Lewis Lakes, Wyoming*; Research Completion Report for Project #1443-IA-15709-9013; National Park Service, Yellowstone National Park: Mammoth, WY, USA, 2004; Available online: https://www.researchgate.net/publication/331744804 (accessed on 10 June 2020).
111. Munro, A.R.; McMahon, T.E.; Ruzycki, J.R. Natural chemical markers identify source and date of introduction of an exotic species: Lake trout (*Salvelinus namaycush*) in Yellowstone Lake. *Can. J. Fish. Aquat. Sci.* **2005**, *62*, 79–87. [CrossRef]
112. Stapp, P.; Hayward, G.D. Estimates of predator consumption of Yellowstone cutthroat trout (*Oncorhynchus clarkii bouvieri*) in Yellowstone Lake. *J. Freshw. Ecol.* **2002**, *17*, 319–329. [CrossRef]
113. Koel, T.M.; Kerans, B.L.; Barras, S.C.; Hanson, K.C.; Wood, J.S. Avian piscivores as vectors for *Myxobolus cerebralis* in the Greater Yellowstone Ecosystem. *Trans. Am. Fish. Soc.* **2010**, *139*, 976–988. [CrossRef]
114. Koel, T.M.; Mahony, D.L.; Kinnan, K.L.; Rasmussen, C.; Hudson, C.J.; Murcia, S.; Kerans, B.L. *Myxobolus cerebralis* in native cutthroat trout of the Yellowstone Lake ecosystem. *J. Aquat. Anim. Health* **2006**, *18*, 157–175. [CrossRef]
115. Murcia, S.; Kerans, B.L.; Koel, T.M.; MacConnell, E. *Myxobolus cerebralis* (Hofer) infection risk in native cutthroat trout *Oncorhynchus clarkii* (Richardson) and its relationships to tributary environments in the Yellowstone Lake basin. *J. Fish Dis.* **2014**, *38*, 637–652. [CrossRef]
116. Murcia, S.; Kerans, B.L.; MacConnell, E.; Koel, T.M. *Myxobolus cerebralis* infection patterns in Yellowstone cutthroat trout after natural exposure. *Dis. Aquat. Org.* **2006**, *71*, 191–199. [CrossRef] [PubMed]
117. Murcia, S.; Kerans, B.L.; MacConnell, E.; Koel, T.M. Correlation of environmental attributes with histopathology of native Yellowstone cutthroat trout naturally infected with *Myxobolus cerebralis*. *Dis. Aquat. Org.* **2011**, *93*, 225–234. [CrossRef]
118. Alexander, J.D.; Kerans, B.L.; Koel, T.M.; Rasmussen, C. Context-specific parasitism in *Tubifex tubifex* in geothermally influenced stream reaches in Yellowstone National Park. *J. N. Am. Benthol. Soc.* **2011**, *30*, 853–867. [CrossRef]
119. Stewart, K.P. Use of Otolith Microchemistry to Identify Yellowstone Cutthroat Trout and Lake Trout Natal Origins and Movement Patterns in Yellowstone Lake, Wyoming. Master's Thesis, Montana State University, Bozeman, MT, USA, 2016.
120. Syslo, J.M.; Guy, C.S.; Arnold, J.L.; Koel, T.M.; Ertel, B.D. *Standard Operating Procedures for Distribution Netting in Yellowstone Lake*; 2010–2012 Final Report; USGS, Montana Cooperative Fishery Research Unit. Yellowstone National Park, WY, USA, 2014; Available online: https://www.researchgate.net/publication/339697733 (accessed on 10 June 2020).
121. Gresswell, R.E.; Heredia, N.A.; Romine, J.G.; Gutowsky, L.F.G.; Sandstrom, P.T.; Parsley, M.J.; Bigelow, P.E.; Suski, C.D.; Ertel, B.D. Identifying movement patterns and spawning areas of lake trout in Yellowstone Lake. *Yellowstone Sci.* **2017**, *25*, 66–69. Available online: https://www.nps.gov/articles/identifying-movement-patterns-and-spawning-areas-of-lake-trout-in-yellowstone-lake.htm (accessed on 10 June 2020).
122. Williams, J.R.; Guy, C.S.; Koel, T.M.; Bigelow, P.E. Targeting aggregations of telemetered lake trout to increase gillnetting suppression efficacy. *N. Am. J. Fish. Manag.* **2020**, *40*, 225–231. [CrossRef]
123. Simard, L.G.; Marsden, J.E.; Gresswell, R.E.; Euclide, M. Rapid early development and feeding benefits an invasive population of lake trout. *Can. J. Fish. Aquat. Sci.* **2019**, *77*, 496–504. [CrossRef]
124. Detjens, C.R.; Carim, K.J. Environmental DNA: A new approach to monitoring fish in Yellowstone National Park. *Yellowstone Sci.* **2017**, *25*, 26–27.
125. Roddewig, M.R.; Churnside, J.H.; Hauer, F.R.; Williams, J.; Bigelow, P.E.; Koel, T.M.; Shaw, J.A. Airborne lidar detection and mapping of invasive lake trout in Yellowstone Lake. *Appl. Opt.* **2018**, *57*, 4111–4116. [CrossRef]

126. Food and Agriculture Organization of the United Nations. *Report on the First Session of FAO Panel of Experts on Integrated Pest Control*; FAO: Rome, Italy, 1968.
127. Dent, D. *Integrated Pest Management*; Chapman and Hall: New York, NY, USA, 1995.
128. Ehler, L.E. Integrated pest management (IPM): Definition, historical development and implementation, and the other IPM. *Pest Manag. Sci.* **2006**, *62*, 787–789. [CrossRef]
129. Flint, M.L.; Van den Bosch, R. *Introduction to Integrated Pest Management*; Plenum Press: New York, NY, USA, 1981. [CrossRef]
130. Peshin, R.; Bandral, R.S.; Zhang, W.; Wilson, L.; Dhawan, A.K. Integrated pest management: A global overview of history, programs, and adoption. In *Integrated Pest Management: Innovation-Development Process, Peshin, R., Dhawan, A.K., Eds.*; Springer: Dordrecht, The Netherlands, 2009; pp. 1–49.
131. Lechelt, J.D.; Bajer, P.G. Modeling the potential for managing invasive common carp in temperate lakes by targeting their winter aggregations. *Biol. Invasions* **2016**, *18*, 831–839. [CrossRef]
132. Weber, M.J.; Hennen, M.J.; Brown, M.L.; Lucchesi, D.O.; St. Sauver, T.R. Compensatory response of invasive common carp *Cyprinus carpio* to harvest. *Fish. Res.* **2016**, *179*, 168–178. [CrossRef]
133. Pearson, J.; Dunham, J.; Ryan Bellmore, J.; Lyons, D. Modeling control of common carp (*Cyprinus carpio*) in a shallow lake–wetland system. *Wetlands Ecol. Manag.* **2019**, *27*, 663–682. [CrossRef]
134. Sawyer, A.J. Prospects for integrated pest management of the sea lamprey (*Petromyzon marinus*). *Can. J. Fish. Aquat. Sci.* **1980**, *37*, 2081–2092. [CrossRef]
135. Christie, G.C.; Goddard, C.I. Sea lamprey international symposium (SLIS II): Advances in the integrated management of sea lamprey in the Great Lakes. *J. Great Lakes Res.* **2003**, *29*, 1–14. [CrossRef]
136. Johnson, N.S.; Yun, S.-S.; Thompson, H.T.; Brant, C.O.; Li, W. A synthesized pheromone induces upstream movement in female sea lamprey and summons them into traps. *Proc. Natl. Acad. Sci. USA* **2009**, *106*, 1021–1026. [CrossRef]
137. Ferreri, P.C.; Taylor, W.W.; Hayes, D.B. Evaluation of age-0 survival and its effect on lake trout rehabilitation in the Michigan waters of Lake Superior. *J. Great Lakes Res.* **1995**, *21*, 218–224. [CrossRef]
138. Cox, B.S.; Guy, C.S.; Fredenberg, W.A.; Rosenthal, L.R. Baseline demographics of a non-native lake trout population and inferences for suppression from sensitivity-elasticity analyses. *Fish. Manag. Ecol.* **2013**, *20*, 390–400. [CrossRef]
139. Bernhart, B.; Blackwood, C.; Gale, R.; Held, E. *Yellowstone National Park Lake Trout Crisis: Report Prepared to Fulfill the Requirements of ME 404/ChE 411, Senior Design*; College of Engineering, Montana State University: Bozeman, MT, USA, 2004. [CrossRef]
140. Gross, J.A.; Farokhkish, B.; Gresswell, R.E.; Webb, M.A.H.; Guy, C.S.; Zale, A.V. *Techniques for Suppressing Invasive Fishes in Lacustrine Systems: A Literature Review*; Final Report for Project RM-CESU H1200040001; National Park Service, Rocky Mountains Cooperative Ecosystem Studies Unit: Yellowstone National Park, WY, USA, 2010; Available online: http://files.cfc.umt.edu/cesu/NPS/MSU/2008/08Zale_YELL_trout%20embryos_lit%20review.pdf (accessed on 10 June 2020).
141. Brown, P.J.; Guy, C.S.; Meeuwig, M.H. A comparison of two mobile electrode arrays for increasing mortality of lake trout embryos. *N. Am. J. Fish. Manag.* **2017**, *37*, 363–369. [CrossRef]
142. Doepke, P.D.; Koel, T.M.; Guy, C.S.; Poole, A.S.; Thomas, N.A.; Zale, A.V. Lake trout suppression alternatives to gillnetting. *Yellowstone Sci.* **2017**, *25*, 70–73. Available online: https://www.nps.gov/yell/learn/ys-25-1-lake-trout-suppression-alternatives-to-gillnetting.htm (accessed on 10 June 2020).
143. Thomas, N.A. Evaluation of Suppression Methods Targeting Non-Native Lake Trout Embryos in Yellowstone Lake, Yellowstone National Park, Wyoming, USA. Master's Thesis, Montana State University, Bozeman, MT, USA, 2017.
144. Thomas, N.A.; Guy, C.S.; Koel, T.M.; Zale, A.V. In-situ evaluation of benthic suffocation methods for suppression of invasive lake trout embryos in Yellowstone Lake. *N. Am. J. Fish. Manag.* **2019**, *39*, 104–111. [CrossRef]
145. Poole, A.S.; Koel, T.M.; Thomas, N.A.; Zale, A.V. Suppression of invasive lake trout embryos by fish carcasses and sedimentation in Yellowstone Lake, Yellowstone National Park. *N. Am. J. Fish. Manag.* (under review).
146. Koel, T.M.; Arnold, J.L.; Bigelow, P.E.; Detjens, C.R.; Doepke, P.D.; Ertel, B.D.; MacDonald, D.J. *Native Fish Conservation Program, Yellowstone National Park: Report 2015–2018*; YCR-2019-04; National Park Service, Yellowstone Center for Resources: Yellowstone National Park, WY, USA, 2019. Available online: https://home.nps.gov/yell/learn/upload/2015-2018-Fish-Report-_Web.pdf (accessed on 10 June 2020).

147. Poole, A.S. Evaluation of Embryo Suppression Methods for Nonnative Lake Trout in Yellowstone Lake, Yellowstone National Park, Wyoming, USA. Master's Thesis, Montana State University, Bozeman, MT, USA, 2019.
148. Dwyer, W.P. Effect of lowering water temperature on hatching time and survival of lake trout eggs. *Prog. Fish-Cult.* **1987**, *49*, 175–176. [CrossRef]
149. Bronte, C.R.; Selgeby, J.H.; Saylor, J.H.; Miller, G.S.; Foster, N.R. Hatching, dispersal, and bathymetric distribution of age-0 wild lake trout at the Gull Island shoal complex, Lake Superior. *J. Great Lakes Res.* **1995**, *21*, 233–245. [CrossRef]
150. Mullins, M.S. Biology and Predator Use of Cisco (*Coregonus artedi*) in Fort Peck Reservior, Montana. Master's Thesis, Montana State University, Bozeman, MT, USA, 1991.
151. Cherry, T.L.; Shogren, J.F. Invasive species management for the Yellowstone Lake ecosystem: What do visitors think? *Yellowstone Sci.* **2001**, *9*, 10–15. Available online: https://www.nps.gov/yell/learn/upload/YS_9_2_sm.pdf (accessed on 10 June 2020).
152. National Park Service. Organic Act 16 USC 1—4. ch. 408, 39 Stat. 53511. 25 August 1916. Available online: https://www.nps.gov/parkhistory/online_books/fhpl/nps_organic_act.pdf (accessed on 10 June 2020).
153. Sax, D.F.; Stachowicz, J.J.; Brown, J.H.; Bruno, J.F.; Dawson, M.N.; Gaines, S.D.; Grosberg, R.K.; Hastings, A.; Holt, R.D.; Mayfield, M.M.; et al. Ecological and evolutionary insights from species invasions. *Trends Ecol. Evol.* **2007**, *22*, 465–471. [CrossRef] [PubMed]
154. Brown, J.H.; Sax, D.F. An essay on some topics concerning invasive species. *Austral Ecol.* **2004**, *29*, 530–536. [CrossRef]
155. Morgan, L.A.; Shanks, P.; Lovalvo, D.; Pierce, K.; Lee, G.; Webring, M.; Stephenson, W.; Johnson, S.; Finn, C.; Schulze, B.; et al. The floor of Yellowstone Lake is anything but quiet! New discoveries in lake mapping. *Yellowstone Sci.* **2003**, *11*, 15–30. Available online: https://www.nps.gov/yell/learn/upload/YS_11_2_sm.pdf (accessed on 10 June 2020).
156. Pianka, E.R. On r- and K-Selection. *Am. Nat.* **1970**, *104*, 592–597. [CrossRef]
157. Reznick, D.; Bryant, M.J.; Bashey, F. r- and K-selection revisited: The role of population regulation in life-history evolution. *Ecology* **2002**, *83*, 1509–1520. [CrossRef]
158. Sibly, R.M.; Barker, D.; Denham, M.C.; Hone, J.; Pagel, M. On the regulation of populations of mammals, birds, fish, and insects. *Science* **2005**, *309*, 607. [CrossRef]
159. Martin, N.; Olver, C.H. The lake charr, *Salvelinus namaycush*. In *Charrs: Salmonid Fishes of the Genus Salvelinus*; Balon, E., Ed.; Kluwer Boston, Inc.: Hingham, MA, USA, 1980; pp. 205–277.
160. Westley, P.A.H.; Fleming, I.A. Landscape factors that shape a slow and persistent aquatic invasion: Brown trout in Newfoundland 1883–2010. *Divers. Distrib.* **2011**, *17*, 566–579. [CrossRef]
161. Gunn, J.M. Spawning behavior of lake trout: Effects on colonization ability. *J. Great Lakes Res.* **1995**, *21*, 323–329. [CrossRef]
162. MacArthur, R.H.; Wilson, E.O. *The Theory of Island Biogeography*; Princeton University Press: Princeton, NY, USA, 1967.
163. Eklöv, P.; Svanbäck, R. Predation risk influences adaptive morphological variation in fish populations. *Am. Nat.* **2006**, *167*, 440–452. [CrossRef]
164. Suski, C.D.; Cooke, S.J. Conservation of aquatic resources through the use of freshwater protected areas: Opportunities and challenges. *Biodivers. Conserv.* **2007**, *16*, 2015–2029. [CrossRef]
165. Hedges, K.J.; Koops, M.A.; Mandrak, N.E.; Johannsson, O.E. Use of aquatic protected areas in the management of large lakes. *Aquat. Ecosyst. Health Manag.* **2010**, *13*, 135–142. [CrossRef]
166. Parker, S.R.; Mandrak, N.E.; Truscott, J.D.; Lawrence, P.L.; Kraus, D.; Bryan, G.; Molnar, M. Status and extent of aquatic protected areas in the Great Lakes. *George Wright Forum* **2017**, *34*, 381–393.
167. Ricker, W.E. Stock and recruitment. *J. Fish. Res. Board Can.* **1954**, *11*, 559–623. [CrossRef]
168. Hilborn, R.; Walters, C.J.; Ludwig, D. Sustainable exploitation of renewable resources. *Ann. Rev. Ecol. Syst.* **1995**, *26*, 45–67. [CrossRef]
169. Rose, K.A.; Cowan Jr, J.H.; Winemiller, K.O.; Myers, R.A.; Hilborn, R. Compensatory density dependence in fish populations: Importance, controversy, understanding and prognosis. *Fish Fish.* **2001**, *2*, 293–327. [CrossRef]
170. Abrams, P.A. When does greater mortality increase population size? The long history and diverse mechanisms underlying the hydra effect. *Ecol. Lett.* **2009**, *12*, 462–474. [CrossRef]

Review

A Decade in Review: Alaska's Adaptive Management of an Invasive Apex Predator

Kristine Dunker [1,*], Robert Massengill [2], Parker Bradley [3], Cody Jacobson [3], Nicole Swenson [4], Andy Wizik [5] and Robert DeCino [2]

1 Alaska Department of Fish and Game, Anchorage, AK 99518, USA
2 Alaska Department of Fish and Game, Soldotna, AK 99669, USA; robert.massengill@alaska.gov (R.M.); robert.decino@alaska.gov (R.D.)
3 Alaska Department of Fish and Game, Palmer, AK 99645, USA; parker.bradley@alaska.gov (P.B.); cody.jacobson@alaska.gov (C.J.)
4 Tyonek Tribal Conservation District, Anchorage, AK 99501, USA; nswenson@tyonek.com
5 Cook Inlet Aquaculture Association, Kenai, AK 99611, USA; awizik@ciaanet.org
* Correspondence: kristine.dunker@alaska.gov; Tel.: +907-267-2889

Received: 6 March 2020; Accepted: 15 April 2020; Published: 21 April 2020

Abstract: Northern pike are an invasive species in southcentral Alaska and have caused the decline and extirpation of salmonids and other native fish populations across the region. Over the last decade, adaptive management of invasive pike populations has included population suppression, eradication, outreach, angler engagement, and research to mitigate damages from pike where feasible. Pike suppression efforts have been focused in open drainages of the northern and western Cook Inlet areas, and eradication efforts have been primarily focused on the Kenai Peninsula and the municipality of Anchorage. Between 2010 and 2020, almost 40,000 pike were removed from southcentral Alaska waters as a result of suppression programs, and pike have been successfully eradicated from over 20 lakes and creeks from the Kenai Peninsula and Anchorage, nearly completing total eradication of pike from known distributions in those areas. Northern pike control actions are tailored to the unique conditions of waters prioritized for their management, and all efforts support the goal of preventing further spread of this invasive aquatic apex predator to vulnerable waters.

Keywords: suppression; eradication; rotenone; fishery restoration; northern pike; salmon

1. Introduction

Management of biological invasions pose significant conservation challenges across the globe. Invasions of aquatic species, particularly freshwater fishes, can be among the most arduous to manage due to factors such as habitat complexity, impacts to co-occurring species, and conflicts between ecological and socioeconomic needs. Due to global anthropomorphic translocations of fishes, native freshwater fish communities rarely resemble those provided by nature [1,2]. In countless waters, many of these events have resulted in the introduced species becoming invasive in their novel environments [3,4]. Notable examples in North America include Asian carp *Hypopthalmichthys spp.* Bleeker 1860 in the Mississippi River drainage, northern snakehead *Channa argus* Cantor 1842 in the Chesapeake Bay watershed, and widespread prolific illegal introductions of game fishes (i.e., bass, trout, walleye). An iconic freshwater game fish that has been widely introduced outside its native range in North America is the northern pike *Esox lucius* Linneaus 1758, and its introductions are currently threatening native fish assemblages in numerous drainages [5].

Northern pike, hereafter pike, have a native Holarctic circumpolar distribution including northern Europe, Asia, and North America generally above 40° latitude. Illegal introductions of pike have expanded their range in several countries in Europe and Africa, southwestern British Columbia, and throughout the American west including southcentral Alaska. Though most of Alaska falls within the native range for pike, Southcentral Alaska does not have any natural populations [6,7]. Pike generally occupy relatively shallow vegetated lakes, flooded wetlands, low-gradient rivers and backwater sloughs [8].

Pike are opportunistic apex predators that are primarily piscivorous but will also prey on small mammals, waterfowl, amphibians, and invertebrates. Where pike are not a native species, they have the capacity to both directly and indirectly alter freshwater fish communities [9–11], especially in waters providing optimal pike spawning and rearing habitat [5]. An effect repeatedly documented following pike introduction is the population-level loss of economically vital fish species [12,13]. Negative ecological and economic impacts such as these classify pike as an invasive species [14] in waters outside its native range [5].

There are a variety of factors that contribute to the invasion success of pike such as its trophic adaptability [15,16] broad physiochemical tolerances [17–19], high fecundity [19], ability to achieve high populations densities [20,21], and popularity as a fishing commodity [22–24]. Throughout the American west, several non-native populations of pike are now well-established. Ramifications of these introductions include complicating endangered species recovery efforts in the Colorado River basin [25], threatening conservation efforts for salmon populations in the Columbia River basin [26], contributing to declines in bull trout *Salvelinus confluentus* Suckley 1859 and westslope cutthroat trout *Oncorhynchus clarki lewisi* Girard 1856 populations in Montana, Washington, and Idaho [27,28], and causing the extirpation and declines of many salmonid populations in southcentral Alaska [5]. The extent of the pike invasion in southcentral Alaska and the resulting consequences to native fish populations in the region have been some of the most extensive in their invaded range. Therefore, the focus of this review will be on the impacts of invasive pike in southcentral Alaska and the on-going adaptive management approaches applied to mitigate the damages.

Data presented in this review originate from a variety of sources spanning from published studies to agency reports, thus making the latter more broadly available to the scientific community. Salmon populations in Alaska are assessed in a variety of ways including aerial enumeration surveys, weir and fishwheel counts, sonar estimates, creel surveys, gillnet surveys, foot stream surveys, guide and commercial logbook records, permit reports, and post-season angler surveys. For all trends reported in this review, cited reports include detailed summaries of the methods employed. Pike are most commonly captured through variations of gillnet surveys, and their abundance patterns are measured through temporal Catch Per Unit Effort (CPUE) comparisons or population estimates (i.e., Peterson mark-recapture methods). In most cases, direct impacts of pike on salmonids have been documented through pike diet investigations, bioenergetics models, and fish assemblage comparisons in waters before and after pike establishment. The collection of work presented in this review documents a decade of progress in managing one of Alaska's most challenging biological invasions.

2. Ecological Role of Pike in Alaska

Pike are a native species in northern and western Alaska but do not naturally occur south or east of the Alaska Mountain Range except for a small, isolated, remnant population near Yakutat [29,30] (Figure 1). Natural pike distribution in Alaska is largely the result of geologic barriers during the Late Pleistocene when the majority of southern Alaska was glaciated [31,32]. For approximately 11,000 years, freshwater fish assemblages in drainages in southcentral (SC) Alaska, hereby defined as the region south of the Alaska Mountain Range, developed in the absence of this aquatic apex piscivore [33,34]. Anecdotal accounts suggest that in the late 1950s, pike were first transported by an angler from the Minto Flats in Interior Alaska to the Yentna River drainage in the Susitna River basin. Subsequent unauthorized introductions and pike movements through open waters resulted in their establishment in over 120 lakes and rivers in the Susitna drainage, Knik Arm drainage, Anchorage vicinity, northern Kenai Peninsula and west Cook Inlet drainages from the Threemile to the Lewis Rivers. Pike populations in SC appear to be genetically distinct from native populations in Alaska and exhibit low genetic diversity indicative of small founding populations [35].

Figure 1. Native and invasive ranges of pike in Alaska. Native range in dashed area; invasive range in solid areas.

Pike occupy a top predator niche in all waters they occur in whether native or non-native [9]. In their native range, pike naturally play a pivotal top-down role in shaping freshwater fish assemblages in shallow low-flow habitats with abundant macrophytes [36]. Examples of natural pike-dominated systems in Alaska include the Minto Flats near Fairbanks, the Dall River tributary of the Yukon River, and the Innoko River in western Alaska [37,38]. In SC Alaska, there is a plethora of similar lowland habitat that naturally functions as vital rearing habitat for Chinook salmon *Oncorhynchus tshawytscha* Walbaum 1792, coho salmon *O. kisutch* Walbaum 1792, and rainbow trout *O. mykiss* Walbaum 1792, among other species [35,39,40]. In many of these waters, pike predation on juvenile salmon and trout has led to localized extirpations of their populations [13,16,41]. Pike diet investigations have demonstrated a propensity for predation on soft-rayed fusiform fishes before other, more energetically expensive, taxa such as sticklebacks *Gasterosteus spp.* Linnaeus 1758 and slimy sculpin *Cottus cognatus* Richardson 1836, waterfowl, small mammals, or conspecifics are depredated [15,16,33,42].

There are approximately 70 lakes in SC Alaska that have documented reduced or extirpated salmonid populations [Alaska Department of Fish and Game (ADFG) unpublished]. Perhaps the most significant case study in SC Alaska waters is that of Alexander Creek, the southernmost western tributary of the Susitna River. Historically Alexander Creek supported a Chinook salmon stock that was vital to the local economy and was one of the most popular recreational fisheries in the Susitna basin. Northern pike were illegally introduced to Alexander Lake at the headwaters of the 84,726-hectare watershed in the 1960s, but it took decades for pike to move and establish throughout the 66 km river corridor and connected waters [43]. In this system [39,40,44] and others in SC Alaska [45], long-distance pike movements over a short duration appear to be uncommon, and pike tend to display a high affinity for macrophyte beds, littoral regions, and outlet streams of individual lakes such that it can take decades for new populations to fully occupy suitable habitat within a drainage [44]. In the late 1980's anglers in Alexander Creek began catching pike [46], but they were not detected in lower Alexander Creek until the late 1990s [43]. Prior to this, Chinook salmon abundance followed similar patterns to those in nearby rivers (i.e., Talachulitna River; Figure 2). By the early 2000's, after pike established in lower Alexander Creek, Chinook salmon abundance precipitously decreased below sustainable levels [43,47]. Exacerbating matters, Chinook salmon in Alaska have been in a period of low productivity since 2007 [48]. However, Chinook abundance in Alexander Creek experienced substantially greater decreases relative to proximate rivers without pike (Figure 2), while angler catches for pike increased (Figure 3).

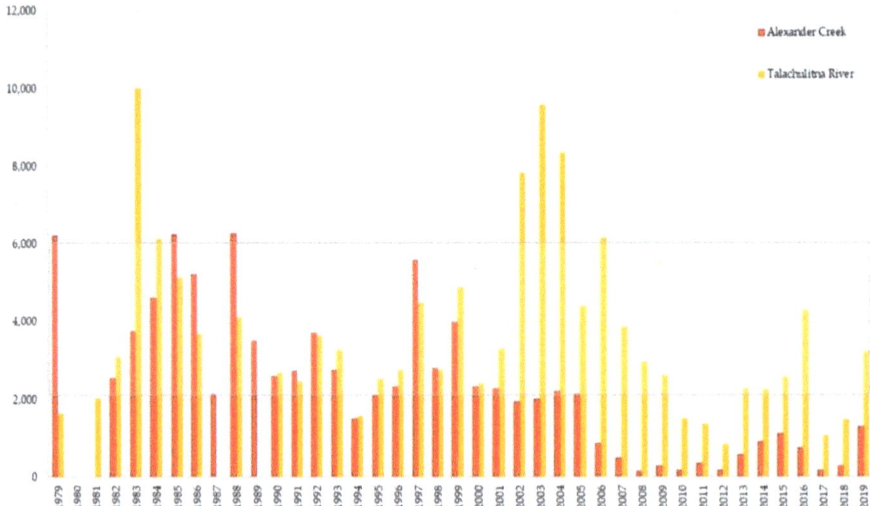

Figure 2. Alexander Creek (pike present) and Talachulitna River (pike not present) aerial survey counts for Chinook salmon escapement. The shaded area shows the escapement goal range (sustainable abundance range) for Alexander Creek Chinook salmon. The escapement goal range for the Talachuitna River is 2200 to 5000 Chinook salmon.

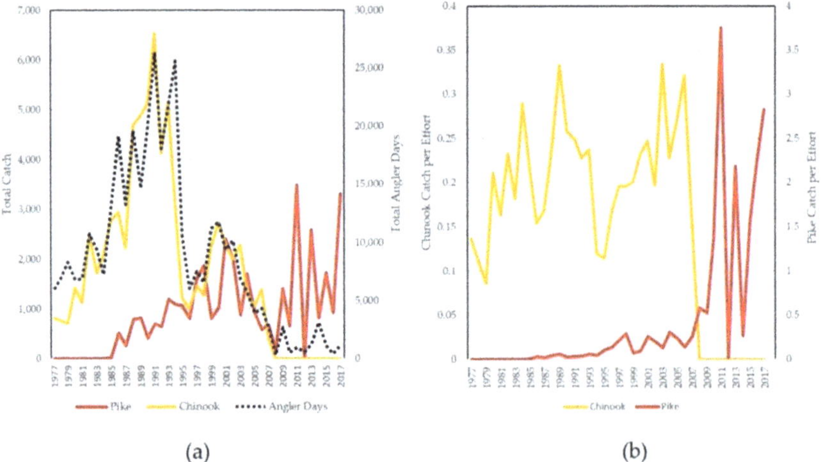

Figure 3. (**a**) Total angler catches of Chinook salmon and Pike in Alexander Creek (1977–2017) based on Statewide Harvest Survey estimates [47,49] and (**b**) Catch per angler effort (angler day) for Chinook salmon and pike in Alexander Creek (1977–2017). Note: Sport fishing for Chinook Salmon was closed in 2008 due to their low returns (Figure 2).

Predation by northern pike is considered the most significant driver of Chinook salmon abundance declines in Alexander Creek [12,16,43]. Pike diet investigations in Alexander Creek began in 2011, about a decade after pike established in the lower river, and their diets reflected the distribution of spawning salmonids, which decreased with distance upstream. In lower Alexander Creek where juvenile salmon were still present but rare, rearing salmon dominated pike diets. Further upstream, where juvenile salmonids had been extirpated, pike diets were dominated by Arctic lamprey *Lethenteron camtschaticum* Tilesius 1811, Pacific lamprey *Entosphenus tridentatus* Richardson 1836 and slimy sculpin, illustrating pike's trophic adaptability following depletion of targeted prey sources [16]. In other SC pike waters such shifts in diet have been shown to eventually deplete secondarily targeted fish species such as threespine sticklebacks *Gasterosteus aculeatus* Linnaeus 1758 [33,34] and eventually become dominated by aquatic macroinvertebrates [41,50]. Bioenergetics models demonstrated that pike could consume up to 1.10 metric tons of juvenile salmonid prey in Alexander Creek annually, which far exceeds the salmonid prey base in the system [12]. This predicted the complete loss of the Chinook salmon stock without management intervention to reduce pike abundance and predation.

The impacts of pike across all SC Alaska waters has been highly variable. This is generally explained by the degree of habitat complexity and connectivity across invaded waters [5,10,16,36]. Where pike are native in western Alaska, they co-occur with the world's largest sockeye salmon *O. nerka* Walbaum 1792 stocks in Bristol Bay. However, large sockeye drainages like the Wood-Tikchick, for example, are enormous systems with deep expansive lakes, high velocity streams, and marshy lowlands. Pike are abundant in littoral regions, marshy lake outlets, and flooded wetland habitats, but are rare in pelagic waters of deep lakes and turbid glacial rivers [51–53]. In drainages like this with a high degree of habitat heterogeneity, salmon may be spatially or temporarily vulnerable to pike predation during parts of their life cycle, but they largely avoid predation otherwise; therefore, the population-level implications for salmon populations are more negligible [16,41]. In drainages where pike are native and have more homogenous habitat with favorable conditions for pike, pike naturally make up a large component of the ichthyofauna in those waters [37,38,54,55].

Where pike are not naturally occurring in SC Alaska, there is similar diversity in habitat throughout the region with areas that are both highly suitable and unsuitable for pike. The Susitna Basin is approximately 6,500,000 hectares, and much of the watershed includes high-velocity clear or turbid glacial rivers, particularly in the eastside Susitna River tributaries. In these waters, pike are unlikely to establish abundant populations nor have significant population-level impacts on salmonid populations because of habitat segregation that mitigates their predation risk. Despite this, the mainstem Susitna River or the Kenai River on the Kenai Peninsula, which do not provide optimal pike habitat, can be used as movement corridors for pike to spread to more favorable habitats. This has been the case in many westside Susitna River tributaries. Westside Susitna streams tend to be lower-gradient, and the surrounding landscape is an extensive interconnected mosaic of lakes, ponds, creeks, marshes and floating bog. Habitat conditions for pike in several westside tributaries are much more favorable, and consequently, pike abundances in these drainages are higher. However, in some cases, westside tributaries with more variable habitat like the Deshka River have high localized pike predation on salmonids, but salmon populations persist because there are areas of the river where salmonids can avoid predation [16]. In contrast, in systems like Alexander Creek that lack predation refugia and have complete habitat overlap between salmon and pike, predation impacts are far greater [16]. This same effect was also observed in over 20 shallow lakes on the Kenai Peninsula [13,56–60]. In waters where habitat conditions favor pike, multiple native fish populations have been lost. Presently, pike remain restricted to only a proportion of their available habitat in SC Alaska, but many drainages and salmon populations remain highly vulnerable to pike invasion [35].

3. Management Approaches

Given the impacts to native fish populations that have already been incurred in SC Alaska and the potential for further impacts, management efforts have been underway to mitigate the damages. Most invasive pike management activities are conducted by or are in consultation with the Alaska Department of Fish and Game (ADFG) and directed through management plans [61,62]. Funding for invasive pike management is limited; therefore, priorities must be identified. Most invasive pike management projects are prioritized using an ADFG-developed scoring matrix designed to ensure that proposed efforts will maximize restoration benefits to impacted fisheries and prevent further invasion of pike to vulnerable waters (Appendix A). The primary functional areas of invasive pike management in SC Alaska include population suppression, eradication, outreach and angler involvement, and research.

3.1. Population Suppression

Population suppression is a common strategy employed for invasive fish management when eradication of an entire population is not feasible [2,25,63,64]. In SC Alaska there are several pike suppression programs in westside Susitna River waters where management is necessary to mitigate impacts to native fisheries, but the spatial extent and reinvasion potential makes successful eradications unlikely with current methods. Current pike suppression programs in the Susitna basin take place in Alexander Creek, Chelatna Lake, Whiskey Lake, Hewitt Lake, Shell Lake and in the Threemile Lake complex and Chuitbuna Lake on the west side of Cook Inlet (Figure 4) [44].

As discussed earlier, the Alexander Creek drainage is one of the most heavily impacted systems in the Susitna Basin, and salmon populations there were predicted to be completely lost without intervention [12]. To prevent this, ADFG conducts an annual program that began in 2011 to reduce pike abundance in the optimal pike habitat of side-channel sloughs along Alexander Creek (Figure 5a). The primary objective is to bolster salmon productivity in the system to sustainable levels [65] by reducing pike predation on juvenile salmon. Each May, during the pike spawning period, gillnets (1.8 m by 35.6 m, variable mesh 1.9 cm–5.1 cm) are systematically fished in approximately 60 sloughs. From 2011–2018, sloughs were fished until they achieved 80% reduction in pike catch from their first day's catch [43]. Beginning in 2019, this method was simplified to fish each slough for three days unless pike continue to be caught, in which case, the slough continues to be netted until pike catches cease [66]. Netting efforts are divided between 2–3 field crews. Due to the remoteness of this area, field crews remain on site for the duration of the season which typically ends by 1 June each year. All pike netted are measured and dissected to study diet patterns, age structure, and sex ratio [43,66]. Co-occurring species distributions and indices of their abundance are monitored through pike stomach contents, gillnet bycatch, and mid-summer minnow trap surveys [43,66]. Annual Chinook salmon returns are monitored through aerial index surveys during the peak of the run that have been flown since 1979 [47] (Figures 2 and 5b). Between 2011 and 2013, pike movements were investigated via radio telemetry to determine if pike at the headwaters of the system in Alexander Lake were a consistent source of fish moving downstream into the creek [44].

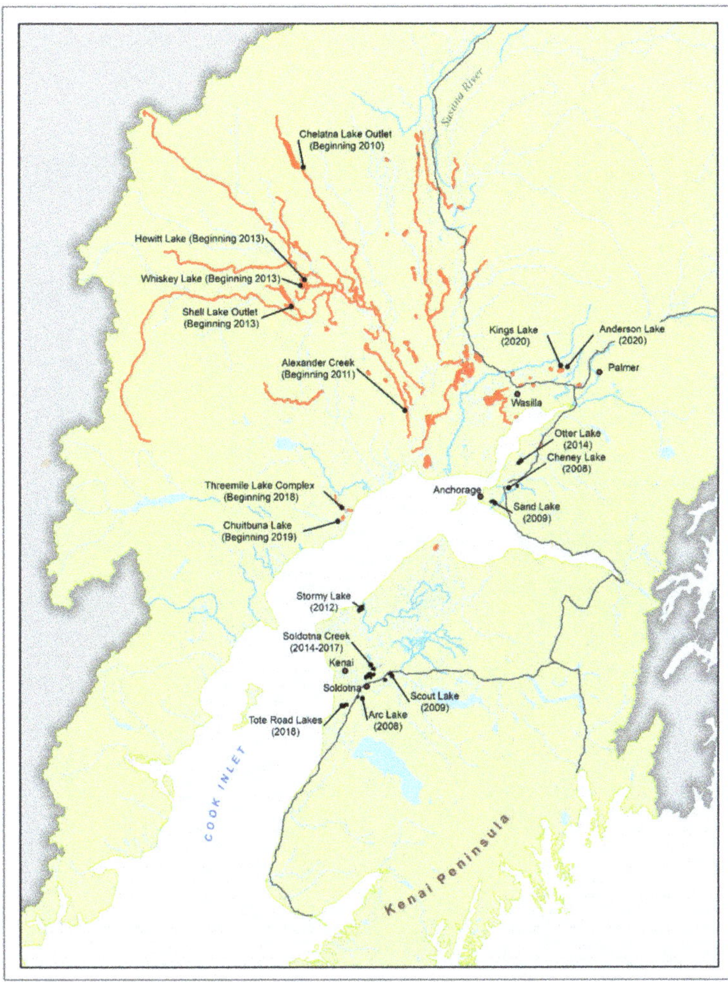

Figure 4. Drainages with pike populations (red). Pike suppression locations noted by year projects were initiated. Pike eradication locations noted by year project was completed.

Between 2011 and 2019, 20,446 pike were removed from Alexander Creek sloughs. During the first three years of suppression between 3000 and 4000 pike were removed annually, but annual pike removal has trended downward since, with less than 1000 pike removed in 2019 (Figure 5b). With the exception of the period between 2016 and 2018, Chinook abundance has trended upward since pike suppression began (Figure 5b). The years between 2016–2018 are difficult to assess because Chinook runs statewide were depressed [48]. However, 2019 saw the largest return of Chinook salmon in Alexander Creek in almost 15 years [65] (Figure 5b). In addition to positive signs with Chinook returns in the system, Chinook distribution has reestablished throughout the entire creek corridor. As discussed earlier, prior to pike suppression, spawning Chinook and juveniles were restricted to the lowest river reach from approximately RM 20.5 to the confluence with the Susitna River (Figure 5c,d). Annual minnow trap surveys, pike stomach analyses, and aerial index surveys confirmed a gradual return of adult Chinook salmon to former spawning grounds and juvenile Chinook salmon reoccurring all the way up to the Alexander Lake outlet by 2014 [43] (Figure 5c,d). In addition, other species like Arctic grayling *Thymallus arcticus* Pallas 1776 and rainbow trout *O. mykiss* Walbaum 1792 have increased in abundance as indicated by gillnet bycatch in recent years. Pike diets in Alexander Creek remain heavily dominated by fish, but despite pike being the dominant fish species in the system, cannibalism was observed in less than 1% of captured pike [43]. During the pike movement investigation in this system, radio-tagged pike displayed a high degree of site fidelity to their tagging locations [44]. Only 7% of tagged pike in Alexander Lake were observed downstream during the study, and all these fish were dispatched in suppression gillnets. At the time, this was interpreted as a positive sign that the objective of increasing salmon productivity in the Alexander system could be accomplished without suppression in Alexander Lake, which would have been a far more costly endeavor. However, in 2014, a new aquatic invasive species, *Elodea canadensis* Babington 1848 was discovered in Alexander Lake. The elodea infestation has rapidly increased and today covers over 90% of the lake [Alaska Department of Natural Resources (ADNR) Unpublished]. While the elodea is under active herbicide management with the goal of eradication, it is unknown if pike movement patterns between Alexander Lake and Alexander Creek have been altered as a result of the significant change in habitat structure. Moving forward, with the current uncertainty in movement dynamics of pike in the system, and to facilitate continued positive trends in Chinook salmon abundance, pike suppression in Alexander Lake is likely now warranted.

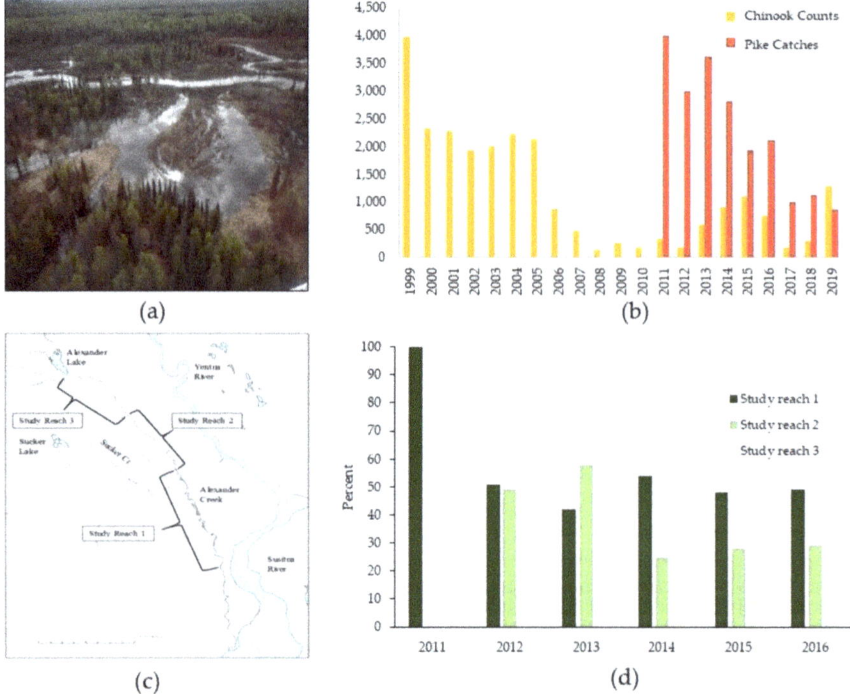

Figure 5. (**a**) A side-channel slough along Alexander Creek, (**b**) 20-year Chinook abundance in Alexander Creek and pike removals by year since suppression began, (**c**) map illustrating study reaches in Alexander Creek (Study Reach 1 = lower Alexander Creek, Study Reach 2 = middle Alexander Creek, Study Reach 3 = Upper Alexander Creek), (**d**) Proportion of juvenile salmon captured in minnow trap surveys in each study reach by year until salmon were again well-distributed throughout the creek corridor (2016).

Pike suppression occurs opportunistically pending funding availability in Shell, Whiskey, Hewitt, and Chelatna lakes and their outlet creeks in a partnership between ADFG and the Cook Inlet Aquaculture Association (CIAA). All these lakes are significant sockeye salmon producers for the Susitna drainage [67]. While some of the lakes, Chelatna in particular, are deep enough to allow habitat segregation between sockeye and pike, sockeye are heavily depredated during both smolt out migrations through outlet streams and juvenile fry recruitment to the pelagic lake habitat. Pike suppression efforts in these lakes have been designed to mitigate these effects [41,68,69].

Natural sockeye production in Shell Lake has declined precipitously since 2007 (Figures 4 and 6). Based on a euphotic volume model, Shell Lake has the estimated potential to contribute up to 10% of the sockeye salmon return to the Susitna River watershed [70]. Between 2006 and 2011, Shell Lake contributed 6.1% of the sockeye salmon return to the Susitna [71]. Due to a notable decline in weir counts of sockeye smolt leaving Shell Lake since 2010 (Figure 6) and a similar trend in adult returns, CIAA initiated pike suppression efforts to reduce predation pressure on smolt. The project included several components including smolt and adult sockeye enumeration, sockeye stocking with Shell Lake brood stock, disease screening, pike suppression, beaver dam modifications, and evaluating the effect of pike suppression on sockeye and pike abundances in the lake.

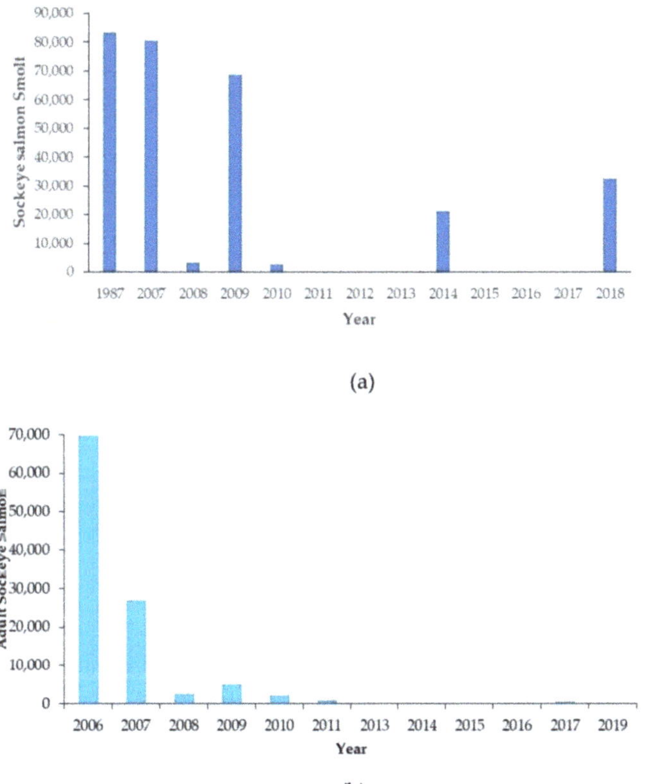

Figure 6. (**a**) Sockeye smolt counts in Shell Lake in 1987 and from 2007–2018 (**b**) Weir counts for adult sockeye in Shell Lake from 2006–2011, 2013–2017, and 2019.

CIAA collected eggs from adult sockeye salmon in Shell Lake during 2012, 2016, and 2017. The fertilized eggs were reared in Trail Lakes Hatchery to the smolt stage and released back into Shell

Lake in 2014, 2018, and 2019. The 2014 smolt release occurred prior to major pike suppression and resulted in the outmigration of only 25% of the released smolt. Diet data taken from captured pike in 2014 showed that the northern pike in Shell Lake stopped preying on most other prey items and focused heavily on smolt when they were available [71–73]. Between 2014 and 2019 CIAA removed 5296 pike from Shell Lake with gillnets (2.5 cm bar gillnets, 2.4 m by 15.2 m) and the number of stocked sockeye smolt emigrating from the lake increased to 71% [73]. Since suppression began, over 7050 pike have been removed from Shell Lake with gillnets reducing the catch-per-unit-effort over that time period from 0.39 pike/gillnet h in 2012 to 0.06 pike/gillnet h in 2019. Bioenergetics models have estimated that this decrease in the pike population reduced the consumption of salmonids by approximately 81% for the period between 2012–2016 [68].

Pike suppression with gillnets (1.8 m by 22.9 m, variable mesh 2.5 cm–7.6 cm) and fyke nets was conducted in Chelatna Lake between 2010–2012 and 2017–2019 by ADFG and intermittently by CIAA between 2009–2016. Like Shell Lake, Chelatna is a significant sockeye producer for the Susitna River but also supports populations of Chinook, coho, chum *O. keta* Walbaum 1792 and pink *O. gorbuscha* Walbaum 1792 salmon. Combined suppression efforts in Chelatna Lake has removed 4335 pike, and CPUE has decreased over time (Figure 7). Pike diet analyses documented that juvenile salmon dominated pike diets during outmigration, sockeye prey size was positively correlated with pike predator size, and 67% of the depredated sockeye smolt were consumed by pike less than 30 cm (fork length). Bioenergetics estimates of sockeye smolt survival resulting from pike suppression activities suggested a potential increase of over 13,000 adult sockeye [74].

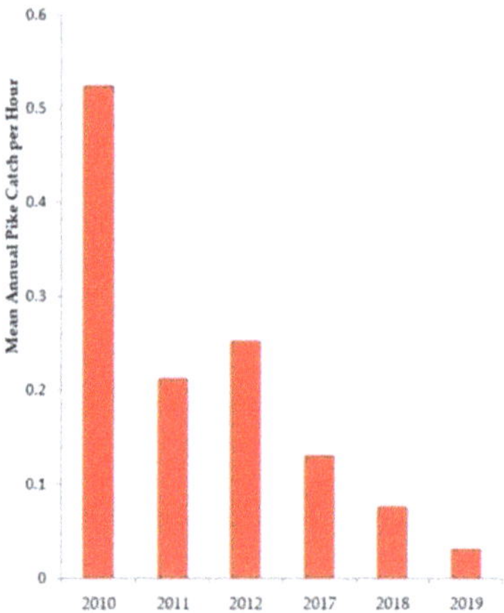

Figure 7. Catch Per Unit Effort (CPUE) of pike removed during ADFG pike suppression in Chelatna Lake 2010–2012 and 2017–2019.

In the Yentna River drainage, pike suppression is conducted in Hewitt and Whiskey lakes and the Hewitt-Whiskey Creek outlet that connects these lakes to the Yentna River (Figure 4). Since Yentna River water backs up into Hewitt Lake during peak summer flows, Hewitt and Whiskey lakes provide rearing habitat for juvenile sockeye salmon produced by adults that spawn in sloughs and tributaries of the Yentna River drainage upstream of these two lakes. Approximately 17–26% of Yentna River sockeye (~48,000–62,000) were estimated to spawn in the upper river based on radio-telemetry investigations [75–77]. Production estimates from a euphotic volume model indicate that Hewitt and Whiskey lakes can support juveniles produced by approximately 51,000 spawners [70]. Since monitoring began in these two lakes, fall fry sockeye salmon abundances in Hewitt Lake declined from 1,105,773 in 2005, to 213,353 in 2006 and 56,161 in 2007 [78]. Sockeye salmon smolt abundances at Whiskey Lake declined from 15,832 in 2012 to 2922 in 2013 and 1395 in 2014 [79]. During this time, pike (>250 mm length) abundances were estimated as 1046 (95% CI: 743–1531) in Hewitt Lake, 3419 (2569–4550) in Whiskey Lake and 3496 (1854–6593) in the Hewitt-Whiskey Creek outlet for a total estimated population of approximately 8000 pike [80]. The relatively high density of northern pike confined to the Hewitt-Whiskey Creek outlet suggests that juvenile salmon suffer high predation mortality while migrating through this system. The goal of current pike suppression is to reduce the pike population in the Hewitt-Whiskey Lake system by 80% over three years. In total, 5087 northern pike from Whiskey Lake were removed between 2012–2019, reducing CPUE from a high of 2.6 pike/gillnet h in 2012 to 0.45 pike/gillnet h in 2019. In Hewitt Lake, 1712 pike have been removed during this period reducing the CPUE from 1.75 pike/gillnet h in 2012 to 0.34 pike/gillnet h in 2019. Assuming a combined population of 8000 northern pike over 250 mm and an average daily individual consumption of one juvenile salmon per day [80] over a summer (150 days), this project is predicted to increase survival of juvenile salmon by approximately 955,000 over three years that would otherwise be depredated by pike. Project success will be evaluated based upon whether hydroacoustic survey estimates indicate the abundance of juvenile sockeye salmon rearing in the system increases by about 1,000,000 by the fall of 2020 [ADFG Unpublished]. The 2018 hydroacoustic surveys estimated populations of juvenile sockeye salmon in Hewitt Lake totaled about 367,000 sockeye fry and, in 2019, the populations of fry increased to approximately 568,000 [ADFG Unpublished].

Finally, in 2018 and 2019, new pike suppression programs were initiated in the Threemile Lake complex and Chuitbuna Lake, respectively, on the west side of Cook Inlet [81,82] (Figures 4 and 8a). These areas are significant because they are at the invasion front of pike in this region. It is presently unknown if these pike populations are the result of illegal anthropomorphic introductions or migrations through Cook Inlet from the Susitna River. Pike are tolerant of low-salinity conditions, and in other parts of their native range, such as the Baltic and Caspian Seas, estuarine habitats can be important migratory corridors [17,83]. In SC Alaska, commercial salmon set-netters occasionally report catching pike in Cook Inlet [ADFG Unpublished]. It is highly feasible that this is now a mode of pike dispersal to west Cook Inlet and, potentially, Kenai Peninsula waters. To investigate this hypothesis further, a new research program between ADFG and the University of Alaska is beginning to analyze pike otolith chemistry to determine origin and utilization of Cook Inlet as a travel corridor [83,84].

The west side Cook Inlet pike suppression sites were conducted in partnership between ADFG the Tyonek Tribal Conservation District (TTCD) and the Native Village of Tyonek (NVT) to increase capacity for pike suppression in a remote region of pike's invaded range in SC Alaska. Alliances between the state, federal agencies, tribal organizations, NGOs, and universities greatly increase resources available for invasive pike suppression and expand the scope of capabilities in the region.

The Threemile and Chuitbuna projects were set up with mark-recapture studies to determine baseline population estimates of pike ≥ 300 mm and assist with long-term evaluation of the pike suppression efforts. In the Threemile Lake complex, which includes Threemile Lake, West Threemile Lake, and Upper Lilly Lake, population estimates in 2018 for pike ≥ 300 mm were 1063 ± 102 (95% CI), 45 ± 11, and 221 ± 70, respectively. In a 10-day suppression event, which doubled as the recapture event, netting efforts reduced those populations by 49% to 57%. A 10-day suppression event in the Threemile Lake complex in 2019 removed a similar number of pike as the previous year. In a 2019 mark-recapture population study for Chuitbuna Lake, the population estimate for pike ≥ 300 mm was 150 ± 13. The suppression event, which doubled as the recapture event, removed 80% of the population in four days of netting. Each lake will be reassessed with mark-recapture evaluations every five years to measure long-term suppression success. It is likely that the similar pike catches in the first two years of suppression in the Threemile Lake complex can be explained by recruitment of pike into the 300 mm size class (Figure 8b). As observed previously, it can take a few netting seasons before catch rates substantially decrease [43] (Figure 5b). Similar to all pike suppression efforts described, pike removed from these lakes are dissected for diet analyses, size class, age, and sex ratio. In addition, over 2000 otoliths have been removed from these pike for use in otolith microchemistry research.

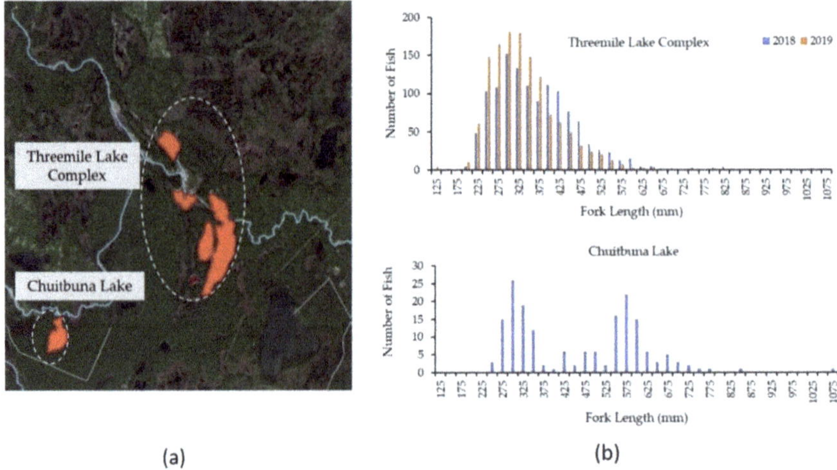

Figure 8. (**a**) Map of west side Cook Inlet pike suppression projects and (**b**) length-frequency distributions of pike removed to date.

3.2. Population Eradication

Although many of the invasive pike management activities in SC Alaska rely on suppression in complex interconnected drainages, the preferred alternative when possible is to eradicate pike populations entirely. This allows for direct restoration of native fish assemblages by removing source populations that could disperse or be introduced to proximate vulnerable waters [85]. At present, there are few management tools that can effectively lead to eradication of invasive fish populations. In rare cases, fish populations from very small lakes can be eradicated with gillnets [86]. In SC Alaska, pike have been eradicated from three lakes on the Kenai Peninsula with intensive winter-long under-ice gillnets (Tiny Lake, Hall Lake and Warfle Lake), but each of these were less than 20 surface hectares, contained low-density populations of less than 30 pike, and juvenile pike catches were low indicating poor recruitment [87]. Besides fortuitous eradications in small lakes with nets, completely draining a waterbody can be another method of fish eradication [85] although this is rarely feasible. This has not been done specifically for pike in SC Alaska, but a pike population in Campbell Lake in Anchorage was eliminated following a municipal project that reduced the water level down to the Campbell Creek thalweg in 2011 [ADFG Unpublished]. The most common method used for invasive fish eradications worldwide is through chemical treatments using piscicides, particularly liquid and powdered formulations of rotenone [88]. Rotenone treatments have been used across the western United States and in SC Alaska for pike eradications [5] (Figure 4).

Rotenone is a naturally occurring compound derived from the roots of tropical plants in the genera *Derris* Graham 1852, *Lonchocarpus* Kunth 1824 or *Tephrosia* Persoon 1807. It has been used for centuries by indigenous cultures in Central and South America to catch fish for food. Rotenone has been used as a piscicide by fishery management agencies in the U.S. since the 1930s to remove unwanted or invasive fish. Currently, rotenone is commercially available as either a wettable-powder or liquid and is registered by the U.S. Environmental Protection Agency (EPA) as a restricted-use pesticide for fish management [89]. Rotenone is lethal to fish because it is readily absorbed through the gills where it instantly enters the blood stream and blocks the biochemical process that allows fish to utilize oxygen during cellular respiration [90]. At the concentrations typically used for invasive fish eradications, the piscicide is not harmful to birds, mammals or adult life stages of amphibians, but is lethal to plankton and macroinvertebrates at varying levels [91]. Rotenone does not enter ground water sources as it is readily bound to organics and, because it is naturally broken down by photolysis and thermal degradation, it generally does not persist long-term in the environment [90]. The piscicidal effects of rotenone can be immediately neutralized with potassium permanganate, and this is typically done when treatment areas include flowing waters.

Decisions to utilize rotenone are strategically made in SC Alaska in waters that will remove important source populations of pike that could spread to nearby vulnerable waters (Appendix A). ADFG completed over 20 rotenone treatments for pike eradication since 2008 (Table 1). To reduce collateral damages, treatments are typically conducted in October just before ice-up. During this time of year, macroinvertebrate populations are seasonally senesced, most piscivorous waterfowl have migrated to wintering areas, and recreational activities for lake residents are less likely to be interrupted. Following all rotenone treatments, water samples are regularly analyzed for degradation rates of rotenone and its byproduct, rotenolone. Under the cold and dark conditions of Alaska's winters, rotenone often remains active throughout the entire ice-covered period [58–60,87]. This is an advantage as it typically ensures sufficient exposure to achieve 100% pike mortality, especially in treatment areas that contain seeps and floating bogs that can be difficult to effectively treat but will freeze solid during the treatment period.

Pike eradication projects for invasive pike in SC Alaska began with small treatments of isolated lakes and later expanded into more complex systems (Table 1). Projects involving rotenone have taken between one to four years to complete. Much of the time investment involves carefully planning and choreographing the treatments, conducting pre-treatment water quality and biological assessments, lake mapping, public scoping, and acquiring state and federal permits (i.e., National Environmental Policy Act. The volume and habitat conditions of each individual waterbody determine the application methods that are employed [58–60,87] (Table 1). Pike eradications in SC Alaska most commonly include rotenone delivery by motorboats equipped with semi-closed pump systems, some specialized for deep water rotenone delivery. Rotenone pumped from an airboat or all-terrain vehicles as well as helicopter spraying, has been used in wetland areas between lakes or along creek corridors. Backpack sprayers, drip stations and rotenone mixture balls are commonly used in small inlet streams, seeps, and difficult-to-treat areas. For treatments requiring deactivation, potassium permanganate is applied at the lake outlets based on flow rates, and sentinel fish downstream are carefully monitored to ensure deactivation is effective [87]. Rotenone treatment success is confirmed through a combination of gillnetting, including under-ice net sets, observations of caged sentinel fish, analytic determination of rotenone concentration achieved and eDNA detection methods [92,93]. Post-treatment, lakes are monitored through routine surveys to ensure pike are not reintroduced [94]. Once rotenone treatments are complete and post-treatment assessments have confirmed successful pike eradication, fisheries are restored to the treated lakes.

Fishery restoration for pike eradication projects is tailored individually based on historical fish assemblages and uses of the treated water body pre pike introduction. In cases where pike have been removed from lakes that were part of ADFG's Statewide Stocking Plan [95], these lakes are re-stocked with hatchery-produced triploid salmonids. In most pike waters with wild fish assemblages that have been treated, pike and stickleback were all that remained [58,59,87]. However, there are two large projects where native fish were present at the time of treatment, and substantial efforts were undertaken to save their populations.

In 2012, Stormy Lake near Nikiski on the Kenai Peninsula was treated to prevent pike from dispersing into the nearby Swanson River drainage (Figure 4). This lake, due to its depth (17 m), had maintained its native Arctic char *Salvelinus alpinus* Richardson 1836 population and other native species like longnose suckers *Catastomus catastomus* Lesueur 1817. As a precaution, gametes were collected from the char to rear in a hatchery and preserve the genetics of that population. As many native fish as possible were collected prior to treatment and held in a net pen in a nearby lake until the rotenone degraded and these fish could be introduced. From there, natural recolonization of native fishes from the Swanson River further re-established native fish assemblages in the lake.

Table 1. Rotenone Applications for Invasive Northern Pike Removal in southcentral (SC) Alaska.

Year	Water Body	Location	Volume (ha-m) [a]	Quantity [b]	Rotenone Concentration mg/L (Target/Actual)	Detoxification Time	Application Method	Species Reintroduced [c]
2008	Arc Lake	Soldotna	17.8	181.7 L CFT	0.050/0.035	8 months	Boat	SS, ST
2008	Cheney Lake	Anchorage	21.6	219.6 L CFT	0.050/0.030	5 months	Boat, Backpack Spray	RT, GR, SS, ST
2009	Scout Lake	Sterling	103.0	700.3 L CFT + 499.0 kg Powder	0.070/0.030	8 months	Boat, Backpack Spray	RT, GR, SS, ST
2009	Sand Lake	Anchorage	140.4	1348 L CFT	0.050/0.030	7 months	Boat	RT, GR, SS, AC
2012	Stormy Lake	Nikiski	858.3	3444.7 L CFT + 3500 kg Powder	0.050/0.048	4 months	Boat (Weighted Hose), Airboat, Backpack Spray, Deactivation	RT, AC, SS, LS
2014	Union Lake	Soldotna	88.7	327.1 L CFT + 504.8 kg Powder	0.050/0.024	8 months	Boat, Backpack Spray	RT, SS, DV, SC, ST
2014	East Mackey Lake	Soldotna	115.6	424 L CFT + 655.4 kg Powder	0.050/0.026	6 months	Boat, Backpack Spray	RT, SS, DV, SC, ST
2014	West Mackey Lake	Soldotna	150.5	564 L CFT + 902.6 kg Powder	0.050/0.024	8 months	Boat, Backpack Spray	RT, SS, DV, SC ST
2014	Derks Lake	Soldotna	56.4	206.3 L CFT + 226.8 kg Powder	0.050/0.024	6 months	Boat, Airboat, Backpack Spray, Deactivation	RT, SS, DV, SC, ST
2015	Otter Lake	JBER	110.2	1400.6 L. CFT + 22.7 kg Powder	0.050/0.024	4 months	Boat, Airboat, Backpack Spray, Drip Stations	RT
2016/17	Sevena Lake	Soldotna	74.0	605.7 L CFT	0.040/0.036	10 days	Boat, Airboat, Backpack Spray	RT, SS, DVSC, ST
2016	Soldotna Creek	Soldotna	-	191.5 L. CFT	0.040/0.036	5 days	Helicopter, Drip Stations, Backpack Sprayers, ATV, Deactivation	All Species Recolonized from Kenai River
2016	Loon Lake	Soldotna	24.4	111.7 L CFT + 65.8 kg Powder	0.040/0.028	8 months	Boat	RT
2018	Hope Lake	Soldotna	50.1	418.3 L CFT	0.040/0.018	3 months	Boat, Backpack Spray	RT, SS, ST
2018	G Lake	Soldotna	35.0	288.4 L CFT	0.040/0.028	3 months	Boat	RT, SS, ST
2018	Crystal Lake	Soldotna	34.2	279.7 L CFT	0.040/0.040	3 months	Boat	RT, SS, ST
2018	Leisure Lake	Soldotna	15.2	124.2 L CFT	0.040/0.040	3 months	Boat	RT, SS, ST
2018	Leisure Pond	Soldotna	1.4	13.6 L CFT	0.040/0.024	3 months	Boat, Backpack Spray	RT, SS, ST
2018	Fred's Lake	Soldotna	1.9	57.5 L CFT	0.040/0.011	3 months	Boat, Backpack Spray	RT, SS, ST
2018	Ranchero Lake	Soldotna	5.1	42.0 L CFT	0.040/0.024	3 months	Boat, Backpack Spray	RT, SS, ST
2018	CC Lake	Soldotna	33.0	26.9 L CFT	0.040/0.026	3 months	Boat	RT, SS, ST
2020	Anderson Lake [d]	Wasilla	118.4	1211.3 L CFT	0.040/TBD	TBD	Boat	RT
2020	King's Lake [d]	Wasilla	107.6	1100.0 L CFT	0.040/TBD	TBD	Boat	RT

[a] Unit = Hectare-Meters (Imperial conversion = Acre-Feet). [b] CFT—CFT Legumine Liquid Rotenone (Unit = Liters; Imperial conversion = Gallons), Powder—Prentox Prenfish Rotenone Fish Toxicant Powder (Kilograms; Imperial conversion= Pounds). [c] SS—Coho salmon *O. kistutch*, RT—Rainbow trout *O. mykiss*, GR–Arctic grayling *Thymallus arcticus*, AC—Arctic char *Salvelinus alpinus*, DV-Dolly Varden *Salvelinus malma*, LS—Longnose suckers *Catastomus catastomus*, SC—Sculpin spp. *Cottus spp*, ST—Threespine stickleback *Gasterosteus aculeatus*; Wild fish in Bold, otherwise hatchery stock. [d] Planned Treatments for October 2020.

In another project, the entire Soldotna Creek tributary drainage of the Kenai River was treated over the course of four years to prevent pike from dispersing into the Kenai River and establishing in more vulnerable tributary systems like the Moose River. This project included the treatment of seven lakes, over 32 km of flowing waters, and over 200 hectares of wetlands. The drainage was divided into two sections with temporary fish barriers (Figures 4 and 9). The first section included five lakes and connecting streams. These lakes no longer contained wild native game fish and were treated in 2014 except for a closed lake (Loon Lake) that was treated in 2017 immediately after pike were discovered (Table 1). In 2015, native fish from Soldotna Creek and Sevena Lake at the headwaters were intensively trapped and relocated into the then fishless four open lakes that had previously been treated (Table 2). In 2016 Sevena Lake and Soldotna Creek were treated, and Sevena Lake was precautionarily re-treated in 2017 because of the complexity of that area. The native fish relocations seeded fish populations in the drainage lakes, and Soldotna Creek was recolonized within 12 months by all previously occurring species via migration from the Kenai River [87].

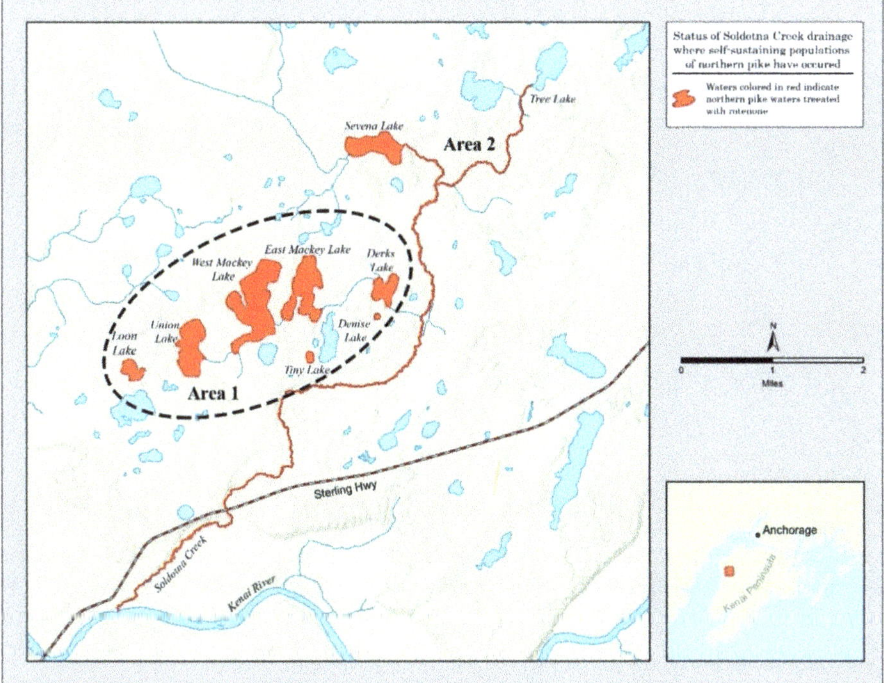

Figure 9. Map of the Soldotna Creek project area for the largest pike eradication project in SC Alaska. This eradication effort was completed between 2014 and 2018. Temporary barriers were installed to separate Area 1 (encircled) from Area 2 (Sevena Lake and Soldotna Creek). The lakes in Area 1 were treated in 2014 (Loon Lake was treated in 2017). During 2015, native fish from Area 2 were rescued and replanted in Area 1. Area 2 was treated in 2016 and 2017, and the barriers were removed to allow fish movement throughout the system.

Table 2. Native fish released into Soldotna Creek Pike Eradication Area 1 lakes during 2015–2018.

Year	Lake	Rainbow Trout [a]	Dolly Varden [b]	Threespine Stickleback [c]	Sculpin Spp. [d]	Coho Salmon [e]	All Species	Salmonids/Hectare [f,g]
2015	Derks	30	161	950	3	6107	7251	68
2016		199	217	3386	229	2452	6483	31
2017		0	0	0	0	38	38	1
	Total	229	378	4,336	232	8597	13,772	100
2015	East Mackey	355	366	5362	960	1396	8439	8
2016		696	484	4103	439	6564	12,286	31
2017		176	436	0	0	2506	3118	13
2018		220	436	0	0	2506	3162	13
	Total	1447	1722	9465	1399	12,972	27,005	65
2015	Union	195	173	3532	183	2173	6256	12
2016		277	407	3563	419	7259	11,925	38
2017		38	130	0	0	604	772	4
2018		35	0	0	0	0	35	1
	Total	545	710	7095	602	10,036	18,988	55
2015	West Mackey	354	437	5553	399	904	7647	4
2016		1088	1034	6401	1062	13,388	22,973	34
2017		203	556	0	0	3374	4,133	9
2018		679	0	0	0	0	679	2
	Total	2324	2027	11,954	1461	17,666	35,432	49
	Grand Total	4545	4837	32,850	3694	49,271	95,197	269

[a] Oncorhynchus mykiss, [b] Salvelinus malma [c] Gasterosteus aculeatus,[d] Cottus spp., [e] Oncorhychus kisutch, [f] Juvenile salmonids were collected from the mainstem of Soldotna Creek and stickleback were collected mainly from Sevena Lake. [g] The majority of fish were collected by minnow trapping; other collection gear used included backpack electrofishing and fyke net traps.

An issue with rotenone projects that is often contentious with the public is concern for piscivorous waterfowl. While these animals are not affected directly by the piscicide, loss of fish prey can displace their populations. As general practice, rotenone projects for pike in SC Alaska now include relocations of sticklebacks in all cases where there are historic records of sticklebacks occurring. This provides a forage base for piscivorous birds and other wildlife in the lakes. It also provides a unique opportunity for evolutionary ecologists to study local adaptations of sticklebacks re-introduced to these lakes [10,96–98]. In several cases, these researchers have been substantial partners in the stickleback translocations.

Since 2008, rotenone treatments for pike eradications predominantly took place in Anchorage and the Kenai Peninsula. Both regions had feasible potential to eradicate invasive pike populations entirely if pike did not spread far beyond their distribution. This was in contrast to the highly interconnected and expansive range of pike in the Susitna Basin. Over the last decade, the primary focus of pike eradication efforts in SC Alaska has been working toward the achievement of this goal, which is uncommon in invasive species management [99]. At the time of writing of this review, both the Anchorage and Kenai Peninsula regions each have one known remaining lake system containing pike populations, and plans are in development for their eradications. If successful, both regions have great potential to eventually be free of invasive northern pike. With this milestone in reach, pike eradication efforts will begin north of Anchorage in two Knik Arm Drainage lakes in 2020 (Table 1, Figure 4).

3.3. Outreach and Angler Engagement

Outreach and angler engagement are interwoven into all pike suppression and eradication efforts and are a critical component of invasive pike management. Communication plans, websites, print materials, magazine articles, social media, media stories, presentations and seminars are all outreach tactics that are commonly used to disseminate information about pike in SC Alaska. For illegal introductions of pike or any other non-indigenous species to cease, public understanding of the ramifications of such actions must be well understood and translate to behavior change. While public opinion on pike in SC Alaska still varies greatly, there is a high degree of public support for invasive pike removals as well as a reluctant understanding that despite being a prized game species, pike in SC Alaska are invasive and must be managed as such. Anglers and the public are encouraged to report any new observations of pike or any other non-native species to a centralized reporting application (http://www.adfg.alaska.gov/index.cfm?adfg=invasivespeciesreporter.main) or by calling a statewide invasive species hotline (1-877-INVASIV). For records of where pike are known to occur, a database and interactive mapping tool for known pike waters is maintained (https://adfg.maps.arcgis.com/apps/webappviewer/index.html?id=ad27ebc052814b66a60d0e52701e64f7&_ga=2.40269538.1172137975.1582067549-1889826575.1579740028).

SC Alaska has some of the most liberalized sport fishing regulations for pike in the country. There is no bag or possession limit for them outside of their native range in the state. Anglers are allowed use of up to five lines while ice fishing, and methods and means not permissible for other species such as bow and arrow and spear can be used for pike. As of 2020, a regulation went into effect requiring anglers to dispatch any pike they catch in waters of the Susitna Basin, Knik Arm, Anchorage, or the Kenai Peninsula. Partnering with anglers to reduce pike abundance through harvest is a strategy that has the potential to both reduce abundance [22,100,101] and provide a means of acquiring observational data on pike in remote locations that are difficult and costly to access.

While bounties are an angler engagement tool implemented in other states [102], to date traditional bounties have not been employed as a pike management strategy in SC Alaska due to funding, complexities of proving the origin of the fish (i.e., native or invasive range), and most importantly, reluctance to create an economic incentive that could motivate additional illegal introductions. However, beginning in 2020, a new winter only ADFG angler incentive program was initiated in Alexander Lake where it was previously noted that pike suppression may now be warranted. Approximately 100 pike in this lake were tagged with passive integrated responder tags (PIT tags) during the summer of 2019. These tags were not visible to anglers. Anglers were encouraged to harvest pike in Alexander

Lake during the winter and bring them into the local ADFG office to be scanned. If an angler had a tag, they received a reward in the form of a $100 gift card and were entered into a drawing for a larger prize at the end of the season. During this first year of the program, approximately 500 pike were harvested, and 13 tags were recovered. This program provided a low-cost avenue for collecting baseline catch and effort data on pike in Alexander Lake, acquiring cleithra and otoliths from the fish for aging and microchemistry research, and facilitated communication and partnership with pike anglers. In addition, Alexander Creek pike suppression crews will be able to scan for tagged pike downstream during annual suppression to continue observing movement patterns of pike in that system. Because eligibility for an award was based on the chance of catching a tagged pike from a known location, there was no motivation for anyone to move pike a result of this program. In the future, this program may be expanded to other pike waters in SC Alaska.

3.4. Population Monitoring and Research

The final tenant of invasive pike management in SC Alaska involves monitoring of pike populations throughout the region and conducting research to learn more about the impacts of pike and effective management tools. Research on pike has been on-going over the last decade and, as has been discussed throughout this review, has included investigations into pike diets and impacts [12,16,50] movement patterns [44,45], population genetics [35], predicting invasion risk [35], developing eDNA tools for pike [92,93,103], and better understanding the degradation process of rotenone [UAA Unpublished]. All these investigations are highly collaborative among ADFG, the University of Alaska, the U.S. Geological Survey, and the U.S. Fish and Wildlife Service as well as local NGOs like the National Fish Habitat Partnerships (NFHP), Kenai Watershed Forum (KWF), CIAA, TTCD, and NVT. Future research will seek to expand alliances with commercial fisherman to acquire samples of pike caught in estuarine waters for otolith microchemistry investigations to learn how pike may be utilizing Cook Inlet for their current dispersal and to develop barrier designs to protect vulnerable drainages from that occurrence. Research will continue to evaluate effectiveness of current pike suppression efforts and look toward the future to determine what new tools and technologies might emerge, such as those in the genetics realm [104–107], that may have future applications for adaptive pike management in SC Alaska.

4. Conclusions

In conclusion, invasive northern pike in SC Alaska have had complex, and in many cases, severe consequences for native fish populations. Pike are certainly not the only factor responsible for salmon declines in the state, but in some cases, they are a substantial part of that story. In an age of climate change and deteriorating ecological conditions, progressive action to reduce stressors facing salmon and other native fish species in Alaska's freshwaters is imperative. Alaska is fortunate to not suffer many of the invasive species impacts that are rampant elsewhere, but there is no guarantee that good fortune will continue. As has been well documented with pike, invasive species are a threat to Alaska's fragile salmon habitats, and pike in SC is one factor that can be mitigated with effective and well prioritized adaptive management.

In this past decade of invasive pike management, significant strides were made toward pike eradication on the Kenai Peninsula and the Anchorage areas. Thousands of pike have been removed from large drainages in the Susitna Basin resulting in increased survival of rearing salmon. Over the last decade close to $5 Million has been spent on these efforts, but the economic value of fisheries in the state is far greater and the cultural value is immeasurable. The invasion and subsequent control efforts for pike in SC Alaska are the most spatially expansive in the world for this species making both the problem and the collaborative program to address it unique. However, there are other locations with similar challenges with invasive populations of pike. Among the most significant examples in the western Unites States are the eastern Columbia River Basin in Washington State and the Yampa River in Utah and Colorado. Significant efforts are underway in these locations to protect native species from pike predation and prevent pike from expanding their ranges. In the Columbia River, this is particularly important as downstream expansion of pike is anticipated to affect anadromous salmon populations just as they have in SC Alaska. In this regard, there is great benefit for western states with invasive pike challenges to collaborate so that successful methods and technologies can be broadly applied. Over the last decade, Alaska's invasive pike program has contributed to a greater understanding of predation impacts on salmonids and other native fish, helped develop enhanced detection capabilities (i.e., eDNA), pioneered largescale pike suppression (i.e., Alexander Creek), and completed over 20 successful eradications of invasive pike populations.

Moving forward, in this next decade of invasive pike management in SC Alaska, the goal will be to complete pike eradication on the Kenai Peninsula and in Anchorage, develop effective strategies and barriers to prevent future introductions, focus eradication efforts in northern Cook Inlet drainage waters where feasible, implement a standardized monitoring program for pike waters in the Susitna basin and use those data to develop effective strategies for continued pike suppression in the region. Ultimately, the primary objective in this coming decade, as it was in the last, will be to prevent pike from spreading beyond their current distribution and causing further damage to Alaska's fisheries.

Author Contributions: Conceptualization, K.D., R.M., P.B., C.J. ,N.S., A.W. and R.D; methodology, K.D, R.M., P.B., C.J., N.S., A.W, and R.D; formal analysis R.M., P.B., R.B; investigation K.D., R.M., P.B., C.J, N.S., A.W., R.D; resources K.D., P.B., C.J., R.M., N.S., A.W. R.D.; writing—original draft preparation K.D; writing—review and editing R.M., P.B., C.J., N.S., A.W. and R.D.; visualization K.D., R.M. N.S.; supervision K.D.,R.M., P.B., N.S., A.W., R.D.; project administration K.D., N.S., R.D.; funding acquisition K.D., N.S., R.D. All authors have read and agreed to the published version of the manuscript.

Funding: Projects discussed in this manuscript were funded by a variety of sources, primarily the Alaska Sustainable Salmon Fund (Grant #s 44609, 44613, 44617,44626, 44807, 44910, 44168, 44353, 44365, 51003, 52013), the Aquatic Nuisance Species Task Force (701818G496), Joint-Base Elmendorf Richardson (COOP 14-010), Matanuska-Susitna Borough (2016-00002088, 2016-00002091), U.S. Fish and Wildlife Service (FWS70181AJ), the Bureau of Indian Affairs (A18AV00629), and State Wildlife Grants (T-10-12).

Acknowledgments: The authors wish to offer their sincere gratitude to Adam Sepulveda with the U.S. Geological Survey Rocky Mountain Science Center. Sepulveda has been a mentor to all authors and a critical partner and collaborator on invasive pike research in SC Alaska. The authors also wish to thank Peter Westley, Andres Lopez, Jeffrey Falke, Matthew Wooller, and Patrick Tomco with the University of Alaska for continued collaboration. Special thanks to Chase Jalbert for this graduate research on pike genetics patterns and invasion potential in SC Alaska. Tammy Davis is the statewide Invasive Species Coordinator for the state of Alaska and is an integral member of the team and a constant support for invasive pike efforts in SC Alaska. Also, the authors sincerely thank David Rutz, ADFG Sport Fish Division Director, for his passionate support of invasive pike efforts and tremendous field assistance. Progress made with invasive pike suppression and eradication in SC Alaska would not be possible without the substantial field and management contributions by the following individuals: Timothy McKinley, Tom Vania, Matt Miller, James Hasbrouck, Jack Erickson, Robert Begich, Colton Lipka, Samantha Oslund, Samuel Ivey, Jerry Strait, Eric Hollerbach, Jillian Jablonski, Christy Cincotta, Justin Trenton, Jessica Johnson, Emily Heale, Savanna Van Diest, Gerrit Van Diest, Tyler Henegan, Kristopher Butler, Cole Deal, RJ Lister, Jordan Couture, Chris Razink, Adrian Baer, Keegun Egelus, Chuck Brazil, Bob Pence, Rip Bliss, and Anne Bliss. Biometric assistance was provided by Ben Buzzee and Adam Craig, and GIS assistance by Skip Repetto and Jason Graham. Most of all, the authors thank their supportive family and friends who endure their frequent absences so the work of protecting Alaska's native fish from invasive pike can be accomplished.

Conflicts of Interest: The authors declare no conflict of interest. Funders of the projects discussed had no role in project design; in the collection, analyses, or interpretation of data; in the writing of the manuscript, or in the decision to publish the results.

Appendix A

Table A1. ADFG Sport Fish Division Invasive Northern Pike Project Prioritization Matrix.

Questions/Criteria	Priority Level (Low, Medium, High)	Weighted** Score	Enter 1 or 0 (Yes = 1) (No = 0)	Score Value
Recreational Fisheries				
Pre-pike introduction, historic fishing level in the water body was low (≤200 days)	Low	1		0
Pre-pike introduction, historic fishing level in the water body was medium (>200–≤1000 days)	Medium	5		0
Pre-pike introduction, historic fishing level in the water body was high (>1000 days)	High	10		0
Is a goal of the project to r restore opportunity for a non-pike fishery?	Medium	5		0
Is the water body currently and/ or formerly stocked by ADFG?	Medium	5		0
Have stocking levels in the water body been altered because of pike presence?	Medium	5		0
Pike Impacts				
Do pike in the water body directly threaten a wild fishery in that water?	High	10		0
Do pike in the water body threaten a wild fishery that is in close proximity (five miles or less)?	High	10		0
Have regulations for wild sport fisheries exceeding 1000 angler-days been restricted because of pike in this waterbody?	High	10		0
Have wild sport fisheries receiving under 1000 angler-days of effort been restricted because of pike in this waterbody?	Low	1		0
Have available data indicated that a wild fish population has been eliminated associated with pike presence?	High	10		0
Have available data indicated that a wild fish population has been impacted associated with pike presence?	Medium	5		0
Has the public indicated concern over the pike population in this water body?	Medium	5		0
Do pike represent >50% of the catch in a netting survey or from other available data?	High	10		0
Do pike represent 25–50% of the catch in a netting survey or from other available data?	Medium	5		0
Do pike represent <25% of the catch in a netting survey or from other available data?	Low	1		0
Does the area management biologist expect a negative impact to a sport fishery associated with pike in this waterbody?	High	10		0
Does the area management biologist expect an imminent loss of a wild stock associated with pike in this drainage?	Very High	30		0
Does the area management biologist associate pike with an inability to meet an escapement goal in this drainage?	High	10		0
Does this water body contain a Board of Fisheries-stock of yield or management concern?	High	10		0
Eliminating pike in this project area removes the pike threat in the entire management area	Very High	30		0
Education and Outreach				
Are there opportunities to use this project as an educational outreach tool to increase public awareness?	High	10		0
Are we demonstrating a pike control strategy to stakeholders?	Medium	5		0
Does the project foster public understanding and awareness of invasive species	Medium	5		0
Has there already been stakeholder input desiring a project of this nature?	High	10		0
Habitat Significance				
Is the project area within an open system?	High	10		0
If successful, can this project prevent pike distribution throughout the drainage?	Very High	30		0
Is the project area within an anadromous system?	High	10		0
Are wild, resident fish species present?	Medium	5		0
If in a closed system, does the project have the potential to reduce native fish populations?	Medium	−5		0
Is the project designed to improve habitat for threatened or endangered species populations?	High	10		0
Project Area Characterization (*Type 1—Suitable pike habitat, Type 2—Marginally suitable habitat, Type 3—Poor pike habitat*)				

Table A1. *Cont.*

Questions/Criteria	Priority Level (Low, Medium, High)	Weighted** Score	Enter 1 or 0 (Yes = 1) (No = 0)	Score Value
(Lakes/Wetlands)				
Is it a Type 1 Lake or wetland—Eutrophic and primarily shallow (≤15 feet) with abundant vegetation throughout?	High	10		0
Is it a Type 2 Lake—Mesotrophic and primarily deep (>15 feet) with vegetation covering 50% or more of the lake?	Medium	5		0
Is it a Type 3 Lake—Oligotrophic and primarily deep (>15 feet) with wither sparse or no aquatic vegetation)?	Low	1		0
(Rivers and Streams)*				
Is the waterbody primarily Type 1—Low stream slope (0.0–0.5%) with abundant vegetation and is capable of supporting rearing coho (i.e., Moose River, Alexander River)?	High	10		0
Is the waterbody primarily Type 2 with some type 1—Moderate stream slope (0.51–2.0%) with semi-permanent woody debris and back-waters sloughs and is capable of supporting rearing coho (i.e., Deshka River)?	Medium	5		0
Is the waterbody primarily Type 2—Moderate stream slope (0.51–2.0%) with semi-permanent woody debris and is capable of supporting rearing coho (i.e., Campbell Creek)?	Medium	5		0
Is the waterbody a combination of Types 2 & 3—High stream slope (>2.0%) and slow back-water sloughs capable of supporting rearing coho (i.e., Willow Creek?)	Medium	5		0
Is the waterbody primarily Type 3—Clear, high stream slope (>2.0%) with few slower back-waters (i.e., the Little Susitna River)?	Low	1		0
Is the waterbody exclusively Type 3—High stream slope (>2.0%) with extensive glacial turbidity (i.e., Klutina River)?	Low	1		0
Cultural Significance				
Are native cultural activities (i.e., fish camps, etc.) threatened by pike in this water body?	High	10		0
Does the project benefit subsistence fisheries?	High	10		0
Is a goal of the project to provide economic benefits for citizens, communities, or industries?	High	10		0
Does the area manager believe that user groups are negatively affected by the pike presence?	High	10		0
Economic Impacts				
Has the area manager received input from local businesses/property owners that they are experiencing a negative financial effect from pike in this water body?	High	10		0
Does the project protect commercially important species?	High	10		0
Research				
Will project goals, objectives, and tasks within the project plan strive to improve understanding of pike behavior or distribution in local waters?	Medium	5		0
Do project goals strive to improve understanding of control or eradication techniques for pike?	High	10		0
Will the project include a follow-up assessment to measure the effects of the management action?	Medium	5		0
Do project goals strive to quantify fishery/economic losses resulting from invasive pike?	Medium	5		0
Feasibility				
Does the water body have public access?	High	10		0
Is the water body on the road system?	Low	1		0
Is it technically feasible that the pike population could be permanently removed or contained if the project is implemented?	High	10		0
Is there a history of reintroductions of pike in this area?	High	−10		0
Is the project designed to achieve program goals within the funding period?	Low	1		0
Does the project achieve long term program goals within a decade of the funding period? (i.e., reestablish wild fish populations)	High	10		0
Can the project begin when funding is received?	Medium	5		0
The goal of the project is to have a measurable, positive outcome for fisheries.	High	10		0

Table A1. *Cont.*

Questions/Criteria	Priority Level (Low, Medium, High)	Weighted** Score	Enter 1 or 0 (Yes = 1) (No = 0)	Score Value
Permitting and Inter-Agency Cooperation				
Does the project provide opportunities to partner or collaborate with other agencies or organizations?	Low	1		0
Is the NEPA process required for this project?	Low	−1		0
Is there reason to believe there would be a conflict with an existing coastal, watershed or restoration plan?	High	−10		0
ADFG Significance				
Is the project programmatically/ scientifically aligned with ADF&G's mission in the Sport Fish Division Strategic Plan?	High	10		0
SCORE				

*** Rearing Coho share the same habitat requirements as northern pike (Rutz 2006) and are a good indicator of pike habitat suitability and potential overlap. ** Low = 1, Medium = 5, High = 10. Note: If project includes work in both categories, score both. Otherwise, score one or the other only.

References

1. Havel, J.E.; Kovalenko, K.E.; Thmoaz, S.M.; Amalifitano, S.; Kats, L.B. Aquatic invasive species: Challenges for the future. *Hydrobiologia* **2015**, *750*, 147–170. [CrossRef] [PubMed]
2. Britton, J.R.; Gozlan, R.W.; Copp, G.H. Managing non-native fish in the environment. *Fish Fish.* **2011**, *12*, 256–274. [CrossRef]
3. Ricciardi, A.; Hoopes, M.F.; Marchetti, M.P.; Lockwood, J.L. Progress toward understanding the ecological implacts of non-native species. *Ecol. Monogr.* **2013**, *8*, 263–282. [CrossRef]
4. Simberloff, D. Invasive species. In *Conservation Biology for All*; Sodhi, N.S., Ehrich, S.R., Eds.; Oxford University Press: New York, NY, USA, 2007.
5. Dunker, K.J.; Sepulveda, A.; Massengill, R.; Rutz, D. The northern pike, a prized native but disastrous invasive. In *Biology and Ecology of Pike*; Skov, C., Nilsson, P.A., Eds.; CRC Press: Boca Raton, FL, USA, 2018; pp. 356–398.
6. Lever, C. *Naturalized Fishes of the World*; Academic Press: San Diego, CA, USA, 1996.
7. Welcomme, R.L. *International Introductions of Inland Aquatic Species*; TP No. 294; Food and Agriculture Organization of the United Nations: Rome, Italy, 1988; pp. 112–113.
8. Inskip, P.D. *Habitat Suitability Index Models: Northern Pike*; FWS/OBS 82/10.17; United States Fish and Wildlife Service: Daphne, AL, USA, 1982.
9. Persson, A.P.A.N.; Brönmark, C. Trophic interactions. In *Biology and Ecology of Pike*; Skov, C., Nilsson, P.A., Eds.; CRC Press: Boca Raton, FL, USA, 2018; pp. 185–211.
10. Heins, D.C.; Knoper, H.; Baker, J.A. Consumptive and non-consumptive effects of predation by introduced northern pike on life history traits in threespine stickleback. *Evol. Ecol. Res.* **2016**, *17*, 355–372.
11. Byström, P.; Karlsson, J.; Nilsson, P.A.; Van Kooten, T.; Ask, J.; Olofsson, F. Substitution of top predators:effects of pike invasion in a subarctic lake. *Freshw. Biol.* **2007**, *52*, 1271–1280. [CrossRef]
12. Sepulveda, A.J.; Rutz, D.S.; Dupuis, A.W.; Shields, P.A.; Dunker, K.J. Introduced northern pike consumption of salmonids in Southcentral Alaska. *Ecol. Freshw. Fish* **2014**. [CrossRef]
13. McKinley, T. *Survey of Northern Pike in Lakes of Soldotna Creek Drainage, 2002*; Special Publication No. 13-02; Alaska Department of Fish and Game: Anchorage, AK, USA, 2013.
14. Beck, K.G.; Zimmerman, K.; Schardt, J.D.; Stone, J.; Lukens, R.R.; Reichard, S.; Randall, J.; Cangelosi, A.A.; Cooper, D.; Thompson, J.P. Invasive species defined in a policy context: Recommendations from the federal invasive species advisory committee. *Invasive Plant Sci. Manag.* **2008**, *1*, 414–421. [CrossRef]
15. Nilsson, P.A.; Eklöv, P. Finding food and staying alive. In *Biology and Ecology of Pike*; Skov, C., Nilsson, P.A., Eds.; CRC Press: Boca Raton, FL, USA, 2018; pp. 9–31.
16. Sepulveda, A.J.; Rutz, D.S.; Ivey, S.S.; Dunker, K.J.; Gross, J.A. Introduced northern pike predation on salmonids in Southcentral Alaska. *Ecol. Freshw. Fish* **2013**. [CrossRef]
17. Jacobsen, L.; Engström-Öst, J. Coping with environments; vegetation, turbidity and abiotics. In *Biology and Ecology of Pike*; Skov, C., Nilsson, P.A., Eds.; CRC Press: Boca Raton, FL, USA, 2018; pp. 32–61.
18. Craig, J.F. *Pike: Biology and Exploitation*; Chapman & Hall: London, UK, 1996.
19. Raat, A.J.P. *Synopsis of Biological Data on the Northern Pike Esox Lucius Linneaeus, 1758*; FAO Fisheries Synopsis No. 30 Rev. 2; Food and Agriculture Organization of the United Nations: Rome, Italy, 1988.
20. Haugen, T.O.; Vøllestad, L.A. Pike population size and structure: Influence of density-dependent and density-independent factors. In *Biology and Ecology of Pike*; Skov, C., Nilsson, P.A., Eds.; CRC Press: Boca Raton, FL, USA, 2018; pp. 123–163.
21. Mann, R.H.K. The numbers and production of pike (*Esox lucius*) in two Dorset rivers. *J. Anim. Ecol.* **1980**, *49*, 899–915. [CrossRef]
22. Arlinghaus, R.J.A.; Beardmore, B.; Diaz, A.M.; Hühn, D.; Johnston, F.; Klefoth, T.; Kuparinen, A.; Matsumura, S.; Pagel, T.; Pieterek, T.; et al. Recreational piking—Sustainably managing pike in recreational fisheries. In *Biology and Ecology of Pike*; Skov, C., Nilsson, P.A., Eds.; CRC Press: Boca Raton, FL, USA, 2018; pp. 288–336.
23. Kuparinen, A.H.L. Northern pike commercial fisheries, stock assessment and aquaculture. In *Biology and Ecology of Pike*; Skov, C., Nilsson, P.A., Eds.; CRC Press: Boca Raton, FL, USA, 2018; pp. 337–355.
24. McMahon, T.E.; Bennet, D.H. Walleye and northern pike: Boost or bane to northwest fisheries. *Fisheries* **1996**, *21*, 6–13. [CrossRef]

25. Zelasko, K.A.; Bestgen, K.R.; Hawkins, J.A.; White, G.C. Evaluation of a long-term predator removal program: Abundance and population synamics of invasive northern pike in the Yampa River, Colorado. *Trans. Am. Fish. Soc.* **2016**, *145*, 1153–1170. [CrossRef]
26. Harvey, S.; Bean, N. *Non-Native Fish Suppression Project, May 2018–April 2019 Annual Report*; Project Number 2007-149-00; Kalispel Tribe of Indians: Usk, WA, USA, 2019; p. 68.
27. Walrath, J.D.; Quist, M.C.; Firehammer, J.A. Trophic ecology of nonnative northern pike and thier effect on conservation of native westslope cutthroat trout. *N. Am. J. Fish. Manag.* **2015**, *35*, 158–177. [CrossRef]
28. Muhlfeld, C.C.; Bennett, D.H.; Steinhorst, R.K.; Marotz, B.; Boyer, M. Using bioenergetics modeling to estiamte consumption of native juvenile salmonids by nonnative northern pike in the Upper Flathead River system, Montana. *N. Am. J. Fish. Manag.* **2008**, *28*, 636–648. [CrossRef]
29. Mecklenburg, C.W.; Mecklenburg, T.A.; Thorsteinson, L.K. *Fishes of Alaska*; American Fisheries Society: Bethesda, MD, USA, 2002; p. 1037.
30. Morrow, J.E. *The Freshwater Fishes of Alaska*; Northwest Publishing Company: Anchorage, AK, USA, 1980; p. 248.
31. Oswood, M.W.; Reynolds, J.B.; Irons, J.G., III; Milner, A.M. Distributions of freshwater fishes in ecoregions and hydroregions of Alaska. *J. N. Am. Benthol. Soc.* **2000**, *19*, 405–418. [CrossRef]
32. Seeb, J.E.; Seeb, L.W.; Oates, D.W.; Utter, F.M. Genetic nvariation and postglacial dispersal of populations of northern pike (*Esox lucius*) in North America. *Can. J. Fish. Aquat. Sci.* **1986**, *44*, 556–561. [CrossRef]
33. Haught, S.; von Hippel, F.A. Invasive pike establishment in Cook Inlet Basin lakes, Alaska: Diet, native fish abundance and lake environment. *Biol. Invasions* **2011**, *13*, 2103–2114. [CrossRef]
34. Patankar, R.; von Hippel, F.A.; Bell, M.A. Extinction of a weakly armoured threespine stickleback (Gasterosteus aculteatus) population in Prator Lake, Alaska. *Ecol. Freshw. Fish* **2006**, *15*, 482–487. [CrossRef]
35. Jalbert, C.S. Impacts of a Top Predator (Esox lucius) on Salmonids in Southcentral Alaska: Genetics, Connectivity, and Vulnerability. Master's Thesis, University of Alaska Fairbanks, School of Fisheries and Ocean Sciences, Fairbanks, AK, USA, 2018.
36. Spens, J.; Ball, J.P. Salmonid or nonsalmonid lakes: Predicting the fate of northern boreal fish communities with hierarchical filters relating to a keystone piscivore. *Can. J. Fish. Aquat. Sci.* **2008**, *65*, 1945–1955. [CrossRef]
37. Baker, B. *Fishery Management Report for Recreational Fishereies in the Tanana River Management Area, 2017, Alaska*; Fishery Management Report No. 18-33; Alaska Department of Fish and Game: Anchorage, AK, USA, 2018; p. 92.
38. Stuby, L. *Fishery Management Report for Sport Fisheries in the Yukon Management Area, 2017, Alaska*; Fishery Management Report No. 18-30; Alaska Department of Fish and Game: Anchorage, AK, USA, 2018; p. 62.
39. Rutz, D.S. *Movements, Food Availability and Stomach Contents of Northern Pike Selected Susitna River Drainages, 1996–1997.*; Fishery Data Series No. 99-5; Alaska Department of Fish and Game: Anchorage, AK, USA, 1999; p. 78.
40. Rutz, D.S. *Seasonal Movements, Age, and Size Statistics, and Food Habits of Upper Cook Inlet Northern Pike during 1994 and 1995*; Fishery Data Series No. 96-29; Alaska Department of Fish and Game: Anchorage, AK, USA, 1996; p. 65.
41. Glick, W.J.; Willette, T.M. *Relative Abundance, Food Habits, Age, and Growth of Northern Pike in 5 Susitna River Drainage Lakes, 2009–2012*; Fishery Data Series No. 16-34; Alaska Department of Fish and Game: Anchorage, AK, USA, 2016; p. 55.
42. Eklöv, P.; Hamrin, S.F. Predatory efficiency and prey selection: Interactions between pike, *Esox lucius*, perch *Perca fluvialtilis* and rudd *Scardinus erythrophthalmus*. *Oikos* **1989**, *56*, 149–156. [CrossRef]
43. Rutz, D.S.; Bradley, P.T.; Jacobson, C.; Dunker, K.J. *Alexander Creek Northern Pike Suppression, 2011–2018, Alaska*; Fishery Data Series No. 20-17; Alaska Department of Fish and Game: Anchorage, AK, USA, 2020; p. 34.
44. Rutz, D.S.; Dunker, K.J.; Bradley, P.T.; Jacobson, C. *Movement Patters of Northern Pike in Alexander Lake*; Fishery Data Series No. 20-16; Alaska Department of Fish and Game: Anchorage, AK, USA, 2020; p. 50.
45. Massengill, R. *Stormy Lake Northern Pike Distribution and Movement Study to Inform Future Eradication Efforts*; Special Publication No. 17-17; Alaska Department of Fish and Game: Anchorage, AK, USA, 2017; p. 58.
46. Mills, M.J. *Alaska Statewide Sport Fish Harvest Studies*; Federal Aid in Fish Restoration, Annual Performance Report 1985–1986, Project F-10-1(27) RT-2; Alaska Department of Fish and Game: Juneau, AK, USA, 1986.
47. Oslund, S.; Ivey, S.; Lescanec, D. *Area Management Report for the Sport Fishereis of Northern Cook Inlet, 2017–2018*; Fishery Management Report No. 20-04; Alaska Department of Fish and Game: Anchorage, AK, USA, 2020; p. 322.
48. ADFG Chinook Salmon Research Team. *Chinook Salmon Assessment and Research Plan*; Special Publication No. 13-01; Alaska Department of Fish and Game: Anchorage, AK, USA, 2013; p. 62.

49. Romberg, W.J.; Raffety, I.; Martz, M. *Estimates of Participation, Catch, and Havest in Alaska Sport Fisheries*; Fishery Data Series No. In Prep; Alaska Department of Fish and Game: Anchorage, AK, USA, 2020; in prep.
50. Catchcart, C.N.; Dunker, K.J.; Quinn, T.P.; Sepulveda, A.J.; von Hippel, F.A.; Wizik, A.; Young, D.B.; Westley, P.A.H. Trophic plasticity and the invasion of a renowned piscivore: A diet synthesis of northern pike (*Esox lucius*) from the native and introduced ranges in Alaska, U.S.A. *Biol. Invasions* **2019**, *21*, 1379–1392. [CrossRef]
51. Chihuly, M.B. Biology of the Northern Pike, *Esox lucius* Linneaus, in the Wood Lakes System of Alaska, with Emphasis on Lake Alegnagik. Master's Thesis, University of Alaska Fairbanks, College of Environmental Sciences, Fairbanks, AK, USA, 1979.
52. Dye, J.; Wallendorf, M.; Naughton, G.P.; Gryska, A.D. *Stock Assessment of Northern Pike in Lake Aleknagik, 1998–1999*; Fishery Data Series No. 02-14; Alaska Department of Fish and Game: Anchorage, AK, USA, 2002; p. 18.
53. Wuttig, K.G. *Estimated Abundance of Northern Pike in Harding Lake, 2012*; Fishery Data Series No. 15-39; Alaska Department of Fish and Game: Anchorage, AK, USA, 2012; p. 32.
54. Russell, R. *A Fisheries Inventory of Waters in the Lake Clark National Monument Area*; Alaska Department of Fish and Game: Anchorage, AK, USA, 1980; p. 207.
55. Schwanke, C.J. *Abundance, Length Composition, and Movement of the Northern Pike Populations in Long Lake of the Chulitna River Dainage, 2007–2010*; Fishery Data Sereies No. 12-23; Alaska Department of Fish and Game: Anchorage, AK, USA, 2012; p. 32.
56. Massengill, R. *Control Efforts for Invasive Northern Pike on the Kenai Peninsula, 2005–2006*; Fishery Data Series No. 10-05; Alaska Department of Fish and Game: Anchorage, AK, USA, 2010; p. 35.
57. Massengill, R. *Control Efforts for Invasive Northern Pike Esox lucius on the Kenai Peninsula, 2007*; Fishery Data Series No. 11-10; Alaska Department of Fish and Game: Anchorage, AK, USA, 2011; p. 49.
58. Massengill, R. *Control Efforts for Invasive Northern Pike on the Kenai Peninsula, 2009*; Special Publication No. 14-11; Alaska Department of Fish and Game: Anchorage, AK, USA, 2014; p. 84.
59. Massengill, R. *Control Efforts for Invasive Northern Pike on the Kenai Peninsula, 2008*; Special Publication No. 14-12; Alaska Department of Fish and Game: Anchorage, AK, USA, 2014; p. 74.
60. Massengill, R. *Stormy Lake Restoration: Invasive Northern Pike Eradication, 2012*; Special Publication No. 17-18; Alaska Department of Fish and Game: Anchorage, AK, USA, 2017; p. 98.
61. ADFG Southcentral Northern Pike Control Committee. *Management Plan for Invasive Northern Pike in Alaska*; Alaska Department of Fish and Game: Anchorage, AK, USA, 2007; p. 62.
62. Fay, V. *Alaska Aquatic Nuisance Species Management Plan*; Alaska Department of Fish and Gmae: Juneau, AK, USA, 2002; p. 116.
63. Syslo, J.M.; Guy, C.S.; Bigelow, P.E.; Doepke, P.D.; Ertel, B.D.; Koel, T.M. Response of non-native lake trout (*Salvelinus namaycush*) to 15 years of harvest in Yellowstone Lake, Yellowstone National Park. *Can. J. Fish. Aquat. Sci.* **2011**, *68*, 2132–2145. [CrossRef]
64. Syslo, J.M.; Guy, C.S.; Cox, B.S. Comparison of harvest scenarios for the cost-effective supression of lake trout in Swan Lake, Montana. *N. Am. J. Fish. Manag.* **2013**, *33*, 1079–1090. [CrossRef]
65. Cook Inlet Staff. *Alexander Creek King Salmon Stock Status and Action Plan, 2020*; Report to the Board of Fisheries; Alaska Department of Fish and Game: Anchorage, AK, USA, 2020; p. 29.
66. Bradley, P.; Jacobson, C.; Dunker, K. *Operational Plan: Alexander Creeknorthern Pike Suppression 2019–2021*; Regional Operational Plan ROF.SF.2A2019.04; Alaska Department of Fish and Game: Anchorage, AK, USA, 2019; p. 31.
67. King, B.E.; Walker, S.C. *Susitna River Sockeye Salmon Fry Studies, 1994 and 1995*; Alaska Department of Fish and Game: Anchorage, AK, USA, 1997; p. 58.
68. Courtney, M.B.; Schoen, E.R.; Wizik, A.; Westley, P.A.H. Quantifying the net benefits of suppression: Truncated size structure and consuption of native salmonids by invasive northern pike in an Alaska lake. *N. Am. J. Fish. Manag.* **2018**, *38*, 1306–1315. [CrossRef]
69. DeCino, R.; Willette, T.M. *Chelatna Lake Northern Pike Suppression Project 2017–2019*; Regional Operational Plan ROP.CF.4K2017.11; Alaska Department of Fish and Game: Soldotna, AK, USA, 2017; p. 22.
70. Tarbox, K.E.; Kyle, G.B. *An Estimate of Adult Sockeye Salmon (Oncorhynchus Nerka) Production, Based on Euphotic Volume, for the Susitna River Drainage, Alaska*; Alaska Department of Fish and Game: Anchorage, AK, USA, 1989; p. 59.

71. Wizik, A. *Shell Lake Sockeye Salmon Progress Report 2015*; Cook Inlet Aquaculture Association: Soldotna, AK, USA, 2015; p. 49.
72. Wizik, A.; Schoen, E.; Courtney, M.; Westley, P. *Shell Lake Sockeye Salmon Progress Report 2017*; Cook Inlet Aquaculture Association: Soldotna, AK, USA, 2017; p. 63.
73. Wizik, A. *Shell Lake Sockeye Salmon Progress Report 2018*; Cook Inlet Aquaculture Association: Soldotna, AK, USA, 2018; p. 59.
74. DeCino, R.; Willette, T.M. *Chelatna Lake Pike Suppression*; Fishery Data Series No. In Prep; Alaska Department of Fish and Game: Anchorage, AK, USA, 2020; in prep.
75. Yanusz, R.J.; Merizon, R.A.; Willette, T.M.; Evans, D.G.; Spencer, T.R.; Raborn, S. *Inriver Abundance and Spawning Distribution of Susitna River Sockeye Salmon Oncorhynchus Nerka, 2006*; Fishery Data Series 07-83; Alaska Department of Fish and Game: Anchorage, AK, USA, 2007; p. 68.
76. Yanusz, R.J.; Merizon, R.A.; Willette, T.M.; Evans, D.G.; Spencer, T.R. *Inriver Abundance and Istribution of Spawning Susitna River Sockeye Salmon Oncorhynchus Nerka, 2007*; Fishery Data Series No. 11-19; Alaska Department of Fish and Game: Anchorage, AK, USA, 2011; p. 50.
77. Yanusz, R.J.; Merizon, R.A.; Willette, T.M.; Evans, D.G.; Spencer, T.R. *Inriver Abundance and Spawning Distribution of Spawning Susitina River Sockeye Salmon Onchorhynchus Nerka, 2008*; Fishery Data Series No. 11-12; Alaska Department of Fish and Game: Anchorage, AK, USA, 2011; p. 44.
78. DeCino, R.; Willette, T.M. *Susitna River Drainage Lakes Pelagic Fish Estimates, Using Split-Beam Hydroacoustic and Midwater Trawl Sampling Techniques, 2005–2008*; Fishery Data Series No. 14-47; Alaska Department of Fish and Game: Anchorage, AK, USA, 2014; p. 58.
79. Smukall, M. *Whiskey Lake Salmon Smolt Progress Report, 2014*; Cook Inlet Aquaculture Association: Soldotna, AK, USA, 2015; p. 33.
80. Smukall, M. *Northern Pike Investigations Final Report (2012–2014)*; Cook Inlet Aquaculture Association: Soldotna, AK, USA, 2015; p. 69.
81. Dunker, K.; Bradley, P.; Jacobson, C. *Threemile Lake Invasive Northern Pike Population Assessment*; Regional Operational Plan SF.2A.2018.11; Alaska Department of Fish and Game: Anchorage, AK, USA, 2018; p. 38.
82. Bradley, P.; Jacobson, C.; Dunker, K. *Operational Plan: Threemile Lake Invasive Northern Pike Suppression and Chuitbuna Lake Population Assessment*; Regional Operational Plan SF.2A.2020.01; Alaska Department of Fish and Game: Anchorage, AK, USA, 2019; p. 28.
83. Larsson, P.; Tibblin, P.; Koch-Schmidt, P.; Engstedt, O.; Nilsson, J.; Nordahl, O.; Forsman, A. Ecology, evolution, and management strategies of northern pike populations in the Baltic Sea. *Ambio* **2015**, *44*, 451–461. [CrossRef] [PubMed]
84. Walther, B.D.; Limburg, K.E. The use of otolith chemistry to characterize diadrmous migrations. *J. Fish Biol.* **2012**, *81*, 796–825. [CrossRef]
85. Britton, J.R.; Brazier, M.; Davies, G.D.; Chare, S.I. Case studies on eradicating the Asiatic cyprinid Pseudorasbora parva from fishing lakes in England to prevent their riverine dispersal. *Aquat. Conserv. Mar. Freshw. Ecosyst.* **2008**, *18*, 867–876. [CrossRef]
86. Knapp, R.A.; Matthews, K.R. Eradication of nonnative fish by gillnetting from a small mountain lake in California. *Restor. Ecol.* **1998**, *6*, 207–213. [CrossRef]
87. Massengill, R. *Soldotna Creek Drainage Restoration: Invasive Northern Pike Eradication, 2014–2018*; Special Publication No. In Prep; Alaska Department of Fish and Game: Anchorage, AK, USA, 2020; in prep.
88. Finlayson, B.; Skaar, D.; Anderson, J.; Carter, J.; Duffield, D.; Flammang, M.; Jackson, C.; Overlock, J.; Steinkjer, J.; Wilson, R. *Planning and Standard Operating Procedures for the Use of Rotenone in Fish Management—Rotenone SOP Manual*, 2nd ed.; American Fisheries Society: Bethesda, MD, USA, 2018; p. 176.
89. USEPA. *Reregistration Eligibility Decision for Rotenone*; Document EPA 738-R-07-005; United States Environmental Protection Agency: Washington, DC, USA, 2007; p. 45.
90. Ling, N. *Rotenone—A Review of Its Toxicity and Use for Fisheries Management*; Science for Conservation 211; New Zealand Department of Conseration: Wellington, New Zealand, 2003; p. 40.
91. Dalu, T.; Wasserman, R.J.; Jordaan, M.; Froneman, W.P.; Wey, O.L.F. An assessment of the effect of rotenone on selected non-target aquatic fauna. *PLoS ONE* **2015**, *10*(11), e0142140. [CrossRef]
92. Dunker, K.J.; Sepulveda, A.J.; Massengill, R.L.; Olsen, J.B.; Russ, O.L.; Wenburg, J.K.; Antonovich, A. Potential of environmental DNA to evaluate northern pike (*Esox lucius*) eradication efforts: An experimental test and case study. *PLoS ONE* **2016**, *11*(9), e0162277. [CrossRef]

93. Sepulveda, A.J.; Hutchins, P.R.; Massengill, R.L.; Dunker, K.J. Tradeoffs of a portable, field-based environmental DNA platform for detecting invasive northern pike (*Esox lucius*) in Alaska. *Manag. Biol. Invasions* **2018**, *9*, 253–258. [CrossRef]
94. Massengill, R.; Begich, L.R.; Dunker, K. *Kenai Peninsula Invasive Northern Pike Monitoring and Native Fish Restoration*; Regional Operational Plan SF.2A.2020.02; Alaska Department of Fish and Game: Anchorage, AK, USA, 2018.
95. ADF&G. *Statewide Stocking Plan for Sport Fish, 2020–2024*; Alaska Deparment of Fish and Game: Anchorage, AK, USA, 2020; p. 26.
96. Bell, M.A.; Heins, D.C.; Wund, M.A.; von Hippel, F.A.; Massengill, R.; Dunker, K.; Bristow, G.A.; Aquirre, W.E. Reintroduction of threespine stickleback into Cheney and Scout Lakes, Alaska. *Evol. Ecol. Res.* **2016**, *17*, 157–178.
97. Wund, M.A.; Singh, O.D.; Geiselman, A.; Bell, M.A. Morphological evolution of an anadromous threespine stickleback population within one generation after reintroduction to Cheney Lake, Alaska. *Evol. Ecol. Res.* **2016**, *17*, 203–224.
98. Kurz, M.L.; Heins, D.C.; Bell, M.A.; von Hippel, F.A. Shifts in life-history traits of two introduced populations of threespine stickleback. *Evol. Ecol. Res.* **2016**, *17*, 225–242.
99. Parkes, J.P.; Panetta, F.D. Eradication of invasive species: Progress and emerging issues in the 21st century. In *Invasive Speces Management: A Handbook of Principles and Techniques*; Oxford University Press: Oxford, UK, 2009; pp. 44–72.
100. Post, J.R.; Sullivan, M.; Cox, S.; Lester, N.P.; Walters, C.J.; Parkinson, E.A.; Paul, A.J.; Jackson, J.; Shuter, B.J. Canada's recreational fishereis: The invisible collapse? *Fisheries* **2002**, *27*, 6–17. [CrossRef]
101. Pierce, R.B.; Tomcko, C.M.; Schupp, D.H. Exploitation of northern pike in seven small north-central Minnesota lakes. *N. Am. J. Fish. Manag.* **1995**, *15*, 601–609. [CrossRef]
102. Pansko, S.; Goldberg, J. Review of harvest incentives to control invasive species. *Manag. Biol. Invasions* **2014**, *5*, 263–277. [CrossRef]
103. Olsen, J.B.; Lewis, C.J.; Massengill, R.L.; Dunker, K.J.; Wenburg, J.K. An evaluation of target specificity and sensitivity of three qPCR assays for detecting environmental DNA from Northern Pike (*Esox lucius*). *Conserv. Genet. Resour.* **2015**. [CrossRef]
104. Schill, D.J.; Heindel, J.A.; Campbell, M.R.; Meyer, K.A.; Mamer, E.R.J.M. Production of a YY male brook trout broodstock for potential eradication of undesired brook trout populations. *N. Am. J. Aquac.* **2016**, *78*, 72–83. [CrossRef]
105. Schill, D.J.; Meyer, K.A.; Hansen, M.J. Simulated effects of YY-male stocking and manual suppression for eradicating nonnative brook trout populations. *N. Am. J. Aquac.* **2017**, *37*, 1054–1066. [CrossRef]
106. Thresher, R.E. Autocideal technoogy for the control of invasive fish. *Fisheries* **2008**, *33*, 114–121. [CrossRef]
107. Thresher, R.W.; Hayes, K.; Bax, N.J.; Teem, J.; Benfey, T.J.; Gould, F. Genetic control of invasive fish: Technological options and its role in integrated pest management. *Biol. Invasions* **2014**, *16*, 1201–1216. [CrossRef]

Article

Case Studies Demonstrate That Common Carp Can Be Sustainably Reduced by Exploiting Source-Sink Dynamics in Midwestern Lakes

Peter W. Sorensen * and Przemyslaw G. Bajer

Department of Fisheries, Wildlife, and Conservation Biology, University of Minnesota, St. Paul, MN 55108, USA; bajer003@umn.edu

* Correspondence: soren003@umn.edu

Received: 18 October 2020; Accepted: 30 November 2020; Published: 4 December 2020

Abstract: The common carp has been highly problematic in North American ecosystems since its introduction over a century ago. In many watersheds, its abundance appears to be driven by source-sink dynamics in which carp reproduce successfully in peripheral ponds that lack egg/larva micro-predators which then serve as sources of recruits for deeper lakes. This manuscript describes how carp were sustainably reduced in two chains of lakes by disrupting source-sink dynamics in three steps. First, we ascertained whether lakes had problematic densities of carp that could be explained by source-sink dynamics. Second, ways to control recruitment were developed and implemented including: (i) aerating source ponds to reduce hypoxia and increase micro-predator abundance, (ii) blocking carp migration, and (iii) locating and removing adults from sinks using targeted netting guided by Judas fish. Third, we monitored and adapted. Using this strategy, the density of carp in 3 lakes in one chain was reduced from 177 kg/ha to ~100 kg/ha in 3 years and held constant for a decade. Similarly, adult density was reduced from 300 kg carp/ha in 2 lakes in the other chain to 25 kg/ha. Once carp densities were low, aluminum sulfate treatments became reasonable and once conducted, water quality improved.

Keywords: common carp; integrated pest control; source-sink; sustainable; micro-predators; water quality

1. Introduction

The common carp, *Cyprinus carpio*, (hereafter "carp") is one of the world's most invasive fishes and likely to become more as our climate warms [1]. Originally from Eurasia, it was initially moved across Europe by the Romans, and from there was exported across the globe in 1800s [2,3]. It is now found in all continents except Antarctica [2–4]. Although the carp is a valuable sport and food fish in many European countries as well as China [5], it has been stocked and become invasive in Africa, Australia, South America and North America [3]. Adult common carp are highly proficient at feeding in bottom sediments where they damage rooted vegetation, release nutrients and sediments into the water column, and degrade water quality (clarity) in shallow ecosystems [6,7]. This process also often causes dramatic reductions in biodiversity and waterfowl [6,8]. Carp control is a particularly onerous issue in North America [3] where control has traditionally focused on drawing-down and/or poisoning (rotenone) entire lakes and wetlands heavily infested with this species, and then installing barriers to prevent re-infestation [9–12]. However, this approach is extremely expensive and ecologically damaging, not applicable in many locales, and very often not sustainable as carp usually reinvade within 5–10 years [6,10]. New solutions to carp control are sought.

Although the common carp is generally considered to be a large river fish [13], it also does very well in many temperate lakes, especially systems of shallow eutrophic lakes with extensive littoral zones or lakes associated with such areas [6,14]. Adult carp are seasonal spawners and migrate, sometimes great distances, into shallow waters (floodplains, wetlands, ponds or shallow lakes) where females release up 3 million eggs each onto submersed vegetation [15,16]. After hatching, larval and juvenile carp forage on plankton in weed beds for 1–2 years but grow quickly [17–21]. Depending on the local environment, carp mature at 2–5 years of age and can live to be over 60 years of age, often reaching weights of up to 10 kg [2,15]. A single female carp might release 100 s of millions of eggs in her life. In shallow lakes in Midwestern North America, correlations between adult carp density, aquatic plant community composition, and water clarity, show that when adult carp densities reach about 100 kg/ha they drive about a 50% reduction in native plant cover, a value which many accept as a threshold for unacceptable ecological damage, and which almost always is associated with a shift in stable state and cyanobacteria blooms along with decreased in water quality [6]. When carp biomass exceeds 200 kg/ha, aquatic plants are typically almost absent [22,23].

For a population of fish to become highly abundant and invasive, its recruitment rate must on average, exceed its mortality and emigration rates. This can happen in many ways. In many inter-connected glacial lakes in the Midwestern North America, a variety of evidence suggests this often happens because adult carp migrate in the spring into peripheral waters which lack egg and larva micro-predators (fish with small gapes that consume eggs and larvae) because they often "winterkill" (suffer from low oxygen in winter which leads to fish kills) [16]. In many of these situations, young carp are able to survive, grow quickly and leave these nurseries for deeper lakes that do not support self-sustaining populations themselves because they do not winterkill (and thus have native micro-predators). Some shallow sources may also support adults on occasion because they only sporadically winterkill, enabling them to also function as sources. This scenario results in watersheds in which peripheral shallow waterbodies including floodplains, wetlands, ponds and shallow lakes can serve as population "sources", while deeper lakes function as "sinks"—a scenario known as a "source-sink" (Figure 1) [21,23–25]. Although not well studied, source-sink dynamics appear to be common in many watersheds in the North American Midwest, especially in its temperate forested zones where shallow lakes that frequently winterkill are commonly connected, sometimes in chains which flow to rivers or deep lakes [14,16,23]. Importantly, source-sink dynamics create special opportunities for controlling invasive species [25] because their recruitment is spatially restricted to a few locations so they can be easier to manage, as can removal of adults from population sinks.

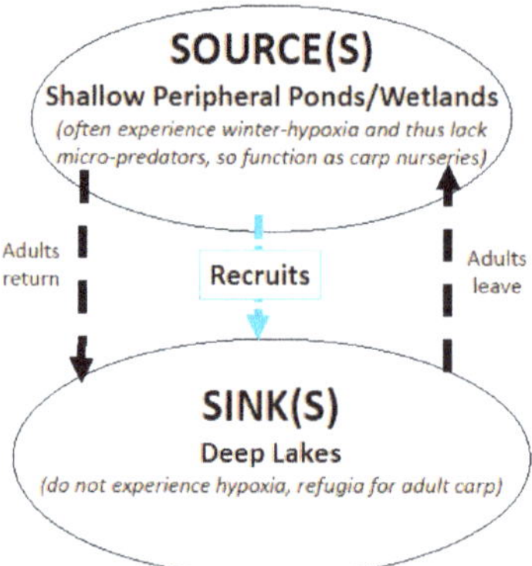

Figure 1. Schematic of the source-sink process hypothesized to determine the abundance of common carp in many Midwestern chains of lakes. The dashed arrows show that most, but not all, adults move between deep lakes and peripheral shallow water bodies to spawn and then return to lakes after the spawning season. Some adults may also occasionally overwinter in shallow waterbodies. Being long-lived, this flexible movement strategy, also known as partial migration, seems to work to the carp's advantage, especially in dynamic and complex temperate ecosystems that experience occasional flooding and winterkill.

Suspicions that adult carp might be superabundant in some lakes in the temperate forest ecoregion of the Midwest arose during the course of a research project conducted by the authors that sought to find the source of young carp in these systems while describing the damage they cause [16]. This work and the results of a dozen or so studies of the questions it addressed [16,18,21,26,27] eventually served as the foundation for long-term carp reduction and control efforts taken over by agency biologists and are described in this manuscript along with our studies. While some of this work has been published (and is referenced), much of it has not been previously published so it is presented here for the first time. The present manuscript describes two decade-long efforts to control carp based on the premise that source-sink dynamics can be sustainably exploited to control carp. In particular, we describe control programs in the two chains of lakes, the Phalen Chain of Lakes and the Riley Chain of Lakes, which have been studied and controlled for the longest period of time. First, we describe our overarching strategy (which we developed over the course of this study), and then how it was implemented, and finally our results.

2. Overarching Strategy and Approach

The overarching goal of Sorensen's and more recently Bajer's research on the common carp has been to develop an understanding of its biology and population dynamics in the Midwest so as to permit development of a sustainable way to reduce and control it. It has been an iterative process which required that we develop many new techniques including the ability to assess the abundance, distribution, and movement patterns of both juvenile and adult carp in our area and then explain them, and the level of ecological damaged caused by this species. During the course of conducting these studies, we developed a stepwise strategy and accompanying set of techniques to assess and control this species based on the observed source-sink dynamics. The present manuscript describes the results of this combined research and management effort in two chains of lakes. While research started in the Riley Chain of Lakes, control was eventually pursued more actively in the Phalen Chain of Lakes so we describe that system first. First, we provide a brief overview of the three steps, some of which were at times conducted concurrently and in an adaptive fashion.

Step 1: Assessing the density and distribution of juvenile and adult carp. We found that before we could consider carp control, we first had to determine the location and number of adult and juvenile carp in the systems, the extent of their populations (carp movement patterns), and whether there were sources and sinks. We developed several techniques including different types of netting, electrofishing, genetic analysis, environmental DNA, and aging to conduct this process. These are described below in the two case studies and related publications.

Step 2: Controlling and reducing the source(s) of juvenile carp while removing adult carp. While developing an understanding of how many carp there were, we realized that we could adopt these assessment techniques and approaches to affect control. Controlling sources required that we develop strategies different from those used for sinks. Many of these were adaptations of assessment strategies (ex. seining used for mark-recapture study were modified to include removal) and conducted as part of assessment itself. Different strategies had to be developed for different locales/carp populations. These are described in the case studies described below.

Step 3. Monitoring and Adapting. Even after carp populations had been reduced and controlled, it was evident that continued systematic monitoring was necessary owing to the high fecundity of this species and how its reproductive success was influenced by variable winter conditions. If/as increases in carp numbers were noted, new techniques to address these increases had to be developed. This adaptive effort was strongly based on assessment techniques and is described below. These data are largely taken from agency records.

3. Case-Studies of Sustainable Carp Control

Since 2006, we have been studying carp biology and developing techniques to use this understanding to reduce carp numbers in several local watersheds. Of these watersheds, the Phalen Watershed and the Riley Watershed, have received the greatest attention so their stories are presented here. As is typical to our region, both watersheds are dominated by a single chain of lakes (inter-connected lakes that drain into rivers) but they do differ in complexity and the number of peripheral shallow waters they drain.

3.1. Case Study 1: The Phalen Chain of Lakes

The Phalen Chain of Lakes is located in the Phalen Subwatershed in Ramsey County, Minnesota, USA (44.9884° N, 93.0545° W; Figure 2) and is similar in size and habitat complexity to many freshwater systems in other glaciated regions of the Upper Mississippi River Basin. It is densely populated and in the late 2000s was experiencing poor water quality, leading biologists in its watershed district (Ramsey Washington Metro Watershed District, RWMWD) to suspect that it might have an overabundance of common carp and to contact Sorensen. This watershed drains 6100 hectares of urban land which has been altered by human development including the draining of wetlands, ditching of creek segments, and construction of storm water ponds (Figure 2). This system contains four relatively deep lakes (surface area: 29–95 ha; maximum depth: 3–28 m) that do not winterkill, and are connected by navigable streams (Table 1). It also contains about half a dozen shallow lakes/ponds, of which Willow Lake, Casey Lake, Markham Pond, and Gervais Mill Pond are the largest (3–21 ha) and most directly connected. All of these shallow lakes/ponds are found in headwater regions of the watershed and are reported to experience periodic low winter oxygen levels and winterkills. At the time of this study, many were highly degraded and received a great deal of nutrient-enriched runoff, especially Casey Lake and Markham Pond.

Figure 2. The Phalen Chain of Lakes.

Table 1. Lakes and ponds in the Phalen Chain of Lakes and their carp as initially assessed in 2009–2010. CPUE (catch per unit effort) for adults is the number of carp caught per hour by electrofishing, while CPUE for juveniles is the number of carp caught per trap-net (see text). Emboldened values are ecologically problematic (i.e., >100 kg/ha). YOY = young-of-the-year carp.

Lake	Size (ha)	Max Depth (m)	Adult Carp (CPUE)	Adult Carp (#)	Adult Carp Density	YOY Carp (CPUE)	Notes
Casey	4.7	6.1	8.5	2585	n.a.	6.2 [b]	Winterkills
Markham	7.0	2.0	8.2	211	n.a.	104 [b]	Winterkills
Willow	21.1	2.5	0	0	n.a.	173 [b]	Winterkills
Kohlman	29.9	2.7	11.2 [a]	1668 [a]	**177 kg/ha** [c]	0 [b]	-
Gervais	94.5	12.5	7.2 [a]	3499 [a]	**177 kg/ha** [c]	0 [b]	-
G Mill Pond	3.0	4.0	n.s.	n.s.	n.s.	n.s.	-
Keller	29.0	2.4	14.7	2103	**177 kg/ha** [c]	0 [b]	-
Round	14.3	2.5	n.s.	n.s.	n.s.	n.s.	-
Phalen	79.6	53	6.0	2504	106 kg/ha [d]	0 [b]	

[a.] Previously published [26]. [b.] Previously published [21]. [c.] Biomass data show the average combined biomass/ha of carp in Kohlman, Gervais and Keller lakes, reflecting our finding that carp readily moved between these three lakes and functioned a single population (Table 2). [d] from 2018. n.s. = not sampled.

Table 2. The absolute number and percentage of total detections of 50 adult carp originally radio-tagged in lakes Gervais, Keller, Kohlman, and Phalen found in 2009 by lake and by season. Emboldened lake names in the left column denote the lake where the carp were originally captured.

Origin Lake	Summer 2009	Fall 2009	Winter 2009–2010	Spring 2010	Summer 2010
			Gervais Lake (Lake Where Detected)		
Kohlman	1 (2%)	0 (0%)	19 (26%)	6 (18%)	4 (15%)
Gervais	35 (71%)	36 (61%)	18 (25%)	16 (47%)	21 (78%)
Keller	13 (27%)	23 (39%)	35 (49%)	12 (35%)	2 (7%)
Phalen	0 (0%)	0 (0%)	0 (0%)	0 (0%)	0 (0%)
			Lake Keller		
Kohlman	0 (0%)	0 (0%)	0 (0%)	0 (0%)	0 (0%)
Gervais	1 (9%)	9 (11%)	3 (33%)	9 (17%)	5 (8%)
Keller	110 (91%)	73 (89%)	6 (67%)	43 (83%)	55 (92%)
Phalen	0 (0)	0 (0%)	0 (0%)	0 (0%)	0 (0%)
			Kohlman Lake		
Kohlman	67 (70%)	56 (84%)	5 (83%)	20 (77%)	36 (68%)
Gervais	21 (22%)	11 (16%)	1 (17%)	5 (19%)	7 (13%)
Keller	8 (8%)	0 (0%)	0 (0%)	1 (4%)	10 (19%)
Phalen	1 (1%)	0 (0%)	0 (0%)	0 (0%)	0 (0%)
			Lake Phalen		
Kohlman	0 (0%)	0 (0%)	0 (0%)	0 (0%)	0 (0%)
Gervais	0 (0%)	0 (0%)	0 (0%)	0 (0%)	0 (0%)
Keller	1 (1%)	26 (30%)	4 (21%)	3 (14%)	12 (20%)
Phalen	82 (99%)	62 (70%)	15 (79%)	18 (86%)	49 (80%)

3.1.1. Step 1: Assessing the Density and Distribution of Adult and Juvenile Carp

In early 2008, we started a research project to develop long-term control of carp in the Phalen Chain of Lakes. We started by concurrently evaluating the relative abundance and distribution of both young-of year (YOY) and adult common carp across the entire chain of lakes. We knew from ongoing work in the Riley Chain of Lakes (see Section 3.2 below), that the easiest and quickest way to obtain an estimate of adult fish abundance, including carp, in our lakes was boat-electrofishing [27,28]. Each of the 4 deep lakes and all of the smaller lakes/ponds we could assess were evaluated in this manner at least once following a well-established protocol which used at least four 20-min transects conducted across each entire lake in 2009. This initial analysis discovered more than 1000 adult carp to be present in the two smallest lakes (<30 ha; lakes Keller and Kohlman), over 2000 individuals in 80 ha Lake Phalen, and finally over 4000 in 95 h Lake Gervais (Table 2). Markham Pond and Casey Lake also had modest populations of small adult carp whose numbers could not be estimated because they were too small to boat-electrofish. Notably, no adult carp were caught in other lakes and ponds. Next, to determine whether adult carp in the main lakes functioned as a single population (i.e., a self-contained group of animals), their movement patterns were described for two years using radio-telemetry following techniques being developed in the Riley Chain of Lakes [16]. Briefly, 10 adult carp were captured in each lake using boat electrofishing and surgically implanted with radio-transmitter tags (model F1850; Advanced Telemetry Systems; Isanti, MN, USA) in April and May 2009. An additional 10 carp were implanted with radio-transmitters and released in June 2009 in Lake Keller. The distribution of these radio-tagged carp was then monitored by surveying the entire system using a small boat and a low frequency loop antenna 1 to 4 times a month for 1.5 years. Monthly detections were then tabulated to assess the presence of fish and percent of all detections also calculated. We found that while those adult carp originally tagged in Lake Gervais tended to move to both Keller and Kohlman lakes in the summer, they returned to Lake Gervais to overwinter and never traveled to Lake Phalen. Carp tagged in Lake Keller tended to enter lakes Gervais and Kohlman but also overwintered in Lake Gervais, although four entered Lake Phalen (one eventually returned). Carp tagged in Lake Kohlman also tended to overwinter Lake Gervais. Tagged adult carp were observed trying to enter Gervais Mill Pond and Markham Pond in 2008–2009 but water levels were too low to permit passage that spring. Finally, only one carp tagged in Lake Phalen left that lake. Based on the results of this radio telemetry study, two populations of adult common carp were identified: a Kohlman–Gervais–Keller population and a Lake Phalen population (Table 2).

Once we knew the relative abundance and movement patterns of adult carp, we sought to determine precise population and density estimates. We focused on lakes Kohlman, Keller, and Gervais as these lakes both appeared to have the highest densities of adult carp while also functioning as a single population (almost all fish overwintered in a single location, Lake Gervais). To estimate the abundance of adult carp we used mark-recapture estimates, the details of which have already been published [28,29]. Briefly, we used radio-tagged carp ("Judas fish") to help us locate aggregations of over-wintering carp in Lake Gervais [29]. A single large aggregation was seined under-ice using a 500 m net which was pulled through the ice after being positioned around the aggregation. A total of 3537 adult carp were captured, marked with numbered t-bar tags and released. The following summer, electrofishing surveys were performed in these lakes to evaluate the ratio of marked fish to unmarked fish in the population. In total, 160 adult carp were captured of which 20 were previously marked. The average weight of these carp was 3.4 ± 0.3 kg. Using a Chapman-modified Lincoln-Petersen mark-recapture equation, the carp population size (of the three lakes) was estimated at 8041 ± 1563 individuals with an average biomass of 177 ± 35 kg/ha, in excess of what we have previously calculated to be ecologically damaging [6]. Numbers and biomass of carp present in each of these deep lakes were then calculated based on the relative distribution of the radio-tagged carp found in each lake in the summer (Tables 2 and 3).

While assessing adult carp, we commenced efforts to determine the source(s) of young-of-year (YOY) carp (<200 mm) [30]. Trap-netting was our primary tool as it was proving useful in the Riley Chain of Lakes [26] although we supplemented it with electrofishing in ponds. Trap-netting, a type of fyke net which is anchored to the shore and strung to trap, is commonly used by fisheries management agencies including the Minnesota Department of Natural Resources (MN DNR) to sample small fish (adopted by us for YOY carp in littoral zones [26,30]. Trap-netting is especially effective in shallow waters where boats cannot reach but it is a considerable amount of work so only a limited areas of lake shores can typically be sampled each year so it not highly quantitative. Trap-netting also does not sample large carp (>200 mm) well. Examining historic trap-netting data, we found that the MN DNR had sampled lakes Kohlman, Gervais, Keller and Phalen 10 times between 1999–2009 and only once caught more than 1 YOY carp per trap-net (Table 3), and Casey and Markham ponds a total of 4 times, twice capturing 25 or more YOY carp (or greater) per trap-net. Given the paucity of data, we started to collect our own trap-net data. Following established procedures [26], each lake in the Phalen Chain was sampled using 5 trap-nets (9 mm bar) set at evenly spaced intervals around the perimeter of each lake for a 24 h period in the falls of 2009 and 2010 (when YOYs were large enough to catch). In both 2009 and 2010, large numbers of YOY carp were captured in all three shallow ponds (Willow, Casey and Markham Ponds; see below), but no YOY carp were captured in any of the four deep lakes (lakes Kohlman, Gervais, Keller, and Phalen) which also had many bluegill sunfish (*Lepomis macrochirus*) (CPUE ranged up to 100), an important micro-predator of carp eggs and larvae in this region but also one susceptible to winterkill [26]. Markham Pond had especially high numbers of YOY carp in both 2009 and 2010 (Catch per unit effort (CPUEs) of 47 and 104) while Lake Willow had a CPUE of 174 in 2009. None of these shallow ponds had appreciable numbers of bluegill sunfish but some had goldfish (*Carassius auratus*) and black bullhead catfish (*Ameiurus melas*). Osborne [30] conducted a mark-recapture study of YOY carp in Lake Casey and Markham Pond in 2010 and estimated the former to contain 12,703 YOY carp, and the latter, 34,782 YOY [30]. Several dozen sexually-mature adult carp (<500 mm, 2–4 years in age) were also sampled in both Casey Lake and Markham Pond using boat electrofishing which was performed as part of the recapture process [30].

Table 3. Common carp control efforts and results in the Lake Phalen Chain of Lakes (previously published data is referenced, the rest is from various data reports to the RWMWD).

Date	Removal Method	# Carp Caught	# Carp Removed
Winter 2010–11	Gervais Winter Seine-1 [a]	3537 [b]	100
Winter 2010–11	Gervais Winter Seine-2	1509	1505
Winter 2010–11	Gervais Winter Seine-3	930	732
Winter 2011–12	Gervais Winter Seine	73	71
Fall 2012	Kohlman Boxnet	630	628
Winter 2012–13	Gervais Winter Seine	827	825
Spring 2011–2013	Spawning Block/Trap	381	351
Summer 2013	Keller Electrofishing	8	8
Summer 2013	Gervais Electrofishing	97	97
Summer 2013	Kohlman Electrofishing	5	5
Fall 2013	Keller Boxnet	446	446
Fall 2013	Gervais Boxnet	192	192
Summer 2013	Drawdown Casey, Markham	-	100^+
Summer 2014	Drawdown Casey, Markham	-	100^+
Winter 2014–2015	Aerate Casey, Markham	-	-
Fall 2015	Gervais Boxnet	302	302
Fall 2016	Gervais Boxnet *	324	324 *

[a]: The lake was seined three times in 2011 (1, 2, 3). [b]: Previously published [29]. * By the fall of 2016, a grand total of 5926 adult carp had been removed of the 7270 estimated present in lakes Kohlman, Gervais, and Keller (81%). The carp biomass was then 34 kg/ha across the 3 lakes, lakes Kohlman, Gervais and Keller (when combined and averaged).

Next, to examine recruitment patterns across time we sampled and aged adult carp. A total of 127 adult carp were captured using boat-electrofishing in Lakes Kohlman, Gervais, and Keller, and then measured and weighed [21]. Following established procedures [16], their asterisci otoliths were removed and the number of annuli counted by three readers [16,31]. We found that recruitment has been very sporadic with distinct peaks in 1988–1962, 1965, 1991, 2000, 2009 (Figure 3). The oldest carp was 64 years old with the average carp being about half that age. Finally, to confirm the sources of these carp, fin samples from adults across the large lakes as well as YOY carp collected from Lake Casey and Markham Pond were analyzed for genetic variation at 12 microsatellite DNA loci (see [21] for full detail and data). This analysis described two genetic populations (strains "A" and "B") in Lakes Kohlman, Gervais and Keller, and a third which was a hybrid of A and B. Over 80% of the adult carp that were less than 30 years old (the majority of the population), were either Strain A or the hybrid, while older fish were all strain B. The same ratio of strain A to hybrid was also found to characterize YOY captured in Casey Lake and Markham Lake, with strain B appearing to come from an unknown historical source at the western edge of the watershed which might have been Gervais Mill Pond [21]. Together, these data strongly suggested that over the past 30 years, YOY had been entering (recruiting to) lakes Kohlman, Gervais and Keller from the peripheral shallow systems as two separate genetic stocks, with Casey Lake and Markham Pond dominating recently. Notably, anecdotal observations suggested that adult carp could reach these two ponds in years when water levels were high from Lake Kohlman while young carp could almost always leave because the stream was nearly always open [21,27].

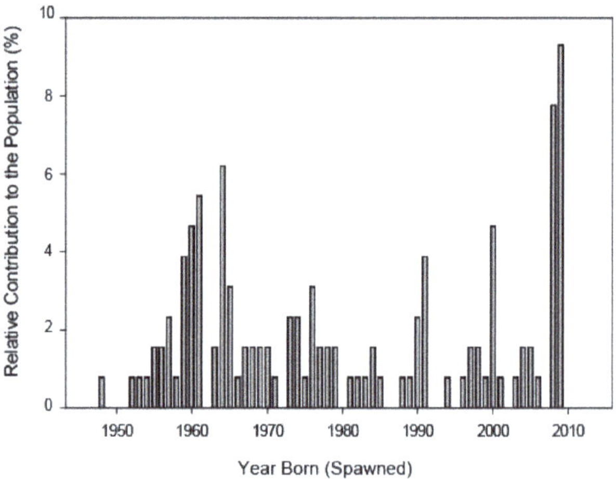

Figure 3. Age structure of adult carp collected from lakes Kohlman, Gervais and Keller. Data from [21] which were re-analyzed for this figure to show their percent contribution to the population. Fish were captured and aged in 2011.

3.1.2. Step 2: Controlling the Number of Adult Common Carp while Controlling Recruitment/Sources

When we realized that there were only a few sources of carp in the Kohlman–Gervais–Keller population which functioned as a sink, we started to develop methods to reduce adults and reduce/control carp recruitment (see below). Adult carp removal commenced in 2009 as part of our effort to collect mark-recapture population estimates (see above). We took three approaches. First, as described above, we used Judas fish to find aggregating carp in the winter and then seined them using under-ice netting. We removed 3233 adult carp in the winter of 2010–2011 (3 hauls) from Lake Gervais where most carp aggregated (Table 3, [29]). Winter-netting was later repeated in the winters of 2011–2012 (825 carp removed) and 2012–2013 (no carp removed) (Table 3). Because under-ice netting is difficult / expensive and its success was waning as carp captures decreased (carp seemingly remembered and avoided the seine net), we shifted to another strategy also being developed in Riley Chain of Lakes [29,32,33] (see Section 3.2) in which adult carp are lured (trained) to food placed on top of a woven box-net (25 m x 25 m) placed on the bottom and pulled when carp are present. This versatile technique was especially successful in late summer when carp were feeding and 2233 adult carp were removed in this fashion from lakes Kohlman, Gervais and Keller in the summers of 2012–2016 (Table 3). In some lakes, a few (n = 110) adult carp were also removed using boat electrofishing in 2013 but this was not very efficient. Finally, we captured and removed 351 adult carp during their spring spawning runs into Kohlman Creek using a stream trap [27]. Annual capture patterns in the creek suggested that this creek was selected by adult carp to access their natal waters [27]. By the fall of 2016, a total of

5926 carp had been removed from this population, or about 80% of the population, and active removal efforts ceased, as values suggested densities were well below 100 kg/ha (Table 3).

While starting to remove adult carp, we undertook several efforts to eliminate carp recruitment in Markham, Casey and Willow. First, we blocked springtime access of adult carp to these ponds by building a fish barrier and trap in Kohlman Creek in the spring of 2010 (see section above). This effort was only moderately successful and its value seemed marginal, especially because we discovered a number of small adult carp overwintering in Markham, Casey and Willow [30]. Second, we attempted to drawdown Lake Casey and Markham Pond in late fall to force a winterkill and carp eradication. This process took three years, because these ponds had natural groundwater seepages, which made drawdowns difficult. Finally, in 2013 we succeeded achieving a complete drawdown and eradication of carp. We followed that with springtime stocking of native micro-predators (bluegill sunfish) and installed a small floating electric water-jet aeration system in both waterbodies to maintain the stocked predator fish. We did not have to address Willow Lake because an aeration system already in it failed, leading to the demise of all of its fish including carp. Trap-net surveys showed these efforts to be successful as YOY were never captured in these locations again (Table 4).

Table 4. Average number of Young-of-Year (YOY) carp caught per trap-net (CPUE) in the Phalen Chain of Lakes.

Year	Lakes						
-	Casey	Markham	Willow	Kohlman	Gervais	Keller	Phalen
1999	0 D	0	-	0 D	0 D	0 D	0 D
2000	-	-	-	-	-	-	-
2001	25 D	-	-	-	-	-	-
2002	-	-	-	-	-	-	-
2003	-	-	-	-	-	-	-
2004	-	-	-	0.33 D	-	-	0 D
2005	-	-	-	1.3 D	0.56 D	0.5 D	-
2006	-	-	-	-	-	-	-
2007	28 D	-	-	-	-	0 D	-
2008	-	-	-	-	-	-	-
2009 S	6.2 *	104 *	173 *	0 *	0 *	0 *	0 *
2010	3.8	47	5.4 W	0	0	-	-
2011	2	0.8	0	0	0	-	-
2012	8	0.7	0	0	0	-	-
2013	0	68.8 DD	0	0	0	-	-
2014	-DD	-DD	-	-	-	-	-
2015	0	0	-	1	-	-	0
2016	0	0		0			
2017	-	-	-	0	-	-	-
2018	0	0	-	0	-	0	0

D = MN DNR online sampling records. S = Start of our study. * Previously published and also shown in Table 2 [21]. W = Winter-killed. DD = Pond drawn-down.

3.1.3. Step 3: Monitoring Carp Populations and Adapting New Management Approaches If/As Needed

While in the process of removing adult carp and controlling YOY, we commenced a monitoring program to confirm that the population was being reduced as expected. Water quality was also assessed each summer and an aluminum sulfate treatment ("alum"; a chemical treatment which binds phosphorous in the sediments) was performed in Lake Kohlman to increase water quality (see Section 3.1.4). Every year a boat electrofishing survey was conducted in all three problematic lakes following our established protocols, while trap-nets were also set each fall across all lakes following protocols to assess YOY carp abundance (Table 4). Because we did not know the precise movement patterns of adult carp between lakes on a weekly basis, and they functioned a single population, we averaged the CPUE for these lakes, weighting abundance estimates for lake size. During the course of monitoring we decided to remove additional adult carp using box-nets in 2014–2016 as well as with boat electrofishing, which was not very efficient as a removal tool (Table 4). Aeration in both outlying ponds continued along with trap-net surveys. These surveys failed to catch any YOY carp and confirmed that native micro-predators remained abundant.

3.1.4. Summary: Carp, Water Quality, and Current Status

During the course of this 10 year study/ management effort, three major sources of young carp were controlled, and nearly 6000 adults removed. This integrated effort caused the carp biomass in lakes Kohlman–Gervais–Keller to fall from an average of about 177 kg/ha to a minimally-damaging level of about 40 kg/ha (mark-recapture) in 2016 from which it rose to ~100 kg/ha in 2018 (as estimated by a single electrofishing survey). In particular, electrofishing showed a drop in average adult carp CPUE (across the three lakes) from 9.41 in 2009 to 4.03 in 2014 (Figure 4). While CPUE estimates in 2018 suggested a possible slight increase in adult density, recent anecdotal reports of juvenile captures in Lake Gervais suggest this likely was caused by a few adult carp entering Gervais Mill Pond to spawn during high water events, (a possibility suggested by the genetic analyses [21]). This possible secondary source was not seemingly been identified earlier when water levels were lower, and in any case appears minor. In the future, measuring environmental DNA, or eDNA [34] might be a very useful tool for identifying possible nursery areas like this because water sampling is much easier than trapping fish. In any case, efforts are now underway to block carp from accessing Gervais Mill Pond. Meanwhile, fisheries surveys conducted by the MN DNR during this time have shown that Lake Gervais has witnessed a possible increase in gamefish from 29.2 CPUE by gill net in 2011 to 71.6 in 2017 (MN DNR). Importantly, an alum treatment was performed in Lake Kohlman to reduce benthic phosphorous release concurrent with carp removal effort while using carp exclosures to evaluate the effects of carp on the benthos and alum. It showed that carp foraging activity greatly disrupts sediments to a depth of over 13 cm [35], enough to greatly impair the benefits of alum treatment unless the carp are removed (as they were). Notably, a substantial (50%) reduction in total summer phosphorous (TP) was measured in Lake Kohlman after this joint alum treatment/carp removal. Furthermore, this lake's water clarity doubled and remains high a decade later (Figure 5). Increased native submersed plant cover has also been measured in this shallow lake, as is common after carp removal [6–8]. Similar improvements in springtime plant cover were noted in Lake Gervais after carp removal, and while some improvement has also been noted in its springtime water clarity (data not shown), this improvement has not persisted into the summer in this lake which was also not yet been treated with alum because of high cost (Figure 5). Trends in Lakes Keller were similar to those in Lake Gervais (https://www.rwmwd.org/projects/keller-lake/) as it too was been treated with alum. In sum, our carp control project has been a success: most of watershed's carp population has been sustainably and meaningfully reduced, many of its ponds are improved, as is fishing, as well as water quality especially where alum treatments occurred. Furthermore, the cost has been reasonable, and no permanent barrier or poisons were needed in this important urban system. While our carp control scheme will need to be maintained into the future and ideally a new effort made to find secondary source(s) of recruits, Ramsey-Washington Metro Watershed District is fully and officially committed to this effort as it has realized substantial benefits (personal communication, RWMWD).

3.2. *Study 2. The Riley Chain of Lakes*

The Riley Chain of Lakes is located in Hennepin and Carver Counties, Minnesota, USA (44.8361° N, 93.5231° W; Figure 6). This watershed drains 3394 hectares of urban land through a chain of 4 interconnected lakes (surface area: 35–118 ha; maximum depth: 3–15 m) and a shallow marsh (Rice Marsh Lake) which are connected by a shallow stream (Riley Creek) and drain into the Minnesota River. All lakes except for Riley and Ann are known to occasionally winterkill, with Rice Marsh Lake being the most prone to this phenomenon, especially in recent decades (Lake Susan used to winterkill up until the early 1990s but an aeration system installed in 1992 has prevented winterkills since). The hydrology of this watershed has also been altered by human development (Table 5) and was suffering from impaired water quality in 2006 when we were approached by its watershed district, Riley Purgatory Bluff Creek Watershed District (RPBCWD), to determine if it had an overabundance of common carp and to develop solutions. Some of this work preceded that reported above in the Phalen Chain of Lakes which we showed first because it demonstrates the role of recruitment control more clearly. The work followed the same steps outlined above for the Phalen Chain of Lakes.

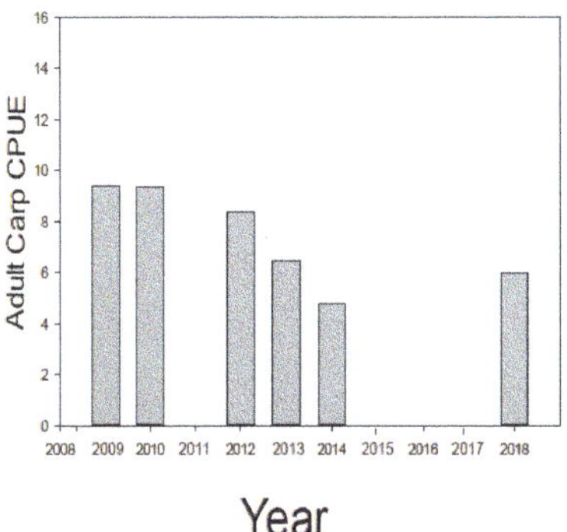

Figure 4. Average number of adult carp caught electrofishing per hour (CPUE) across lakes Kohlman, Gervais and Keller (data are combined and averaged). The study started in 2009.

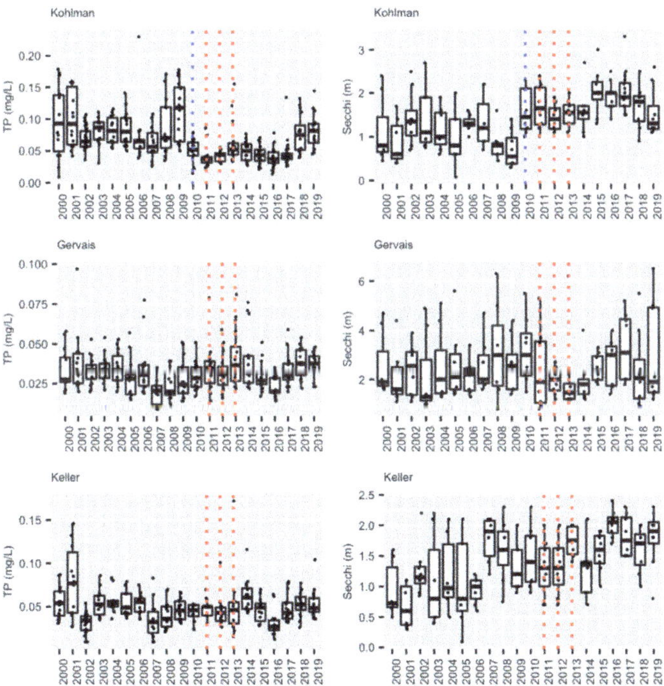

Figure 5. Summer (June–August) epilimnion water quality values (TP, or Total Phosphorous, and Secchi Depth (clarity in m)) for lakes Kohlman, Gervais and Keller. The blue dotted line shows when the alum treatment occurred [35] while the dotted red lines show years of carp removal in the Kohlman-Gervais-Keller system of lakes (see text).

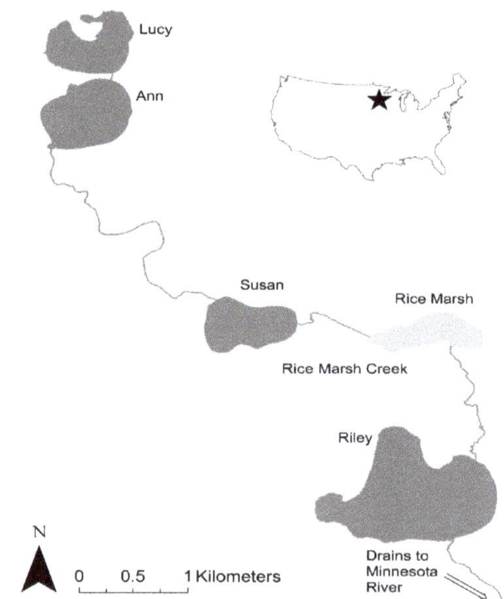

Figure 6. The Riley Chain of Lakes.

Table 5. Lakes in the Riley Chain of Lakes and their carp in 2008–2009. Data on carp densities are described in the text (emboldened values are for adult densities that are ecologically problematic).

Lake	Size (ha)	Max Depth(m)	Adult Carp (CPUE)	Adult Carp (#)	Adult Carp (kg/ha)	YOY Carp (CPUE)	Notes
Lucy [a]	34.6	6.0	8.15 [b]	808 [c]	70 kg/ha [d]	n.s.	winterkills
Ann	44.1	13.7	3.57 [e]	662 [c]	66 kg/ha [d]	0	-
Susan	35.1	5.1	17.26 [b]	4181 [c]	**307/kg/h** [e]	0.4	aerated
RiceMarshL	40.0	3.0	3.0	178 [c]	16 kg/h	0	winterkills
Riley	118.8	14.7	12.16 [b]	6419 [c]	**176 kg/ha** [d]	0	-

[a]: Lake Lucy winterkilled in 2012 and most fish died and population temporarily dropped to 20 kg/ha, before returning to 60–70 kg/ha seemingly because of adult immigration. [b]: Boat electrofishing 2008 and 2010 (Lucy), some of these data are published [28,29]. [c]: Published, population size extrapolated from mark-recapture study [28]. [d]: Biomass densities extrapolated from population size [28] and fish size. [e]: Previously reported [36].

3.2.1. Step 1: Assessing the Density, Abundance, and Distribution of Adult and Juvenile Carp

We started this project in 2006 by evaluating the relative abundance, distribution, and movement patterns of adult common carp across the entire chain of lakes. First, we sought to determine whether adult carp in Riley and Susan (which pilot studies showed to have the most carp) functioned as a single population or several distinct populations, and to what extent carp from those lakes used the shallow lake/marsh in-between as spawning site. To determine this, we examined the movement patterns of approximately two dozen adult carp caught in each lake and also in Rice Marsh Lake, which we implanted with radio-transmitters ATS F1850; [16] and tracked for two years. We found that adult carp originally tagged in Lake Susan had a strong tendency to enter Rice Marsh Lake in the spring-summer to spawn and then return, while carp tagged in Lake Riley also tended to move into Rice Marsh Lake and then return and rarely entered Susan [37] (fully summarized in Table 6). A few fish appeared to be resident in Rice Marsh Lake, but they died when this marsh winterkilled in early 2007. No carp moved to lakes other than Lake Susan, Lake Riley, and Rice Marsh Lake while spawning was observed in all three lakes each spring. Based on both these movement data and boat electrofishing surveys, we concluded that this chain of lakes contained two major (and ecologically damaging) populations of adult carp: one centered in Lake Riley where some of the adult carp moved in and out of this system in spring to spawn in Rice Marsh Lake, and the other in Lake Susan which did the same. We then began studying both populations in more detail while conducting management efforts to reduce them below the management threshold. In the summer of 2008, we began mark-recapture analyses to precisely estimate carp abundance and biomass in lakes Susan and Riley. In Lake Susan, we used electrofishing and open water seining to mark and release 101 carp [28] in the summer 2008. We then conducted winter seining guided by radio-telemetry in March 2009. In that event, over 3000 carp were captured and removed, including 79 of the 101 marked carp [28]. This allowed us to estimate that Lake Susan had approximately 4181 adult carp with a biomass of 307 kg/ha in 2008 [28]. In Lake Riley, we also used radio-tagged carp to find aggregations of carp, seined them using an under-ice 500 m net, after which some carp were tagged and released, and then resampled the next summer and winter to calculate abundance using mark-recapture analyses [28,29]. A total of 4440 carp were captured in Lake Riley in January 2009, of which 600 were marked and released. Of the 600 marked carp, 388 were recaptured the following year among 2303 carp that were captured in another winter seine [29]. Overall, mark-recapture analyses showed that the carp population in Lake Riley in 2009 was estimated to be 6491 individuals with a biomass of 176 kg/ha [28,29], in excess of what we had previously calculated to be ecologically damaging (Table 5). The relationship between electrofishing CPUE and adult carp abundance and biomass was calculated, so boat electrofishing alone could be used here and elsewhere including the Phalen Chain to estimate carp abundance [28], see Section 3.1.

Table 6. The number of adult carp radio-tagged in lakes Susan, Rice Marsh, and Riley and their locations in these lakes by season (% = portion of the population they represent; some of these data have been published [37]). Lakes were carp were captured is emboldened.

Original Lake	Spring 2006	Summer 2006	Winter 2006–7	Summer 2007	Winter 2007–8
			Lake Susan (Lake Where Detected)		
Susan	16 (100%)	10 (62%)	15 (88%)	24 (100%)	26 (79%)
RiceM	-	6 (38%)	1 (6%)	0 (0%)	7 (21%)
Riley	-	0 (0%)	1 (6%)	0 (0%)	0 (0%)
			Rice Marsh Lake		
Susan	-	6	1	7 (0%)	0 (0%)
RiceM	15 (100%)	15 (63%)	D	0 (0%)	11 (61%)
Riley	-	3	D	7 (0%)	7 (39%)
			Lake Riley		
Susan	-	0 (%)	0 (0%)	0 (0%)	0 (0%)
RiceM	-	3 (15%)	0 (0%)	0 (0%)	1 (6%)
Riley	19 (100%)	16 (85%)	18 (100%)	24 (100%)	17 (94%)

D = died in Rice Marsh Lake due to winterkill.

While assessing the population of adult carp, we started to assess the source(s) of these fish by examining trap-net data previously collected by the MN DNR. Although sparse (only six samples between 1997–2006), only Rice Marsh Lake was seen to have produced any YOY carp (Table 6). Starting in 2006, we began our own sampling program and set 5 trap-nets traps at evenly spaced intervals around the perimeter of each of the five lakes in the chain for 24-h every fall, similar to MN DNR protocol [26]. Very few (CPUE < 0.5) YOY carp were captured in any lake in 2006. However, on 28 March 2007, a winterkill was reported in Rice Marsh Lake and we found a large number (many hundreds) of juvenile carp, 18 of which were sampled and found to have an average length of 342 ± 10 mm (TL) and be 3 years old. Subsequent trap-netting in 2008, 2009, 2010 and 2011 showed that while Rice Marsh Lake produced a few YOY carp (CPUE = 1), the other lakes produced even smaller numbers of YOY (Table 7). No winterkill events were reported past 2007 in these lakes, but analysis of historical records showed that Lake Susan had winterkilled in 1990 and 1991 (after which it was aerated most winters), while Rice Marsh Lake had in 1997, 2000, and 2004 although these data were incomplete [16]. Lake Riley had no known records of winterkill, likely due to its depth (over 15 m). Ageing analyses and winterkill records then provided the key clue to solving the carp recruitment puzzle in the Susan and Riley systems. Examining the relationship between winterkill occurrence and carp recruitment, we collected 100 carp from Lake Susan and Lake Riley, and aged them using otoliths [16]. This analysis showed that recruitment in Lake Riley was sporadic and had not occurred since 1998 with large events in 1997, 1994, 1991, 1980 and 1955. The median age of carp in Lake Riley was 15 years (i.e., hatched in 1991) while the oldest carp was 51 years old (Figure 7a). In Lake Susan, the largest recruitment events occurred in 1989–1991, with other small events in 2004, 2003, and 1997; the median age was 15, maximum age was 17 (This lake shared large year classes with Riley in 1997 and 1991; Figure 7b). Historical records showed that winterkills had occurred in Lake Susan in 1989–91 (a winter aeration was installed in Susan in 1992 after this event which also explains the absence of older carp in this lake), and then in Rice Marsh Lake in 1997, 2000 and 2004 [16]. Together, these trends seemingly demonstrate that Lake Susan and Rice Marsh Lake had been the primary source of YOY carp in this chain of lakes for decades, and while the aeration system has prevented winterkills (and recruitment) in Lake Susan since 1991, Rice Marsh Lake has continued to serve as a source of carp.

Table 7. Trap-net captures of YOY carp (CPUE) in the Riley Chain of Lakes.

Year	Lakes (CPUE)				
-	Lucy	Ann	Susan	RiceMarsh	Riley
1999	-	-	-	-	0 [D]
2000	0 [D]	0 [D]	-	-	-
2001	-	-	-	-	-
2002	-	-	-	-	-
2003	-	-	0 [D]	-	-
2004	-	-	-	-	-
2005	-	-	-	-	0 [D]
2006 [S]	0 [D]	0 [D]	0	0	0
2007	-	-	0	T	0
2008	-	-	0	0	0
2009	0	0	1.3	0	0
2010	0.8	0.6	1	1	1
2011	0.2	0	0.5	1	1
2012	0.2	0.2	0.8	1	0.8
2013	0.6	0.4	0.8	1	1
2014	0.6	0.4	0.6	0.8	0.8
2015	-	-	0	0	0
2016	-	-	0 [W]	0.2 [W]	0
2017	-	0 [D]	0 [D]	-	-

[S] = start of this study; [D] = MN DNR sampling in previous years; [W] = Watershed District data. T = hundreds of juvenile carp were seen leaving this pond during a winterkill event, 18 of which were later aged (see text).

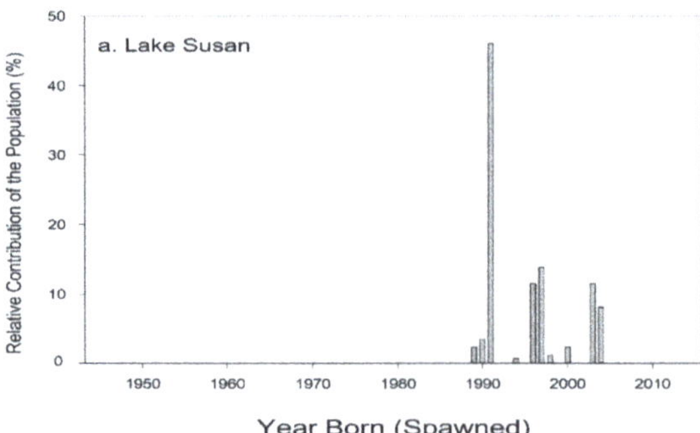

Figure 7. *Cont.*

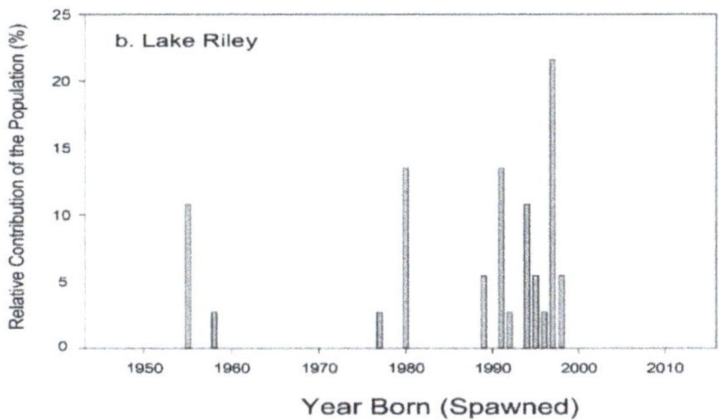

Figure 7. Age structure of adult carp collected from lakes: (**a**) Susan, and (**b**) Riley. Fish were aged in 2009. The Lake Susan data are published [16].

3.2.2. Step 2—Reducing and Controlling Adult Carp while Controlling Recruitment

Efforts to control carp started in 2008 during the mark-recapture studies (Section 3.2.1) and focused on removing adults from sink populations in lakes Susan and Riley, while suppressing recruitment in Rice Marsh Lake, the putative source. (Lake Susan may have only have become a sink in early 1992 when it was first aerated and its internal source of recruits was likely removed.) We focused on winter-seining using radio-tagged Judas fish to remove carp in both Lake Susan and Lake Riley (while also conducting mark-recapture studies see above). A total of 3425 adult carp were removed from Lake Susan in this manner in single haul in March 2009 [28,29], realizing a reduction in biomass of over 80%. Later, a few hundred other adult carp were removed from Lake Susan using a barrier with a trap set in Rice Marsh Creek (Table 8). Meanwhile, a total of 5619 adult carp were removed from Lake Riley in the course of three successful winter-seines starting in 2008/2009, leaving only about 10% of the original population by number and biomass past 2011 (Table 8) [29].

Table 8. Common carp control efforts and results in the Lake Riley Chain of Lakes.

Date	Removal Method	# Carp Caught	# Carp Removed
Winter 2008–2009	Susan Winter Seine	3981	3425
Spring 2009–2011	Spawning block/trap	242	242 [a]
Winter 2008–2009	Riley Winter Seine	4040	2940
Winter 2009–2010	Riley Winter Seine	376	376
Winter 2009–2010	Riley Winter Seine	2303	2303
Spring 2009–2011	Spawning block/trap	21	21 [b]
Fall 2011	Rice Marsh Lake aerated	-	-
Winter 2010–2011	Lucy Winterkills	630	500 [c]

[a]: A total of 3667 adult carp were captured and removed from Lake Susan, leaving in theory 514 of the original 4181 (13%) or 40.8 kg/ha in 2011. [b]. A total of 5619 adult carp were captured and removed from Lake Riley, leaving, in theory, 800 of the original 6419 (12%) or 21 kg/ha in 2011. [c]: Lake Lucy was seined in 2010 and then winterkills in 2011, most carp appeared to die (we examined this lake as a side project intended to benefit lakes upstream of Lake Susan).

Concurrent with adult removal, efforts to suppress recruitment focused on Rice Marsh Lake. In the spring of 2009 barriers with traps were installed at both the outflow of Lake Susan to Rice Marsh Lake and the inflow of Lake Riley to prevent Rice Marsh Lake from serving as a spawning site (this was also used for carp removal, see above). These barriers were maintained for 3 years and demonstrated that about 5–30% of the adult population of Lake Susan left it each spring for Rice Marsh Lake, while 5–10% left Lake Riley for Rice Marsh Lake, and fluctuated yearly with flow, suggesting this movement should be considered a type of a partial migration [27]. In addition, a winter aeration system was established in Rice Marsh Lake in 2011-2012 to prevent winterkills. This system is only operated in the winter and uses a bottom diffuser to aerate most of the lake while minimizing sediment disturbance. Winter oxygen and carp recruitment have been monitored every year since. Oxygen has not dropped below 2 ppm since its installation and no winterkills have been reported.

3.2.3. Step 3—Monitoring Carp Populations and Using Adaptive Management Approaches

While in the process of removing adult carp and suppressing recruitment, we commenced a monitoring program to confirm that the population was being reduced as expected in a sustainable and effective manner. A systematic (i.e., at least every-other year) boat electrofishing survey has been conducted in all lakes [28]. Water quality has also been monitored. An alum treatment was performed in Lake Riley once it was clear that its carp had been sustainably reduced and it was reasonable to expect lasting effects. Meanwhile, yearly autumnal trap-net surveys for YOY carp were initiated and these too showed no increase in recruitment so no new actions have been taken in this system. Native fishes are abundant in all lakes and bluegill sunfish average 53.4 sunfish/net in Rice Marsh Lake.

3.2.4. Summary: Carp, Water Quality, and Current Status

We have been able to sustainably bring carp levels down to non-damaging levels in two lakes in this watershed within several years after removing Rice Marsh Lake as its source of young, and targeting adult carp removal. Briefly, average adult CPUE across Lake Susan has fallen from over 16 carp/h in 2007 to below 2.0 carp/h in 2017, corresponding with a biomass reduction from ~307 kg/ha in 2008 to ~24 kg/ha in 2017 (Figure 8a). Similar results have been seen in Lake Riley where CPUE dropped from just over 12 (76 kg/ha) to under 1, suggesting a density of only ~15 kg/ha in 2017 (Figure 8b). It is possible that this continuous fall has been driven by natural mortality (annual mortality is estimated at 5–15% annually; [16]. Meanwhile only a handful of YOY carp have been captured during this entire time in either Lake Susan, Rice Marsh Lake, or Lake Riley (CPUE < 1; Table 7). Early springtime water clarity has increased by at least two-fold in Lake Susan, while plant cover and species richness has also increased [36]. However, while there have been small decreases in springtime total phosphorous and water clarity in this lake, this improvement has not persisted through the summer [36] (Figure 9), almost certainly because of benthic loading and hypolimnetic mixing unrelated to carp; see [35,36]. Since carp removal, submersed aquatic plant abundance and species richness have also increased greatly in Lake Riley [38,39]. Furthermore, the improvements associated with carp removal and reduction in their foraging activity have inspired the transplantation of submersed native plants into lakes Susan and Riley which have been moderately successful see [38,39]. These improvements enabled an alum treatment in Lake Riley in 2016, which led to a doubling in summer water clarity which has persisted as well a small decrease in total dissolved phosphorous (Figure 9). In sum, we have achieved successful, sustainable carp control by exploiting source-sink dynamics, and this, in turn, has enabled improvements in water quality and plant biodiversity throughout the entire watershed, especially where followed by alum treatment—the effects of which have been sustainable. The administration of RPBCWD remains committed to maintaining carp control indefinitely and is now planning an alum treatment for Lake Susan.

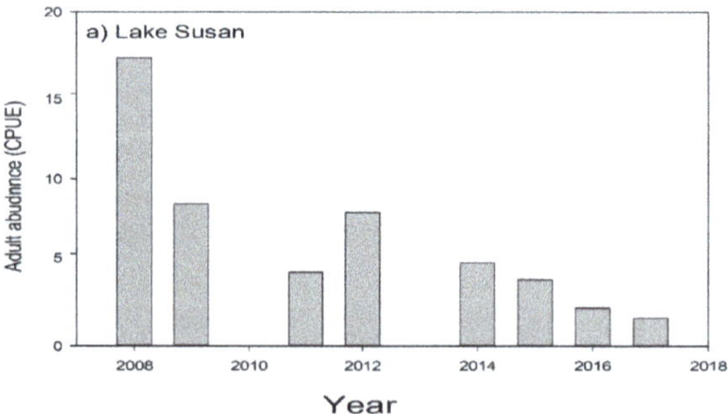

Figure 8. *Cont.*

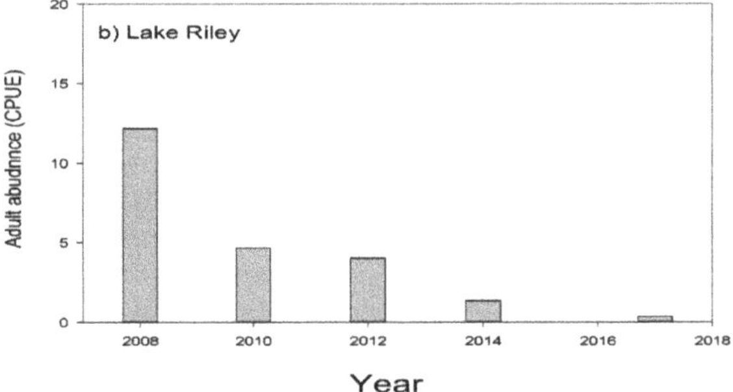

Figure 8. Adult carp CPUE (carp/h) as monitored by boat electrofishing in: (**a**) Lake Susan, and (**b**) Lake Riley.

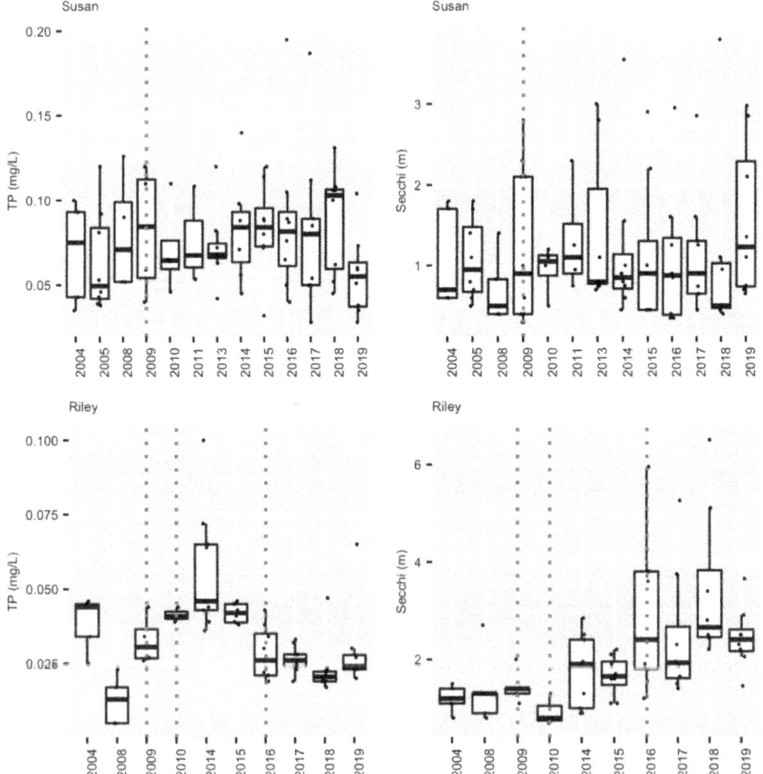

Figure 9. Summer (June–August) epilimnion water quality values (TP or Total Phosphorous, and Secchi depth (clarity)) for lakes Susan and Riley. The blue dotted line shows when the alum treatment occurred while the dotted red lines show years of carp removal in this system of lakes (see text).

4. Discussion

This manuscript describes decade-long case studies of common carp management in seemingly typical chains of lakes in the North American Midwest. Both studies demonstrate that carp abundance was explained by source-sink dynamics and thus exploited in a sustainable manner to reduce carp abundance, enabling lasting improvements in water quality with alum. Both chains experienced substantial declines in carp biomass to levels that are no longer ecologically damaging as result of a combination of targeted removal and recruitment suppression at identified sources. While these systems are similar in size, control has been especially effective in the Riley Chain of Lakes. This is likely because the Phalen Chain of Lakes has many more shallow ponds (over half a dozen vs. a single pond) and thus more sources of carp including some secondary ones. Nevertheless, in both instances carp reduction has been sustainable and adequate to permit successful alum treatment which has provided long-term improvements in submersed plant cover as well as water quality. These case studies appear to be the only known examples of successful invasive fish control that do not involve expensive and damaging use of poisons and barriers. An added benefit of carp removal/control was improved springtime plant communities and some apparent improvement in some game fisheries. It seems likely that our three-step approach based on source-sink dynamics could work in other chains of lakes in the Midwest and perhaps elsewhere to control populations of carp and other invasive fishes whose abundance might also be explained by source-sink dynamics.

The key to our strategy was likely our ability to identify sources of young carp and then suppress them for extended periods. Because of the high fecundity of adult carp and their longevity in Midwestern waters, these efforts needed (and still need) to continue for many years. As mentioned, this was a simpler task in the Lake Riley Chain of Lakes which seemingly had a single source (Rice Marsh Lake) than the Phalen Chain of Lakes which had many (Casey, Markham, Gervais Mill Pond). Nevertheless, because the solution involved restoring ponds in the Phalen Chain of Lakes that had previously been considered problematic, the effort is considered justifiable by the watershed district. It is notable that winter aeration alone was sufficient to control carp recruitment and native fisheries quickly rebounded albeit with some stocking in lakes Casey and Markham. In fact, Lake Casey is now a popular urban fishing lake. Increasingly strong evidence that adult carp return (home) to the source waters also made the task somewhat easier because carp do not appear to readily move to different ponds for spawning if blocked so efforts can be targeted at just a few locations [21,27,40]. In spite of our success, a significant challenge remains in these watersheds because of the longevity of adult carp which requires these efforts and monitoring continue for decades, but similar efforts are already spent for game fish. Lake systems with discrete and discontinuous patterns of recruitment such as found in many lakes of the temperate forest ecoregion of the Midwest are likely easier to control than other systems, such as shallow prairie lakes where recruitment might occur within many regions of highly interconnected main lakes [14,20].

Efficient removal of adult carp was another key to our success. In both our two case studies this involved several factors. One very important factor was our being able to develop good population estimates and couple them with an understanding of how much ecological damage different densities of carp cause in the glacial lakes so that removal targets could be established 100 kg/ha [6]. A second factor contributing to our success was our ability to develop a good understanding of adult carp distributions and movement patterns through the use of radio-tagged Judas carp. Not only did radio-tagged Judas carp provide essential information on where and how adult carp were aggregating under the ice in the winter [29], but also how and when they found food in baited box-nets and likewise how they could be specifically targeted for removal using baited traps [32,33]. Judas fish were also very helpful for locating and targeting carp spawning areas in the spring [6]. A third factor was our development of several different removal techniques that could be applied at different times of year to different types of carp aggregations. Notably, carp removal became progressively more difficult as their numbers declined so under-ice netting which worked very well initially [29], became less useful as carp densities dropped and required that we remove carp using food-induced aggregations [33] as well as spawning movements, the latter of which are unfortunately not highly predictable [27]. Boat electrofishing did not work well for removal because it was labor intensive although it proved to be a reliable method for rapidly and relatively accurately assessing carp abundance and biomass [27]. Finally, monitoring adult population size and distribution during removal efforts was essential. While boat electrofishing worked well in deeper navigable waters, it does not work well in shallow areas and it is possible the environmental DNA (eDNA) could supplement it as it has the advantage of being rapid and easy to use at almost any location and time [34]. Tracking sexually active adult carp could also be improved by the simultaneous deployment of sex pheromone measurement [32] or archival tag receivers. The success of commercial fishing schemes might also be improved with highly targeted approaches which do not rely on selling fish at market price, but reducing propagule pressure for a fee [41]. There is little doubt that all of our monitoring and removal techniques could be improved to drive an even more effective integrated control program for this species.

Several factors likely played into why we were able to eliminate carp recruitment in source ponds and are relevant to how this approach might be pursued by others in other locations. Our reliance on native egg and larval micro-predators was critical. There is little doubt, based on several laboratory and field studies [16,18,42], that bluegill sunfish, a voracious and agile predator with a small gape but high oxygen demand, played a key role eliminating any eggs and surviving larvae in restored aerated ponds and marshes. Nevertheless, several ponds (ex. Markham Pond) had few bluegill sunfish but many bullhead catfish, suggesting that they too, might have been important predators on larval and juvenile carp. A laboratory experiment also supports this possibility [18]. Likely, other species of native fish also serve this function in other locations and habitats. In addition, our efforts to control carp recruitment were likely strongly enhanced by our ability to both block adults from entering these ponds and to then remove the small number of resident adults they contained (i.e., Markham and Casey Ponds). Integrated, multi-faceted responses to invasive species control such as this are warranted. Especially remarkable about our study was how effectively our relatively low-cost strategy using native micro-predators worked: no carp recruitment was measured after winter aeration was installed in any pond. Carp lay their eggs on aquatic vegetation which also serves as habitat for larvae and young [18], and they are extremely vulnerable. In contrast, efforts to control carp by creating fishless wetlands using poisons (rotenone) and then protecting them with barriers [9] may, if not conducted with great care, enhance the long-term success of adults, because if migratory carp do manage to enter these systems (flood, power failure to electric barrier), reproductive success is all but assured. It thus makes sense that if possible (not all systems can support fish year-around) to manage native fishes together with carp, as we did in Lake Casey and Markham Pond. Notably, all carp nursery ponds restored by this study had been previously neglected and had poor water quality because of nutrient loading (poor water clarity, low oxygen, few resident fish), so their restoration was welcomed for its own sake. While it appears from our studies that most juvenile carp generally do not leave nursey ponds until they at least a year old [14,21,43], this issue remains an unknown and warrants study because it could be exploited. In sum, restoration of shallow degraded ponds so that they can support native fishes is clearly a concept worth exploring to control carp.

Our study also sheds new light on the life history of common carp and how it may have evolved to use source-sink dynamics. Like the large rivers in which carp evolved, the chains of lakes we studied are complex and possess large interconnected ponds/wetlands (which resemble flood plains) that are subject to fluctuating physical conditions (winter oxygen, flows and water levels) and thus not amenable to supporting stable populations of resident native micro-predators. It is fascinating that common carp have evolved a life history which simultaneously includes high fecundity, longevity and directed yet flexible movement strategies that allow adults to exploit peripheral wetland habitats which often lack micro-predators in the spring. Their ability to find these spawning/ nursery habitats is also remarkable, as is their ability to exhibit these migratory behaviors when appropriate (high flow years). Being long-lived, they only need to be successful on rare occasions. We hypothesize that these movement patterns may reflect a type of homing, or ability to remember and preferentially return to natal areas when appropriate (at times of high water), thus explaining how and why they select certain ponds and not others [27,37,40]. We are not aware of any other species that has coupled a source-sink life history with partial migration as well as homing [44], and it may do much to explain the invasiveness of this species in many regions.

Our study will hopefully provide guidance to others hoping to control carp. Our case studies emphasize the need to develop an understanding of the population dynamics of carp in local areas, especially the sources of young, and to appreciate the need that a long-term commitment is needed given the longevity of this species. We recommend the same three-step process we employed which includes assessment, control, and monitoring. However, the highly adaptive and flexible nature of the carp's life history means that its population dynamics are likely to take different forms in different locations and different approaches to this control scheme may be required elsewhere. Nevertheless, targeting the production of young is always likely to be key. As shown by others [35], we too show that if improved water quality is the objective of carp removal, alum treatment may often be required after carp removal because nutrient input from eutrophic sediments may persist even in the absence of carp, but that if alum is then applied it can work well and in a sustainable manner. Added benefits of carp control include increased aquatic plant cover [39] and likely aquatic biodiversity, but we did not assess the later possibility.

Finally, a fascinating and important question is whether source-sink dynamics might commonly explain high densities of common carp and perhaps other invasive fish species, and thus be commonly exploitable in ways similar to those used by us. Many hundreds of species of fish employ migratory life histories which include occasional seasonal use of spawning areas and nurseries [44], and thus could also be considered to be employing a type of source-sink dynamics. In addition to common carp, several of these species are highly invasive including rainbow trout (*Oncorhynchys mykiss*), brown trout (*Salmo trutta*), and sea lamprey (*Petromyzon marinus*). Aside from the sea lamprey, which uses migratory pheromones to select spawning/nursery streams [45], all of these fish home so they have only a few sources. Indeed, recent efforts to control lake trout in western lakes by targeting embryos on their home spawning grounds are similar in concept [46]. Carp reproductive success seems to vary greatly with region and local ecosystem, perhaps because the presence of native micro-predators for their eggs/larvae likely varies for several reasons including: native fish biodiversity (ex. Australia), flooding (river floodplains in Eurasia), and/or local/seasonal hypoxia (Midwestern lakes). This variation will make this species easier to control in some regions/ ecosystems than others. The question of what ecological factors determine the causes and sources of common carp recruitment in different ecosystems is thus one that must be actively explored, especially because it is now clear that common carp may be controlled once this question has been answered.

Author Contributions: Conceptualization: P.W.S.; methodology: P.G.B. and P.W.S.; formal analysis: P.W.S. and P.G.B.; writing—original draft: P.W.S.; writing—review and editing: P.W.S. and P.G.B.; project administration: P.W.S.; Funding: P.W.S. All authors have read and agreed to the published version of the manuscript.

Funding: This research was funded by The Minnesota Environment and Natural Resources Trust Fund; The Riley Purgatory Bluff Creek Watershed District and The Ramsey Washington Metro Watershed District.

Acknowledgments: This study is the culmination of the work of many over a decade. The late Bill Oemichen kick-started this research with a personal gift of USD 10,000 to Sorensen with no strings attached. This paper is dedicated to him. Ray Norrgard from the MN DNR was also very encouraging and helped us get initial funding. Since that time, additional funding has been provided from many sources (see above) which have supported many (without promise of practical results) who have dedicated themselves to understanding carp biology and carp control including the following graduate students: Mario Travaline, Jake Osborne, Justin Silbernagel, Nate Banet, Reid Swanson, Justine Dauphinais, Joey Lechelt, Josh Poole; Undergraduate students: Keith Phillips; Technicians: Nathan Berg, Mary Haedrick Newman, Brett Miller; and Postdoctoral fellows: Przemek Bajer Haude Levesque, Hangkyo Lim, Chris Chizinski, Ratna Ghosal, Jessica Eichmiller. Bill Bartodziej, Matt Kocian, Paul Haik, Mike Casanova and Ken Wentzel were also very encouraging. Dave and Anne Florenzano supported and encouraged our early work on Lake Riley. Daryl Ellison from the MN DNR was also very helpful and pointed out the importance of winterkill to local fisheries while also helping us select the Riley Chain of Lakes for our study and providing the permits we needed. Sorensen's visit with the Tasmanian Fisheries Service and the inspiration they offered in the mid-2000s was extremely influential and appreciated. Justine Dauphinais, Ray Newman, Jeff Whitty, and Bill Bartodziej all read initial drafts and provided feedback. Sorensen is deeply grateful to all of these people.

Conflicts of Interest: P.W.S. Sorensen has no conflict. P.G.B. Bajer is the founder and owner of Carp Solutions LLC, a company involved in the development of management strategies for invasive fish. These interests have been reviewed and managed by the University of Minnesota in accordance with its Conflict of Interest policies.

References

1. Britton, J.R.; Cucherousset, J.; Davies, G.D.; Godard, M.J.; Copp, G.H. Non-native fishes and climate change: Predicting species responses to warming temperatures in a temperate region. *Freshw. Biol.* **2010**, *55*, 1130–1141. [CrossRef]
2. Balon, E.K. Origin and domestication of the wild carp, Cyprinus carpio: From Roman gourmets to the swimming flowers. *Aquaculture* **1995**, *129*, 3–48. [CrossRef]
3. Sorensen, P.W.; Bajer, P.G. Carp, common. In *Encyclopedia of Biological Invasions*; Simberloff, D., Rejmanek, M., Eds.; University of California Press: Berkeley, CA, USA, 2011.
4. Kohlmann, K.; Kersten, P. Deeper insight into the origin and spread of European common carp (Cyprinus carpio carpio) based on mitochondrial D-loop sequence polymorphisms. *Aquaculture* **2013**, *376*, 97–104. [CrossRef]
5. Funge-Smith, S. *Review of the State of the World Fisheries Resources: Inland Fisheries*; FAO Fisheries and Aquaculture Circular: Rome, Italy, 2018; 253p.
6. Bajer, P.G.; Sullivan, G.; Sorensen, P.W. Effects of a rapidly increasing population of common carp on vegetative cover and waterfowl in a recently restored Midwestern shallow lake. *Hydrobiologia* **2009**, *632*, 235–245. [CrossRef]
7. Weber, M.J.; Brown, M.L. Effects of Common Carp on Aquatic Ecosystems 80 Years after "Carp as a Dominant": Ecological Insights for Fisheries Management. *Rev. Fish. Sci.* **2009**, *17*, 524–537. [CrossRef]
8. Matsuzaki, S.S.; Usio, N.; Takamura, N.; Washitani, I. Contrasting impacts of invasive engineers on freshwater ecosystems: An experiment and meta-analysis. *Oecologia* **2008**, *158*, 673–686. [CrossRef]
9. Verrill, C.R.; Berry, B.R., Jr. Effectiveness of a lake drawdown and electrical barrier for reducing common carp and bigmouth buffalo abudances. *N. Am. J. Fish. Manag.* **1995**, *5*, 137–141. [CrossRef]
10. Meronek, T.G.; Bouchard, P.M.; Buckner, E.R.; Burri, T.M.; Demmerly, K.K.; Hatleli, D.C.; Klumb, R.A.; Schmidt, S.H.; Coble, D.W. A Review of Fish Control Projects. *N. Am. J. Fish. Manag.* **1996**, *16*, 63–74. [CrossRef]
11. Roberts, J.; Tilzey, R. *Controlling Carp: Options for Australia*; CSIRO Press: Griffith, Australia, 1997; 141p.
12. Weber, M.J.; Hennen, M.J.; Brown, M.L.; Lucchesi, D.O.; Sauver, T.R.S. Compensatory response of invasive common carp Cyprinus carpio to harvest. *Fish. Res.* **2016**, *179*, 168–178. [CrossRef]
13. Balon, E.K. The common carp, Cyprinus carpio: Its wild origin, domestication in aquaculture, and selection as colored nishikigoi. *Guelph Ichthyol. Rev.* **1995**, *3*, 1–55.
14. Bajer, P.G.; Cross, T.K.; Lechelt, J.D.; Chizinski, C.J.; Weber, M.J.; Sorensen, P.W. Across-ecoregion analysis suggests a hierarchy of ecological filters that regulate recruitment of a globally invasive fish. *Divers. Distrib.* **2015**, *21*, 500–510. [CrossRef]
15. McCrimmon, H.R. Carp in Canada. In *Bulletin Fisheries Research*; Fisheries Research Board of Canada: Ottawa, ON, Canada, 1968; 165p.
16. Bajer, P.G.; Sorensen, P.W. Recruitment and abundance of an invasive fish, the common carp, is driven by its propensity to invade and reproduce in basins that experience winter-time hypoxia in interconnected lakes. *Biol. Invasions* **2009**, *12*, 1101–1112. [CrossRef]
17. Billard, R. *Carp: Biology and Culture*; Praxis Publishing: Chichester, UK, 1999.
18. Silbernagel, J.J.; Sorensen, P.W. Direct Field and Laboratory Evidence that a Combination of Egg and Larval Predation Controls Recruitment of Invasive Common Carp in Many Lakes of the Upper Mississippi River Basin. *Trans. Am. Fish. Soc.* **2013**, *142*, 1134–1140. [CrossRef]
19. Koehn, J.D. Carp (*Cyprinus carpio*) as a powerful invader in Australian waterways. *Freshw. Biol.* **2004**, *49*, 882–894. [CrossRef]
20. Lechelt, J.; Bajer, P. Elucidating the mechanism underlying the productivity-recruitment hypothesis in the invasive common carp. *Aquat. Invasions* **2016**, *11*, 469–482. [CrossRef]
21. Dauphinais, J.D.; Miller, L.M.; Swanson, R.G.; Sorensen, P.W. Source-sink dynamics explain the distribution and persistence of an invasive population of common carp across a model Midwestern watershed. *Biol. Invasions* **2018**, *20*, 1961–1976. [CrossRef]
22. Vilizzi, L.; Tarkan, A.S.; Copp, G.S. Experimental evidence from causal criteria analysis for the effects of common carp Cyprinus carpio on freshwater ecosystems: A global perspective. *Rev. Fish. Sci. Aquac.* **2015**, *23*, 253–290. [CrossRef]

23. Bajer, P.G.; Beck, M.W.; Cross, T.K.; Koch, J.; Bartodziej, B.; Sorensen, P.W. Biological invasion by a benthivorous fish reduced the cover and species richness of aquatic plants in most lakes of a large North American ecoregion. *Glob. Chang. Biol.* **2016**, *22*, 3937–3947. [CrossRef] [PubMed]
24. Pulliam, H.R. Sources, Sinks, and Population Regulation. *Am. Nat.* **1988**, *132*, 652–661. [CrossRef]
25. Travis, J.M.J.; Park, K. Spatial structure and the control of invasive alien species. *Anim. Conserv.* **2004**, *7*, 321–330. [CrossRef]
26. Bajer, P.G.; Chizinski, C.J.; Silbernagel, J.J.; Sorensen, P.W. Variation in native micro-predator abundance explains recruitment of a mobile invasive fish, the common carp, in a naturally unstable environment. *Biol. Invasions* **2012**, *14*, 1919–1929. [CrossRef]
27. Chizinski, C.J.; Bajer, P.G.; Headrick, M.E.; Sorensen, P.W. Different migratory strategies of invasive Common Carp and native Northern Pike in the American Midwest suggest an opportunity for selective management strategies. *N. Am. J. Fish. Manag.* **2016**, *36*, 769–779. [CrossRef]
28. Bajer, P.G.; Sorensen, P.W. Using boat electrofishing to estimate the abundance of invasive Common Carp in small Midwestern lakes. *N. Am. J. Fish. Manag.* **2012**, *32*, 817–822. [CrossRef]
29. Bajer, P.G.; Chizinski, C.J.; Sorensen, P.W. Using the Judas technique to locate and remove wintertime aggregations of invasive common carp. *Fish. Manag. Ecol.* **2011**, *18*, 497–505. [CrossRef]
30. Osborne, J. Distribution, Abundance and Overwinter Survival of Young-of-the-Year Common Carp in a Midwestern Watershed. Master's Thesis, University of Minnesota, Minneapolis, MN, USA, 2012, (unpublished).
31. Brown, P.; Green, C.; Sivakumaran, K.P.; Giles, A.; Stoessel, D. Validating otolith annuli for use in age-determination of carp (*Cyprinus carpio* L.) from Victoria, Australia. *Trans. Am. Fish. Soc.* **2004**, *133*, 190–196. [CrossRef]
32. Ghosal, R.; Eichmiller, J.J.; Witthuhn, B.A.; Sorensen, P.W. Attracting Common Carp to a bait site with food reveals strong positive relationships between fish density, feeding activity, environmental DNA, and sex pheromone release that could be used in invasive fish management. *Ecol. Evol.* **2018**, *8*, 6714–6727. [CrossRef]
33. Bajer, P.G.; Lim, H.; Travaline, M.J.; Miller, B.D.; Sorensen, P.W. Cognitive aspects of food searching behavior in free-ranging wild Common Carp. *Environ. Boil. Fishes* **2010**, *88*, 295–300. [CrossRef]
34. Eichmiller, J.J.; Bajer, P.G.; Sorensen, P.W. The Relationship between the Distribution of Common Carp and Their Environmental DNA in a Small Lake. *PLoS ONE* **2014**, *9*, e112611. [CrossRef]
35. Huser, B.J.; Bajer, P.G.; Chizinski, C.J.; Sorensen, P.W. Effects of common carp (*Cyprinus carpio*) on sediment mixing depth and mobile phosphorus mass in the active sediment layer of a shallow lake. *Hydrobiologia* **2015**, *763*, 23–33. [CrossRef]
36. Bajer, P.G.; Sorensen, P.W. Effects of common carp on phosphorus concentrations, water clarity, and vegetation density: A whole system experiment in a thermally stratified lake. *Hydrobiologia* **2014**, *746*, 303–311. [CrossRef]
37. Bajer, P.G.; Parker, J.E.; Cross, T.K.; Venturelli, P.A.; Sorensen, P.W. Partial migration to seasonally-unstable habitat facilitates biological invasions in a predator-dominated system. *Oikos* **2015**, *124*, 1520–1526. [CrossRef]
38. Knopik, J.M.; Newman, R.M. Transplanting aquatic macrophytes to restore the littoral community of a eutrophic lake after the removal of common carp. *Lake Reserv. Manag.* **2018**, *34*, 365–375. [CrossRef]
39. Dunne, M.A.; Newman, R.M. Effect of light on macrophyte sprouting and assessment of viable seedbank to predict community composition. *J. Aquat. Plant Manag.* **2019**, *57*, 90–98.
40. Banet, N. Partial Migration, Homing, Diel Activity, and Distribution of Adult Common Carp across a Large, Model Watershed in the North American Midwest. Master's Thesis, University of Minnesota, Saint Paul, MN, USA, 2016, (unpublished).
41. Colvin, M.E.; Pierce, C.L.; Stewart, T.W.; Grummer, S.E. Strategies to control a common carp population by pulsed commercial harvest. *N. Am. J. Fish. Manag.* **2012**, *32*, 1251–1264. [CrossRef]
42. Poole, J.R.; Bajer, P.G. A small native predator reduces reproductive success of a large invasive fish as revealed by whole-lake experiments. *PLoS ONE* **2019**, *14*, e0214009. [CrossRef]
43. Lechelt, J.; Kocian, M.; Bajer, P. Low downstream dispersal of young-of-year common carp from marshes into lakes in the Upper Mississippi Region and its implications for integrated pest management strategies. *Manag. Biol. Invasions* **2017**, *8*, 485–495. [CrossRef]
44. Chapman, B.B.; Hulthén, K.; Brodersen, J.; Nilsson, P.A.; Skov, C.; Hansson, L.-A.; Brönmark, C. Partial migration in fishes: Causes and consequences. *J. Fish Biol.* **2012**, *81*, 456–478. [CrossRef]

45. Sorensen, P.W.; Fine, J.M.; Dvornikovs, V.; Jeffrey, C.S.; Shao, F.; Wang, J.; Vrieze, L.A.; Anderson, K.R.; Hoye, T.R. Mixture of new sulfated steroids functions as a migratory pheromone in the sea lamprey. *Nat. Chem. Biol.* **2005**, *1*, 324–328. [CrossRef]
46. Koel, T.M.; Arnold, J.L.; Bigelow, P.E.; Brenden, T.O.; Davis, J.D.; Detjens, C.R.; Doepke, P.D.; Ertel, B.D.; Glassic, H.C.; Gresswell, R.E.; et al. Yellowstone Lake ecosystem restoration: A case study for invasive fish management. *Fishes* **2020**, *5*, 18. [CrossRef]

Review

Eradication of the Invasive Common Carp, *Cyprinus carpio* from a Large Lake: Lessons and Insights from the Tasmanian Experience

Jonah L. Yick [1,*], Chris Wisniewski [1], John Diggle [1] and Jawahar G. Patil [2]

[1] Inland Fisheries Service, 17 Back River Road, New Norfolk, TAS 7140, Australia; chris.wisniewski@ifs.tas.gov.au (C.W.); john.diggle@ifs.tas.gov.au (J.D.)
[2] Fisheries and Aquaculture Centre, Institute for Marine and Antarctic Studies, University of Tasmania, Nubeena Crescent, Taroona, TAS 7053, Australia; Jawahar.patil@utas.edu.au
* Correspondence: jonah.yick@ifs.tas.gov.au

Abstract: Common carp (*Cyprinus carpio*, L. 1758) are the most abundant pest fish species in Australia, detrimental to ecosystem integrity and values, and in need of suitable management solutions. In January 1995, this destructive pest was discovered in two large, connected Tasmanian lakes—Lakes Crescent (23 km^2) and Sorell (54 km^2). After an initial assessment, carp were immediately contained to these waters using screens to prevent their escape down-stream, followed by swift legislation to enforce closure of the lakes to the public. Assessment and evaluation of carp numbers occurred throughout the eradication program, with effort focused on Lake Crescent. Beginning with undirected removal, techniques progressively evolved to more sophisticated targeted removal with assistance from biotelemetry, in conjunction with gill netting and electro-fishing. Real-time population estimates and in situ observations resulted in a detailed cumulative understanding of carp population dynamics, behaviour and seasonal habitat choice. This allowed strategic deployment of fences to block access to marshes, and the installation of steel traps within the fences. These gears specifically prevented spawning opportunities, while concurrently capturing mature fish. Following 12 years of adaptive and integrated effort, 7797 carp (fry, juvenile and adult) were captured from Lake Crescent, with the last carp being caught in December 2007. The subsequent 14 years of monitoring has not resulted in the capture of any carp, confirming the successful eradication of carp from Lake Crescent. These management practices have been successfully replicated in the larger Lake Sorell, where 41,499 carp (fry, juvenile and adult) have been removed. It is now estimated that there are few, if any carp remaining. Collectively, the techniques and strategies described here were reliable, and can be applied as a model to control or eradicate pest populations of carp in freshwater lakes elsewhere.

Keywords: common carp; invasive; incursion; alien fish; fyke net; pest fish; Lake Sorell; Lake Crescent; biotelemetry

Citation: Yick, J.L.; Wisniewski, C.; Diggle, J.; Patil, J.G. Eradication of the Invasive Common Carp, *Cyprinus carpio* from a Large Lake: Lessons and Insights from the Tasmanian Experience. *Fishes* **2021**, *6*, 6. https://doi.org/10.3390/fishes6010006

Academic Editor: Peter W. Sorensen

Received: 28 November 2020
Accepted: 18 February 2021
Published: 23 February 2021

1. Introduction

The impacts of invasive species on native species, communities, and ecosystems have been extensively documented [1–4] and recognised as a threat to global biodiversity [1,5], second only to habitat loss. While some aquatic introduced species have had little or no detectable effects, many have been damaging to the environment and human interests [6]. Aquatic invasive species significantly impact economic enterprises such as agriculture and fisheries [7], with freshwater ecosystems particularly at higher risk as they harbour greater biodiversity per surface area than marine and terrestrial ecosystems [6,8,9], and provide more direct human services such as a ready source of potable water.

The invasive common carp *Cyprinus carpio*, is one of the most widely established freshwater fish globally [10,11], and is the most abundant pest fish species in Australia [12]. It was first introduced in 1859, after being released into numerous ponds in Victoria, but never established in the wild [13]. It was not until 1964 when it was released into the Murray River near Mildura, Victoria, that it began to spread throughout Australia [13]. With a distribution of over one million km^2, common carp are now established in all states and territories in Australia, apart from the Northern Territory [12–14]. In addition to competing with native and other desirable fish species for both food and space [13,15,16], carp can inflict major environmental and economic costs by reducing water quality [13,17]. Carp have also been implicated in macrophyte destruction through direct grazing, physical uprooting of plants, increasing water turbidity, enriching nutrient loads, and reducing invertebrate biomass through predation [13,18,19].

Physical eradication of carp was found to be the most cost effective and environmentally benign option in Tasmania, despite the high level of sustained effort required. The aim here is to detail the steps, strategies, and techniques that were successfully implemented to eradicate carp from Lake Crescent, Tasmania. This was initiated by setting up a designated Carp Management Program (CMP) within the Inland Fisheries Service (IFS) Tasmania, as soon as the carp were discovered in 1995. The integrated strategies employed focused on the exploitation of the biological vulnerabilities of carp (including their specific behaviours), and the optimisation of capture methods. The events and strategies of the CMP in the Tasmanian lakes have been previously summarised in Diggle et al. (2012) [20] and Wisniewski et al. (2015) [21], however this current report aims to provide a refined, and more rigorous update of the successful eradication.

2. Common Carp in Tasmania

2.1. *The Discovery of Common Carp in Tasmania*

Carp were first discovered in Tasmania in the mid-1970s, then again in 1980, in several farm dams on the North West coast [20–22]. Given the relatively small size of the dams, all populations were eradicated using rotenone treatments. It wasn't until 28 January 1995, when an angler found the remains of a fish on the banks of Lake Crescent, that a much more serious threat was identified. The fish was handed to the Inland Fisheries Service (IFS) on 30 January 1995, and suspected to be a carp [20].

2.2. *Lakes Crescent and Sorell*

Lake Crescent is a large, shallow, freshwater lake located in the south-east corner of the Tasmanian Central Plateau (Coordinates: Lake Crescent 147°16′ E, 42°18′ S), and is connected with the upstream Lake Sorell [20] (Figure 1). The lakes are situated at the head of River Clyde catchment, a 97 km watercourse which eventually drains into the River Derwent. Water temperatures can range from a low of about 0–2 °C in winter and reach up to 20 °C in summer (Figure A1) [23]. Snow can occur in any month of the year, and frost events are also common [23]. Lake Crescent (23 km^2) is less than half the size of Lake Sorell (54 km^2), with average and maximum depths of 1.5 and 3.8 m, and 2.5 and 4.4 m respectively, at full supply [24,25]. At times of full supply, both lakes have extensive wetland areas that connect to the main lake bodies, which are some of the largest areas of shallow water in Tasmania [26].

The Interlaken Lakeside Reserve in the north-west corner of Lake Crescent is internationally recognised under the Ramsar Convention on wetlands [28]. The wetland supports a diverse assemblage of aquatic macrophytes, including the poorly reserved swamp wallaby-grass (*Amphibromus neesii*) (currently listed as rare under the Tasmanian Threatened Species Protection Act 1995) [29]. Lake Crescent is a turbid-phytoplankton dominated system [29,30], with phytoplankton biomass 10 times that of the neighbouring Lake Sorell [30]. It is also inhabited by numerous endemic fauna, that includes a snail species *Austropyrgus* sp., and the golden galaxias fish (*Galaxias auratus*), which are listed as endangered under the Commonwealth Environment Protection and Biodiversity Conservation Act 1999 [20,21,31,32]. The reserve is an important resting and feeding refuge for many waterbirds, including significant migratory birds. The wetland periodically dries out and when inundated, provides important habitat for macroinvertebrates, frogs and fish [29]. The two lakes also support a recreational brown (*Salmo trutta*) and rainbow trout (*Oncorhynchus mykiss*) fishery, a commercial fishery for the native short-finned eel (*Anguilla australis*), and provides water for domestic use and irrigation, which are all of significant importance to the State's economy [20,27,33,34].

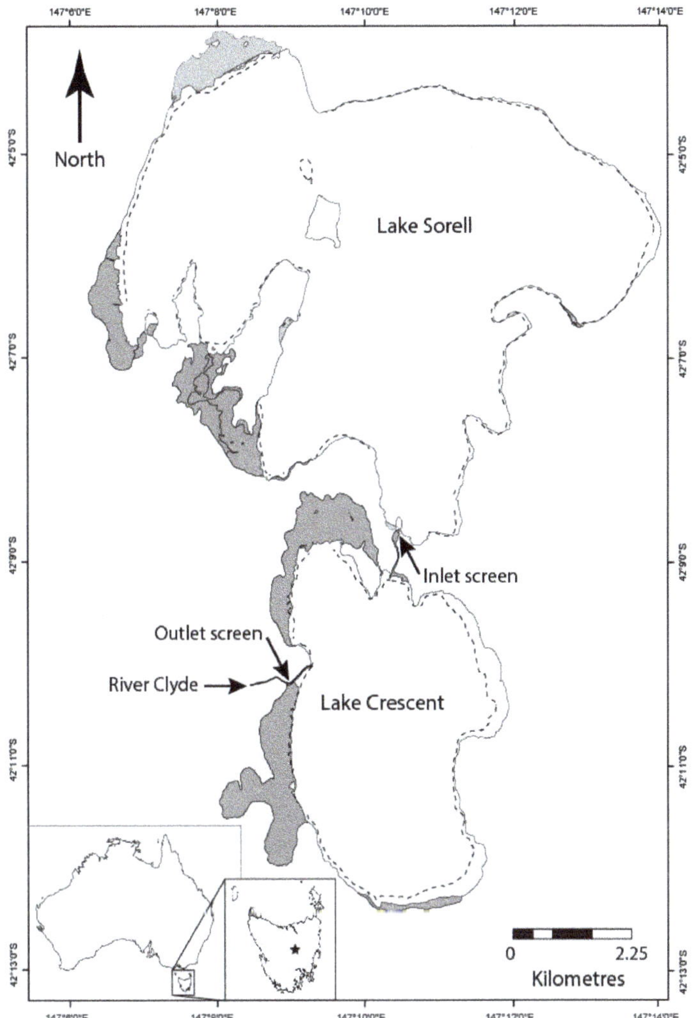

Figure 1. Map of lakes Sorell and Crescent showing the location of the River Clyde, containment screens, wetland areas (shaded grey), and the level of wetland inundation (dotted lines). The integrated management activities contained the carp to these two lakes with eradication achieved in the latter (Adopted from Taylor et al. (2012) [27]). Inset are maps of Australia and Tasmania, with the location of the lakes (star) (Left hand side, bottom).

3. Developing a Strategy and Approach to Control Carp in Lakes Sorell and Crescent

The confirmation of carp in the two largest water bodies on the eastern side of Tasmania's central plateau was of immediate concern to the Tasmanian community at large. This triggered an immediate fisheries management response led by the Inland Fisheries Service, Tasmania.

3.1. Initial Assessment and Response

Backpack electro-fishing surveys were undertaken in Lake Crescent on 1 February 1995, and resulted in multiple carp caught, confirming their presence in this lake. On 6 March 1995, an electro-boat was deployed in Lake Sorell, resulting in the capture of several carp, making it apparent that carp were established in both the lakes [20,21]. Further assessments of both lakes soon determined that Lake Crescent had a relatively large population of carp compared to Lake Sorell. As a result, most of the initial removal effort was focused on Lake Crescent, in an attempt to reduce the population numbers while the density of carp was high. Electrofishing surveys downstream of Lake Crescent were also undertaken annually in the River Clyde catchment, to assess if carp had inadvertently escaped and established in any sections of the river. None were found downstream of Lake Crescent, except for a single carp caught immediately downstream of the Lake Crescent outlet in July 1995 [20]. Electro-fishing boats from New South Wales and Victorian fisheries agencies were procured in February and March 1995, to assist the carp surveys [20,21]. All initial surveys suggested that the carp were contained to lakes Sorell and Crescent. The preliminary fishing effort in the lakes resulted in significant fish-down, giving an indication of the size of the carp population in both lakes, which guided the subsequent management steps.

Following the confirmation of carp in both lakes, a basic containment strategy was implemented, which allowed time to assess the various options [20,21]. Primarily, this involved the prevention of escape or transfer of carp from the Lake Sorell and Lake Crescent populations [22]. Carp could have potentially spread throughout the lower River Derwent via the River Clyde catchment, as far down as Bridgewater (approximately 15 km north of Hobart, and 60 km downstream of the lakes system). Containment was achieved by closing the outlet from Lake Crescent. To prevent the risk of further inadvertent human assisted spread (e.g., by anglers) and/or reintroductions [35], Lake Crescent was closed to the public on 18 February 1995. It remained closed until 2004 when carp numbers were deemed low enough that the risk of transfer from the lake was unlikely [20]. The closure included, but was not limited to, angling, duck shooting, boating, wading, swimming, or any activity that results in contact with the water [22].

3.2. Creation and Implementation of the Adaptive Carp Management Plan

In the absence of a similar scale eradication attempt anywhere else in the world at the time, it was 'a learn as you go' adaptive management strategy from the outset. To begin with, the State Government initiated a joint agency response consisting of a carp task force which later transitioned to a Carp Working Group (CWG) [20–22]. The initial mission of the CWG was to determine the preferred strategy between suppression (i.e., control of carp population numbers) or eradication. Although contained to the Lake Sorell and Lake Crescent systems, a control strategy alone posed long term risk of translocation and spread to other waterbodies. This directly threatened the multimillion-dollar recreational trout fishing industry, as well as a wide range of endemic aquatic flora and fauna of the state. Therefore, containment with the aim of eradication was determined to be the preferred outcome. On 1 June 1995, the CWG were successful in obtaining additional funding from the Tasmanian Government, which was put towards the establishment of a four person CMP team, to begin the carp fish-down process.

For eradication of any pest species to be successful, six criteria need to be achieved, which are detailed in Bomford and O'Brien (1995) [35]. The lakes Crescent and Sorell were soon assessed and rendered "closed systems" with an ability to contain carp. Complemented by sustained support from the government and community, it became clear that eradication was a feasible option [20]. Consequently, the objective of the CMP was to eradicate carp from Tasmanian waters and, in the meantime, to minimise the impact of carp on Tasmania, from economic, recreational and ecological points of view.

The following broad strategies were adopted to achieve this objective [20,21]. They are:

1. Contain carp to the lakes Sorell/Crescent catchment;
2. Reduce the existing carp population;
3. Eradicate carp;
4. Develop and implement a water management plan to ensure water supply for reliant townships and irrigators, while facilitating 1–3 above;
5. Prevent introduction of carp to new water bodies and it's reintroduction to cleared waters (from both inter and intrastate sources) and;
6. Implement legislative and communication strategies to minimise damage to inland fisheries and tourism, while facilitating the above objectives in a timely fashion.

The next obvious step was to determine the most feasible method of achieving eradication. Three main options were considered; draining the lakes, poisoning the lakes and physical removal of carp [20,21]. Due to bathymetrical limitations and the potential adverse environmental impacts, draining or poisoning the lakes were not considered. As a result, an integrated approach with physical removal as the cornerstone was deemed the best option, and implemented at the outset.

Although techniques and strategies evolved with time and growing knowledge, the physical removal remained the central theme. This was complemented by containment, spawning prevention and real-time estimation/calibration of population size. Collectively, these four elements were deemed adequate to address criteria one to three [35] for eradication.

4. Following the Plan: Reducing Carp Abundance in Lake Crescent

The capture and removal of carp in Lake Crescent evolved with time and was reliant on using a range of techniques, alone or in combination. Typically, all size classes of carp were caught concurrently. However, specific techniques were developed to target carp fry (<50 mm fork length) and juveniles (50–250 mm fork length), as part of the recruitment prevention strategy (See Step 3 below).

4.1. Step 1: Containment

First, the two lakes and the upper River Clyde were closed to the public, then containment screens were installed at the outlet of Lake Crescent (Figure 2). To further facilitate transfer of water downstream (for domestic, stock and irrigation use, as well as to restore environmental flows), the outlet structure at Lake Crescent was fitted with an internal screen and the outlet reopened on 24 February 1995 (Figure 2) [21]. The screens (Figure 2) constituted a physical barrier to carp by screening objects to 1 or 5 mm, using interchangeable stainless steel mesh [22]. The 1 mm mesh was used from the start of October until the end of April each year, throughout the carp spawning period, to minimise the risk of eggs or fry escaping containment [20,22]. From May until the end of September, the mesh size was increased to 5 mm or greater reflecting the increased size of carp fry and allowing increased flow releases over the wetter winter months [20,22]. This approach balanced the greater risk of uncontrolled spill/overflow, and the potential escape of fry, juvenile, and adult carp. Observations indicated that all carp fry would outgrow the 5 mm mesh by the end of April [20]. The screens required manual cleaning, an aspect increasingly labour intensive as the flow rates increased [20]. Importantly, this screening strategy was successful in containing carp over a 25-year period. Initially, a course 25 mm grate was installed at the outlet of Lake Sorell on 7 August 1995, to prevent the movement of adult carp from Lake Crescent upstream into Lake Sorell [20]. In 2001, 5 mm screens were installed at the Lake Sorell outlet structure, which completely isolated the lakes from each other, preventing both upstream and downstream movement of carp [22].

Figure 2. The Lake Crescent outlet structure showing the coarse external screens.

Modifications were also implemented around the lake to prevent uncontrolled and unscreened spill/overflow [22]. Specifically, levees were constructed to direct all flood waters through an overflow channel [22]. The channel was screened with 5 mm mesh to prevent the passage of fry, juvenile and adult carp [22]. In addition to the screening infrastructure, an active compliance operation around the lakes was also undertaken, as well as a public awareness program. This was followed by on-going electrofishing surveys in the River Clyde that indicated that carp had not spread downstream, nor established self-supporting populations. These downstream electrofishing surveys, conducted annually since 1995, did not detect any carp populations, confirming the containment strategy was successful, reinforcing the possibility of eradication.

4.2. Step 2: The Carp Fish-Down and Associated Strategies

The CMP began fishing-down carp in February 1995 (Table 1) [20,33]. All quantitative data associated with the eradication has been summarized in Table 1, with detailed analyses of population estimation and seasonal distribution of carp in Lake Crescent described in two separate studies [27,34]. In the early stages of the program there was no knowledge of preferred habitat or fish movement under the local conditions. Therefore initial carp removal strategies were basic, relying on visual observations of fish aggregations and undirected electro-fishing (backpack and boat) or net fishing (gill and seine nets) [20,21]. It was soon identified that for all sized carp to be at risk of capture, a range of gear types were necessary for effective fish-down, and therefore implemented from the outset. Although the gear types evolved and changed with time, they were categorised as primary or secondary techniques based on their catch efficiency for a given season. Specifically, the primary technique refers to the gear type/s that resulted in the highest proportion of carp caught, while the secondary technique refers all other gear types that collectively caught a small proportion of carp, in a given season (Table 1). Specific techniques were developed to target and remove each life stage of carp [20]. For example, the routine collection of length-frequency information derived from capture data provided information on age structure and growth rates of carp in the lake. This information provided insights into gear selectivity that was essential for predicting and deploying suitable gear types (e.g., gill net mesh size) on a regular basis, including forward planning for upcoming fishing seasons, so all cohorts in the population, at a given time, could be fished most effectively [20,36].

Table 1. Summary of techniques and strategies used to eradicate carp from Lake Crescent (1995–2010).

Season *	No. of Carp Caught	Primary Technique **	Secondary Technique ***	Annual Strategy	Description
94/95	234	SN, E		CT, LC, DS	First carp discovered in Lake Crescent on 1 Feb 1995. Carp contained to lakes Sorell and Crescent via the installation of screens at the Lake Crescent outlet. Populations of carp between the two lakes were isolated via the installation of screen structures at the Lake Sorell outlet. Lake Crescent closed to the public. Carp fish-down commenced, with seine nets and electro-fishing accounting for most carp caught.
95/96 †	1300	SN, E	E, GN	CT, LC, DS	Seine nets and electro-fishing effort accounted for most carp caught. Monofilament gill nets starting to be used.
96/97 †	1731	GN+TC+E	FN	CT, LC, DS	Radio transmitter implanted carp deployed for the first time, which were used in conjunction with appropriately sized monofilament gill nets and electro-fishing. This technique accounted for most carp caught this season.
97/98	1547	GN+TC+E	FN	CT, LC, DS	The combination of monofilament gillnets, transmitter carp and electro-fishing accounted for most carp caught.
98/99	1554	GN+TC+E	SN, FN, PN	CT, LC, DS	The combination of monofilament gillnets, transmitter carp and electro-fishing accounted for most carp caught. Population estimate undertaken.
99/00	800	GN+TC, SN, E	FN	CT, LC, DS	The combination of monofilament gillnets, transmitter carp and electro-fishing accounted for most carp caught.
00/01 †	261	GN+TC+E	FN	CT, LC, DS	The combination of monofilament gillnets, transmitter carp and electro-fishing accounted for most carp caught.
01/02	210	GN+TC+E	SN, FN	CT, LC, DS	The combination of monofilament gillnets, transmitter carp and electro-fishing accounted for most carp caught.
02/03	57	GN+TC+E	FN	CT, LC, DS	The combination of monofilament gillnets, transmitter carp and electro-fishing accounted for most carp caught.
03/04	60	GN+TC+E		CT, LC, F, DS	The combination of monofilament gillnets, transmitter carp and electro-fishing accounted for most carp caught. Second population estimate undertaken. Fences with inbuilt traps were installed in front of marshes to prevent access into spawning grounds.
04/05	24	TP	GN, TC, FN, E	CT, F, DS, RS	Traps built into fences accounted for most carp caught. Lake Crescent re-opened to the public. Annual carp recruitment surveys begin.
05/06	11	GN+TC+E	TP, FN	CT, F, DS, RS	The combination of monofilament gillnets, transmitter carp and electro-fishing accounted for most carp caught.
06/07	5	GN+TC+E		CT, F, DS, RS	The combination of monofilament gillnets, transmitter carp and electro-fishing accounted for most carp caught.
07/08	3	GN+TC+E		CT, F, DS, RS	The combination of monofilament gillnets, transmitter carp and electro-fishing accounted for most carp caught. Last carp caught in Lake Crescent in Dec 2007.
08/09	0	GN+TC+E		CT, F, DS, RS	Combination of monofilament gillnets, transmitter carp and electro-fishing effort applied.
09/10	0			CT, F, DS, RS	All remaining active transmitter carp removed from the lake. Carp eradication in Lake Crescent declared.
Total	7797				

GN—Gill Net, E—Electro-fishing (backpack and boat), SN—Seine Net, TC—Transmitter Carp, TP—Trap, FN—Fyke Net, PN—Pound net. CT—Containment Screens, LC—Lake Closure, RS—Recruitment Survey, DS—Downstream Surveys (electrofishing), F—Fence, +—Indicates techniques which are used in combination. *—Season is in line with Australian financial year (1 July to 30 June). **—Primary technique refers to the gear type/s responsible for dominant (highest proportion of carp caught) capture of carp for the season. ***—Secondary technique refers to the gear type/s other than the primary for the season (small proportion of carp caught). †—Indicates a spawning event. Note: All fencing removed in September 2013. Containment screens at outlet of Lake Crescent removed in August 2014. Monitoring of Lake Crescent is on-going post 2010. Downstream and recruitment surveys are still ongoing.

From 1995 to 1996 seine netting was the primary technique, along with electro-fishing and to a lesser extent, gill nets. Due to a higher density of carp at the outset, these fishing techniques proved very efficient, with over 1500 carp caught in the first two years (Table 1). In late 1996, it became apparent that the fish-down efforts employed by the CMP were making a significant impact on the population, as catch per unit effort (CPUE) began dropping, and the capture of carp became increasingly difficult [20]. The CMP began looking into new techniques to improve the effectiveness of locating fish, with the use of biotelemetry adopted in 1997 [37].

Wild male carp were surgically implanted with appropriately sized radio transmitters in their gut cavity and then released back into the lakes [20,21,37]. Adult females were not used as transmitter carp to minimise recruitment risk. It was predicted that transmitter carp would begin mixing back with the wild population in the lakes and hence betray their locations. Transmitter carp were located daily, and were used as indicators of carp activity, movement, distribution and range [27,38]. On 10 March 1997, the first nine adult male transmitter carp were released into Lake Crescent [20,22]. On 9 April 1997 the first transmitter aggregation was targeted and resulted in 202 carp captured [22]. From that point through to 2009, the use of gill nets, transmitter carp (Judas carp), and electro-fishing in combination, became the primary technique for locating and capturing carp, particularly adults (Table 1). The technique was highly effective and resulted in the capture of most carp in the aggregations. The high catch efficiency when using transmitter carp in combination with nets was also observed in studies conducted in the USA [38]. Winter aggregations of carp in three Midwestern lakes were targeted using transmitter carp and seine nets, where 94% of carp in the aggregations were removed [38].

The gill nets used to target adult carp ranged in mesh size from 4 to 7 inches (102 to 178 mm), with the exact specifications of the nets detailed in Walker and Donkers (2003) [36]. The carp were enmeshed in gill nets in three main ways [39]: wedged—held tightly by the mesh around the body, gilled—prevented from backing out of the net by a mesh caught behind the operculum and/or tangled—held by projections, usually spines in the mesh. The susceptibility of carp to entanglement by their spines (dorsal and anal), provided a broader size range of catchability for the gill nets, particularly with the finer monofilament mesh. Gill nets were only set during daylight hours, and were removed from the lake at the end of each day.

The daily movements of transmitter carp were first tracked and monitored from a boat using a radio telemetry receiver attached to a large boat-mounted directional antenna (Figure 3). Typically, the boat was driven parallel to the shore, approximately 250 m from the shoreline. Once a transmitter carp (or group of transmitter carp) was detected in the shallows, care was taken not to spook or startle the aggregation, and the boat was carefully parked on shore, a safe distance away. A small hand-held antennae was then connected to the receiver, and the aggregation was further pin-pointed on foot by wading (two person team). A first gill net (usually the smallest mesh size 2.5 inch) was carefully waded and set around the area of the transmitter carp, while the second person continued to monitor the position of the transmitter carp with the receiver. This net was always set from shore to shore, to ensure all carp in the vicinity of the transmitter carp were enclosed within the net. The secondary gill nets (3, 4, 5, 6 and 7 inch mesh) were set inside the smaller meshed outer net, with increasing (the largest mesh on the inside) mesh size ensuring all sized carp were vulnerable to capture. The areas inside the nets were also systematically electro-fished with the backpack electro-fishers and/or the electro-boat, in order to herd all carp into the nets. The transmitter carp caught in the aggregation were detected first, quickly removed from the net, and released back into the lake to avoid injury and damage. On average, ten active male transmitter carp were maintained in the lake. Re-captured transmitter carp with expired batteries, incidental mortalities, or in poor condition, were replaced with new healthy male transmitter carp [22].

The radio telemetry enabled delineation of inter-seasonal and inter-annual patterns of carp movement and habitat choice in response to changes in lake water level and water temperature (Figure A1) [27]. The resulting knowledge was then used for carp removal, identifying life cycle vulnerabilities, recruitment prevention, and resource rationalisation [20]. Specifically, it facilitated targeted removal of carp by establishing a detailed understanding of carp movement, behaviour and habitat preferences in the lakes [20,27]. Radio telemetry was also critical in locating spawning sites, allowing early intervention through liming or spot poisoning with rotenone.

Figure 3. A CMP work boat with a directional antenna installed (**a**), hooked up to a Lotek telemetry receiver (**b**) for tracking and monitoring transmitter carp.

Population Estimation

It was essential to estimate the population abundance of carp accurately and as frequently as possible, to assist the ongoing fish-down efforts and to allow timely management decisions [20]. This aided planning of fishing effort, including allocation of resources, and to monitor the success of the eradication program [20,33,40]. Although estimating population numbers can be challenging or near impossible in large lakes, a mark recapture method based on a model developed by Petersen (1896) [41], which proved to be accurate [34], was used. Specifically, deployment of a combination of gear types (to minimise capture bias), use of radio transmitter carp to increase capture efficiency, and the adoption of a reverse-Schnabel approach, resulted in an accurate population estimation in Lake Crescent [34]. These estimates assisted in making several informed management decisions which included the effort/time required to eradicate a cohort, the estimated number of fish remaining, and ongoing CPUE based population estimates within the lake. This information was useful in communicating the program outcomes, and garnering sustained socio-political support including funding. More significantly, the estimates were central to support the management and legislative decisions to declare Lake Crescent free of carp, and its reopening to the public for fishing and recreational activities.

5. Step 3: Preventing and Eliminating Recruitment

Preventing and eliminating carp recruitment was a key strategy in the eradication of carp from Lake Crescent. It occurred throughout the fish-down process, either to prevent/capture mature carp accessing spawning habitat, to actively survey for potential spawning events or to respond to spawning events.

Cooler temperature, particularly during winter, resulted in comparatively slower growth, therefore Tasmanian carp matured more slowly than those on mainland Australia [20]. Male carp in Lake Crescent took a minimum of three years to mature (spermiate) and females, four years. However, some females did not mature until seven years of age [20]. The movement and aggregation of mature carp triggered by the urge to spawn made them particularly vulnerable for capture, by both passive gears as well as active targeting [20]. As males matured at least a year earlier, they were susceptible for earlier capture in traps/fyke nets, providing an opportunity to remove most mature males prior to females in the cohort maturing, further aiding the prevention of spawning.

Spawning of the carp is known to occur in spring and summer at temperatures around 17–25 °C, in shallow, wetland areas [22,42–44]. The number of eggs produced by a female can vary depending on size, age and condition [13]. On mainland Australia, it is estimated that a female carp can produce over one million eggs per year [13]. The temperature data collected from Lake Crescent indicates suitable temperatures for viable spawning are present each year, usually from November to February. This restricted window, in part, allowed implementation of a focused and targeted spawning prevention strategy. In Lake Crescent, the carp have also been observed spawning in water temperatures as low as 11 °C [22]. However, there is no evidence of successful recruitment from eggs spawned at such low temperatures [22].

5.1. Spawning Prevention by Blocking Adults

To ensure removal exceeded the rate of population increase [35], spawning prevention was critical. In Lake Crescent, this was initiated in 2004 (Table 1), by installing 200 m of heavy gauge chicken wire mesh (approximately 50 mm in diameter) around prime spawning habitats (shallow, macrophyte rich wetland areas) [22,45]. The fences were also exploited by installing lockable steel traps at vantage points along the fence, such as near drains and creek outlets and inlets (Figure 4) [22,45]. These traps were particularly efficient at capturing mature carp that were attempting to push into the wetlands for spawning, at times of rising lake levels and warming conditions in spring. However, the fences could not completely block access to the wetlands during periods of high lake levels. The efficiency of these traps would peak for a few months of the year (October–November) when the spawning cues were present, as they preferentially removed fish which were reproductively primed, thus preventing both spawning and future recruitment [45]. For example, in the first year of installation, these traps installed within the fences accounted for a large percentage of carp captured, and were the primary technique for the 2004/05 season (Table 1) [45]. Of the 115 carp caught over the 2004/05 season, 63 (55%) were caught in the traps, and mainly over the peak spawning period in October and November [45].

5.2. Egg and Fry Removal

The preferred action plan was to prevent spawning from occurring in the first place, through careful monitoring of environmental cues, in conjunction with monitoring transmitter carp movements, and preventing access to breeding habitat. However, when eggs were detected, a combination of physically removing the affected substrate, in addition to isolating and poisoning the spawning beds, proved to be most effective [20,21]. Carp lay sticky eggs in shallow water amongst submerged macrophytes. When necessary, spawning beds were detected and marked by visual survey, followed by the localized application of hydrated lime (Limil®) to rapidly raise the pH level (above 11.0) and kill developing embryos [20]. All stages of carp are vulnerable to alkaline stress in excess of pH 9.0 [20]. The caustic effect of the reactive lime appeared to have an added detrimental effect on the developing embryos. Following liming, the sites were monitored using visual inspections and fine mesh net sweeps to detect any surviving larvae.

Figure 4. The wire fences that were installed to prevent carp from accessing potential spawning habitat, with lockable traps set at strategic locations. The large expanse of wetlands can be seen behind the fence and trap.

Eight weeks after hatching, the surviving fry began dispersing into the wetlands [20]. The abundance of fry (<50 mm fork length) could be reduced while they were inhabiting the wetlands by dropping water levels (either naturally or artificially). Falling water levels concentrated the young carp into depressions and channels, as the receding water drew them away from the wetlands. These aggregations were located and backpack electro-fished, and/or poisoned by spot rotenone applications [20]. Although the efficacy of these techniques were not quantified, they were collectively effective.

5.3. Juvenile Carp Removal

Once maturing past the fry stage (<50 mm fork length), juvenile carp (defined here as 50–250 mm fork length) tended to disperse out of wetlands into the main lake, and forage around the margins [20]. For the next 15–24 months, carp of this size became vulnerable to capture using fyke nets set in the shallows, and strongly responded to increases in temperature of the shallow margins around the lake. During warm settled weather, carp of this size class pushed into the shallows and formed feeding aggregations [20]. The fyke nets used consisted of single 5 m wing, 800 mm high D shaped entrance, three internal hoops, and 18 mm stretched mesh. Small meshed gill nets (2.5 inch) combined with electro-fishing, also became standard techniques for targeting small carp, or determining whether spawning had occurred [20,22]. The use of seine nets was also effective at catching all sizes of carp (including juveniles), with good catches made when aggregations were detected over soft bottom and shallow water.

5.4. Recruitment Surveys

It was essential to detect the frequency and magnitude of any recruitment events, in order to plan the most appropriate and effective management strategy [20,21], where recruitment is defined as the presence of carp fry (<50 mm fork length). It was particularly important to identify these events as early as possible so appropriate intervention, such as liming of spawned eggs, isolation and/or spot poisoning of nursery grounds, and choice of appropriate capture gear for targeting could be implemented [20]. To detect spawning and recruitment events an ongoing systematic annual recruitment survey was adopted [20]. Specifically, this involved visual inspections of high-risk spawning habitat, fine mesh dip net sweeps, and backpack electro-fishing in areas of known breeding habitat, or where transmitter carp were known to frequent. This was also in conjunction with an annual lake wide fyke net survey, that was used for detecting any cryptic recruitment event/s reliably [20]. Data collected suggested that any carp recruited during the peak spawning period (November–February) would have grown sufficiently large and mobile to be captured by fyke nets set perpendicular to the shore, during the scheduled annual surveys later in the season [20]. These annual recruitment surveys have been ongoing in Lake Crescent since 2004 and have continued despite the last carp capture in 2007 (Table 1). This aspect is likely to remain until carp are completely eradicated from Tasmania.

6. Step 4: Final Eradication and Monitoring

After nine years of intensive fish-down efforts using a range of techniques, the remaining number of carp left in Lake Crescent was estimated to be very low. As a result, Lake Crescent was re-opened to the public in 2004, while fishing efforts continued and were focused on removing the last few carp. The use of male transmitter carp (combined with gill nets and electrofishing) was still the key technique for carp detection and capture. However, the installation of wire fences with steel traps in 2004 also proved to very efficient.

6.1. *Opening the Lake to the Public*

In August 2004, it was determined that the remaining carp population in Lake Crescent was so low that the risk of spread via human activities was unlikely, and therefore the lake was re-opened to the public [45]. Fish-down efforts continued despite opening, while ensuring recreational fishers did not interact with the gears.

6.2. *Continued Fish-Down Efforts*

Over the 2004/05 season, the majority of carp were caught in the fish traps installed within the wire fences, while a smaller number were caught using transmitter carp, electrofishing and gill nets in combination [45]. The installation of these new fences prevented mature carp from accessing the spawning habitat within the marshes. Between 2005 to 2006 there were numerous occasions when transmitter carp were captured by the traps, however no other wild untagged carp were found with them [46]. To entice what may be trap-shy carp that may have evaded capture, chemoattraction traps were deployed. This involved priming an odour-donor female carp with pituitary extract and holding it securely in an enclosure behind a trap. This allowed an odour plume to disperse into the lake, potentially attracting other mature carp [47]. One such trapping event (on 31 October 2005) accounted for 9 fish (all males), including a male transmitter carp, which was quite significant considering only 11 fish were caught from the lake after that event [46]. Moreover, a follow up study later demonstrated that mature carp are particularly chemo-attuned during breeding season and hence susceptible for capture by this technique [48]. This finding supports previous work suggesting that a F prostaglandin-based spawning pheromone released by ovulated carp is attractive [49], and could prove useful in their control [50].

During the 2007/08 season, 15 transmitter carp aggregations were targeted with gill nets and electro-fished, with 14 of these events resulting in the capture of only transmitter carp, indicating there were very few carp remaining in the lake [51]. In December 2007, two transmitter carp were detected in shallow water, in response to a significant rain event, combined with warm, humid weather. This particular aggregation resulted in the capture of two tagged male carp and a single wild female carp, which turned out to be the last carp caught out of Lake Crescent (Figures 5 and 6) [51].

Figure 5. The last carp caught from Lake Crescent.

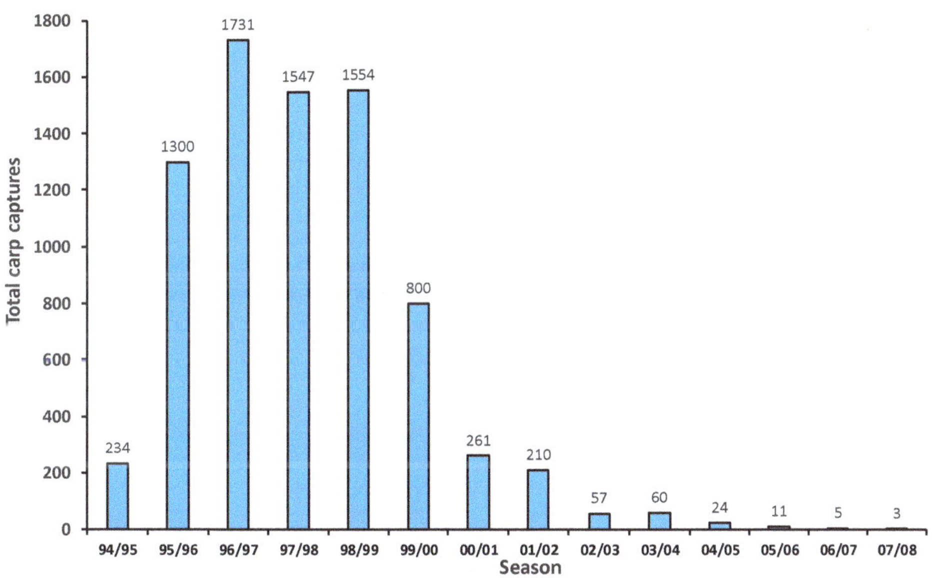

Figure 6. The total number of all carp caught in Lake Crescent from 1995 to 2007, with the last carp caught in 2007.

6.3. Ongoing Monitoring

After this event, transmitter carp were monitored for the next two years, with no other wild fish captured [52,53]. All transmitter carp were removed from the lake in December 2009, followed by months of intensive surveying for recruitment events [53]. Monitoring for recruitment involved electro-fishing, spot rotenone poisoning at known aggregation sites and multiple fyke net surveys around the lake [53]. In addition, commercial eel fishers were permitted to fish Lake Crescent which significantly increased fyke netting effort [53]. Despite the extra effort, no further carp were caught from the lake, verifying the carp free status of Lake Crescent. However, in lieu of an as yet unconfirmed carp free status of the upstream sister Lake Sorell, annual recruitment surveys are ongoing with no carp caught for the last 14 years in Lake Crescent.

7. Summary and Lessons Learned

Despite the challenges of the high fecundity of carp, the large expanse of the lake, and relatively cryptic nature of the threat (i.e., compared to terrestrial incursions), the eradication of carp from Lake Crescent was successful. This is particularly remarkable, given the limitations associated with multiple use conflicts including the need to protect the natural values of the lake and maintain the conservation status of the surrounding wetlands. In all, it took 12 years to eradicate the carp population from Lake Crescent, employing mostly physical removal approaches, whilst retaining the natural, social, and in large part economic values of the lake throughout the 12-year campaign.

One of the many challenges faced in eradicating a pest species from an aquatic habitat is knowing when it is finally achieved. We now know the last wild carp was removed from Lake Crescent in December 2007. Several years of fishing effort was applied after this capture to validate the eradication, with recruitment surveys also used to verify success. Fishing effort and lessons learned have since been applied successfully to the last remaining Tasmanian carp population in Lake Sorell.

The integrated and multifaceted approach was the key feature which resulted in the eradication of carp from Lake Crescent. This involved using a range of gear types and techniques in combination with one another during the fish-down, whilst containing carp to the system and preventing/eliminating recruitment at the earliest opportunity. The need and ability to learn and adapt continuously with changing environmental conditions and capture efficiency with decreasing density, were also crucial. Through the Lake Crescent experience, it was apparent that any given carp cohorts can be fished-down systematically within a window of about seven years. However, the most difficult aspect was blocking spawning, despite having reduced the population. The inability to prevent every spawning event meant the introduction of new cohorts, which prolonged the eradication process.

It is difficult to put an exact monetary value on the eradication of carp from Lake Crescent, as the fish-down in Lake Sorell was also undertaken concurrently, albeit intermittently. However, from inception of the Carp Management Program in 1995, to the declaration of carp eradication from Lake Crescent in 2009, a total cost of AU$ 6.5 million was incurred. As a result of the Lake Crescent experience, extensive improvements were made possible to the gear types and fish-down strategies currently used in the Lake Sorell eradication campaign. This includes gill nets which select for a wide size class of carp, combined with over-night netting. The recruitment prevention strategies in Lake Sorell are also more effective, with the use of polypropylene barrier nets to replace the fences, and various layers of gill net permanently installed behind the barrier nets in the spawning season, for additional security. These improvements would have made a substantial difference if employed in Lake Crescent, especially if they were utilised from the outset, and may have reduced the eradication time frame significantly. Replicating these strategies elsewhere in a similarly invaded habitat is possible, with an estimated eradication time frame of about 7–10 years, with a particularly focused and intensive fishing effort, during the peak spawning season.

At present, over 99% of the original carp population in Lake Sorell have been removed, and it is estimated that there are few, if any carp remaining. Consistent with the approach in Lake Crescent, the control order was lifted in February 2020 allowing Lake Sorell to be re-opened to the public, after 25 years of intermittent closures. The CMP expects to complete the eradication of carp from Lake Sorell in the next two years. The Tasmanian experience clearly demonstrates that an integrated physical removal and containment strategy can be successfully applied to eradicate common carp from large freshwater lakes. With enough resolve, socio-political support and funding, the Tasmanian approach is a viable control technique for application elsewhere.

Author Contributions: Individual authors contributed as follows: Manuscript conceptualization J.L.Y.; program design and methodology J.D., C.W.; data collection J.D., C.W.; writing-original draft preparation J.L.Y., J.G.P.; writing-review and editing J.L.Y., J.G.P., C.W. and J.D. All authors have read and agreed to the published version of the manuscript.

Funding: The Carp Management Program was primarily funded by the Tasmanian State Government, with support from the Federal Government. It was also assisted by a Fisheries Research and Development Corporation (FRDC) grant (Project No. 2000/182).

Institutional Review Board Statement: Not applicable.

Informed Consent Statement: Not applicable.

Acknowledgments: We thank past and present field staff from the Carp Management Program and the Inland Fisheries Service for their input and support. Thanks to Helen O'Neill and Rob Freeman for assisting during the review process. Special thanks also to Professor Peter Sorensen, the guest editor for providing extensive comments and feedback during the review process which substantially improved the manuscript.

Conflicts of Interest: The authors declare no conflict of interest.

Appendix A

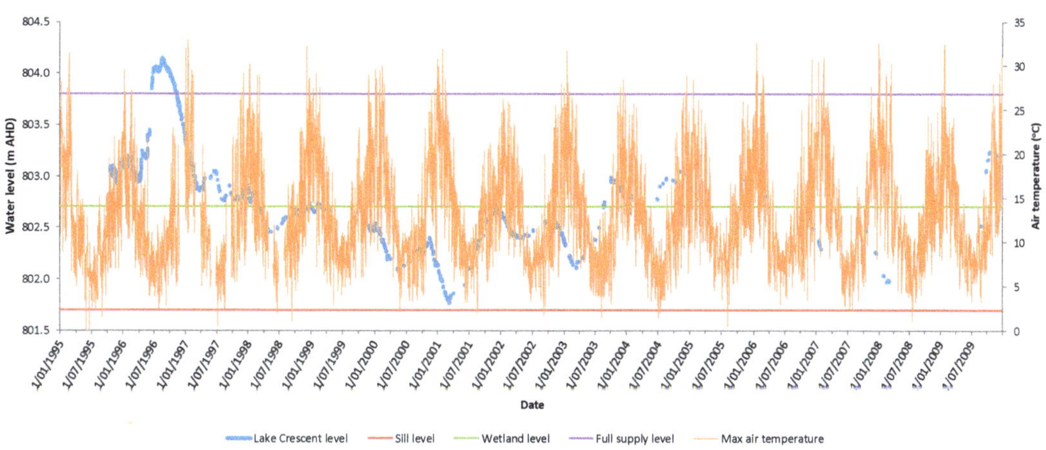

Figure A1. Lake Crescent lake levels from 1995 to 2009, in relation to the full supply, wetland, and sill levels of the lake. Daily maximum air temperature taken from Lake St Clair National Park is also plotted in conjunction (Downloaded from www.bom.gov.au). Note: Lake level data is deficient at certain periods.

References

1. Lowe, S.; Browne, M.; Boudjelas, S.; De Poorter, M. *100 of the World's Worst Invasive Alien Species: A Selection from the Global Invasive Species Database*; The Invasive Species Specialist Group: Auckland, New Zealand, 2000. Available online: http://www.issg.org/pdf/publications/worst_100/english_100_worst.pdf (accessed on 8 April 2020).
2. Simberloff, D. Impacts of introduced species in the United States. *Consequences* **1996**, *2*, 13–22.
3. Clavero, M.; García-Berthou, E. Invasive species are a leading cause of animal extinctions. *Trends Ecol. Evol.* **2005**, *20*, 110. [CrossRef]
4. Elton, C.S. *The Ecology of Invasions by Animals and Plants*, 1st ed.; Methuen & Co. Ltd.: London, UK, 1958.

5. Walker, B.; Steffen, W. An overview of the implications of global change for natural and managed terrestrial ecosystems. *Conserv. Ecol.* **1997**, *1*, 1–17. [CrossRef]
6. Havel, J.E.; Bruckerhoff, L.A.; Funkhouser, M.A.; Gemberling, A.R. Resistance to desiccation in aquatic invasive snails and implications for their overland dispersal. *Hydrobiologia* **2014**, *741*, 89–100. [CrossRef]
7. Lovell, S.J.; Stone, S.F.; Fernandez, L. The economic impacts of aquatic invasive species: A review of the literature. *Agric. Resour. Econ. Rev.* **2006**, *35*, 195–208. [CrossRef]
8. Dudgeon, D.; Arthington, A.H.; Gessner, M.O.; Kawabata, Z.I.; Knowler, D.J.; Lévêque, C.; Naiman, R.J.; Prieur-Richard, A.H.; Soto, D.; Stiassny, M.L. Freshwater biodiversity: Importance, threats, status and conservation challenges. *Biol. Rev.* **2006**, *81*, 163–182. [CrossRef]
9. Balian, E.V.; Segers, H.; Martens, K.; Lévéque, C. The freshwater animal diversity assessment: An overview of the results. *Hydrobiologia* **2008**, *595*, 627–637. [CrossRef]
10. Bajer, P.G.; Sorensen, P.W. Using boat electrofishing to estimate the abundance of invasive common carp in small Midwestern lakes. *North Am. J. Fish. Manag.* **2012**, *32*, 817–822. [CrossRef]
11. Sorensen, P.W.; Bajer, P.G. Carp, common. In *Encyclopedia of Biological Invasions*; Simberloff, D., Rejmanek, M., Eds.; University of California Press: London, UK, 2011; pp. 100–104.
12. Koehn, J.D. Carp (*Cyprinus carpio*) as a powerful invader in Australian waterways. *Freshw. Biol.* **2004**, *49*, 882–894. [CrossRef]
13. Koehn, J.; Brumley, A.; Gehrke, P. *Managing the Impacts of Carp*; Bureau of Rural Sciences: Canberra, Australia, 2000.
14. Winker, H.; Weyl, O.L.; Booth, A.J.; Ellender, B.R. Life history and population dynamics of invasive common carp, *Cyprinus carpio*, within a large turbid African impoundment. *Mar. Freshw. Res.* **2011**, *62*, 1270–1280. [CrossRef]
15. Fletcher, A.R.; Morison, A.K.; Hume, D.J. Effects of carp, *Cyprinus carpio* L., on communities of aquatic vegetation and turbidity of waterbodies in the lower Goulburn River basin. *Mar. Freshw. Res.* **1985**, *36*, 311–327. [CrossRef]
16. Brumley, A.R. Cyprinids of Australasia. In *Cyprinid Fishes: Systematics, Biology and Exploitation*; Winfield, I.J., Nelson, J.S., Eds.; Springer: Dordrecht, The Netherlands, 1991; pp. 264–283.
17. Bajer, P.G.; Sorensen, P.W. Effects of common carp on phosphorus concentrations, water clarity, and vegetation density: A whole system experiment in a thermally stratified lake. *Hydrobiologia* **2015**, *746*, 303–311. [CrossRef]
18. Matsuzaki, S.S.; Usio, N.; Takamura, N.; Washitani, I. Contrasting impacts of invasive engineers on freshwater ecosystems: An experiment and meta-analysis. *Oecologia* **2009**, *158*, 673–686. [CrossRef]
19. Sorensen, P.W.; Bajer, P.G. Case studies demonstrate that common carp can be sustainably reduced by exploiting source-sink dynamics in Midwestern lakes. *Fishes* **2020**, *5*, 36. [CrossRef]
20. Diggle, J.W.; Patil, J.G.; Wisniewski, C. *A Manual for Carp Control: The Tasmanian Model*; Invasive Animals Cooperative Research Centre: Canberra, Australia, 2012.
21. Wisniewski, C.D.; Diggle, J.D.; Patil, J.G. Managing and Eradicating Carp: A Tasmanian Experience. In *New Zealand Invasive Fish Management Handbook*; Collier, K.J., Grainger, N., Eds.; LERNZ (The University of Waikato) and Department of Conservation: Hamilton, New Zealand, 2015; pp. 82–89.
22. IFS. *Inland Fisheries Service, Carp Management Program Report Lakes Crescent and Sorell 1995–June 2004*; Inland Fisheries Service: Hobart, Australia, 2004.
23. DSEWPC. *Interlaken Lakeside Reserve Ramsar Site Ecological Character Description*; Department of Sustainability, Environment, Water, Population and Communities: Canberra, Australia, 2012.
24. Hardie, S.A. *Current Status and Ecology of the Golden Galaxias (Galaxias Auratus)*; Inland Fisheries Service: Hobart, Australia, 2003.
25. Uytendaal, A. Water Clarity in Two Shallow Lake Systems of the Central Plateau, Tasmania, Australia. Honours Thesis, University of Tasmania, Hobart, Australia, 2006.
26. Kirkpatrick, J.B.; Tyler, P.A. Tasmanian wetlands and their conservation. In *The Conservation of Australian Wetlands*; McComb, A.J., Lake, P.S., Eds.; Surrey Beatty & Sons Pty: Sydney, Australia, 1988; pp. 1–16.
27. Taylor, A.H.; Tracey, S.R.; Hartmann, K.; Patil, J.G. Exploiting seasonal habitat use of the common carp, *Cyprinus carpio*, in a lacustrine system for management and eradication. *Mar. Freshw. Res.* **2012**, *63*, 587–597. [CrossRef]
28. UNESCO. *Convention on Wetlands of International Importance, Especially as Waterfowl Habitat*; UNESCO: Paris, France, 1994.
29. Heffer, D.K. *Wetlands of Lakes Sorell and Crescent: Conservation and Management*; Inland Fisheries Service: Hobart, Australia, 2003.
30. Cheng, D.M.H.; Tyler, P.A. Lakes Sorell and Crescent—A Tasmanian paradox. *Int. Rev. Gesamten Hydrobiol. Hydrogr.* **1973**, *58*, 307–343. [CrossRef]
31. Cleary, C. Habitat Preferences of the Endemic Hydrobiid Gastropod. "*Austropyrgus*" sp., and Other Aquatic Gastropods from Lakes Sorell and Crescent, Central Plateau, Tasmania. Honours Thesis, University of Tasmania, Hobart, Australia, 1997.
32. Hardie, S.A. Conservation Biology of the Golden Galaxias (*Galaxias auratus*) (Pisces: Galaxiidae). Ph.D. Thesis, University of Tasmania, Hobart, Australia, 2007.
33. Diggle, J.; Day, J.R.; Bax, N.J. *Eradicating European Carp from Tasmania and Implications for National European Carp Eradication*; Inland Fisheries Service: Hobart, Australia, 2004.
34. Donkers, P.; Patil, J.G.; Wisniewski, C.; Diggle, J.E. Validation of mark-recapture population estimates for invasive common carp, *Cyprinus carpio*, in Lake Crescent, Tasmania. *J. Appl. Ichthyol.* **2012**, *28*, 7–14. [CrossRef]
35. Bomford, M.; O'Brien, P. Eradication or control for vertebrate pests? *Wildl. Soc. Bull.* **1995**, *23*, 249–255.

36. Walker, R.M.; Donkers, P. *An Examination of the Selectivity of Fishing Equipment in Relation to Controlling Carp (Cyprinus carpio) in Lakes Sorell and Crescent*, 2nd ed.; Technical Report No. 2; Inland Fisheries Service: Hobart, Australia, 2011.
37. Macdonald, A.; Wisniewski, C. *The Use of Biotelemetry in Controlling the Common Carp (Cyprinus carpio) in Lakes Crescent and Sorell*, 2nd ed.; Technical Report No. 1; Inland Fisheries Service: Hobart, Australia, 2011.
38. Bajer, P.; Chizinski, C.; Sorensen, P. Using the judas technique to locate and remove wintertime aggregations of invasive common carp. *Fish. Manag. Ecol.* **2011**, *18*, 497–505. [CrossRef]
39. Hamley, J.M. Review of Gillnet Selectivity. *J. Fish. Board Can.* **1975**, *32*, 1943–1969. [CrossRef]
40. Donkers, P.D. *An Investigation into the Abundance of European Carp (Cyprinus carpio) in Lakes Sorell and Crescent*; Technical Report No. 3; Inland Fisheries Service: Hobart, Australia, 2003.
41. Petersen, C.G.J. The yearly immigration of young plaice in the Limfjord from the German Sea. *Rept Dan. Biol Sta* **1896**, *6*, 1–48.
42. McDowall, R.M. *New Zealand Freshwater Fishes: A Natural History and Guide*; Heinemann Reed MAF Publishing Group: Auckland, New Zealand, 1990.
43. Brumley, A.R. Family Cyprinidae—Carps, minnows. In *Freshwater Fishes of South Eastern Australia*, 2nd ed., McDowall, R.M., Ed.; Reed: Chatswood, Australia, 1996; pp. 99–106.
44. Penne, C.R.; Pierce, C.L. Seasonal distribution, aggregation, and habitat selection of common carp in Clear Lake, Iowa. *Trans. Am. Fish. Soc.* **2008**, *137*, 1050–1062. [CrossRef]
45. IFS. *Inland Fisheries Service, Carp Management Program Annual Report Lakes Crescent and Sorell July 2004–June 2005*; Inland Fisheries Service: Hobart, Australia, 2005.
46. IFS. *Inland Fisheries Service, Carp Management Program Annual Report Lakes Crescent and Sorell July 2005–June 2006*; Inland Fisheries Service: Hobart, Australia, 2006.
47. Patil, J.G.; Wisniewski, W. *Hypohysation: A Technique for Deployment of Odour Donor Fish to Control the Common Carp (Cyprinus carpio)*, 2nd ed.; Technical Report No. 5; Inland Fisheries Service: Hobart, Australia, 2011.
48. Adair, B.J.; Purser, G.J.; Patil, J.G. Peripheral olfactory structures and maturity-related crypt receptor neuron kinetics in the olfactory epithelium of carp *Cyprinus carpio* (L.): Implications for carnal vulnerability and pest management. *Mar. Freshw. Res.* **2018**, *69*, 1604–1613. [CrossRef]
49. Lim, H.; Sorensen, P.W. Common carp implanted with prostaglandin F2α release a sex pheromone complex that attracts conspecific males in both the laboratory and field. *J. Chem. Ecol.* **2012**, *38*. [CrossRef] [PubMed]
50. Sorensen, P.W.; Johnson, N.S. Theory and application of semiochemicals in nuisance fish control. *J. Chem. Ecol.* **2016**, *42*, 698–715. [CrossRef] [PubMed]
51. IFS. *Inland Fisheries Service, Carp Management Program Annual Report 2007–2008*; Inland Fisheries Service: Hobart, Australia, 2008.
52. IFS. *Inland Fisheries Service, Carp Management Program Annual Report 2008–2009*; Inland Fisheries Service: Hobart, Australia, 2009.
53. IFS. *Inland Fisheries Service, Carp Management Program Annual Report 2009–2010*; Inland Fisheries Service: Hobart, Australia, 2010.

MDPI
St. Alban-Anlage 66
4052 Basel
Switzerland
Tel. +41 61 683 77 34
Fax +41 61 302 89 18
www.mdpi.com

Fishes Editorial Office
E-mail: fishes@mdpi.com
www.mdpi.com/journal/fishes

www.ingramcontent.com/pod-product-compliance
Lightning Source LLC
LaVergne TN
LVHW070150170726
843515LV00007B/1886
9783036526829